Cryotechniques

in Biological Electron Microscopy

Edited by
R. A. Steinbrecht and K. Zierold

With 108 Figures

Springer-Verlag
Berlin Heidelberg New York
London Paris Tokyo

Prof. Dr. Rudolf Alexander Steinbrecht
Max-Planck-Institut für Verhaltensphysiologie
D-8131 Seewiesen, FRG

Dr. Karl Zierold
Max-Planck-Institut für Systemphysiologie
Rheinlanddamm 201
D-4600 Dortmund, FRG

Cover illustration:
Pressure-frozen and freeze-fractured fresh liver cell
From H. Moor, Fig. 6, p. 186, this volume

ISBN-13: 978-3-642-72817-4 e-ISBN-13: 978-3-642-72815-0
DOI: 10.1007/ 978-3-642-72815-0

Library of Congress Cataloging-in-Publication Data. Cryotechniques in biological electron microscopy. Includes bibliographies and index. 1. Cryobiology–Technique. 2. Electron microscopy–Technique. I. Steinbrecht, Rudolf A. (Rudolf Alexander), 1937– . II. Zierold, Karl, 1944– . QH324.9.C7C79 1986 578'.45 87-23358.

Softcover reprint of the hardcover 1st edition 1987

Typesetting: K+V Fotosatz GmbH, Beerfelden

2131/3130-543210

In lieu of a motto:

... Noch einmal bückt der Meister sich,	... And once again the Master bends,

Doch nicht geschickt erweist er sich.	The key eludes him past amends.
Das Wasser in dem Fasse hier	The water in the barrel here

Hat etwa null Grad Réaumur.	Is close to zero Réaumur.
Es bilden sich in diesem Falle	At this point, scientists advise,

Die sogenannten Eiskristalle ...	The substance tends to "crystallize" ...

From: WILHELM BUSCH (1832–1908)
Eine kalte Geschichte (A Chilly Affair)
English translation by Walter Arndt
from: *The Genius of Wilhelm Busch – Comedy of Frustration,*
University of California Press 1982,
with kind permission of the publishers.

Preface

To preserve tissue by freezing is an ancient concept going back presumably to the practice of ice-age hunters. At first glance, it seems as simple as it is attractive: the dynamics of life are frozen in, nothing is added and nothing withdrawn except thermal energy. Thus, the result should be more life-like than after poisoning, tanning and drying a living cell as we may rudely call the conventional preparation of specimens for electron microscopy. Countless mishaps, however, have taught electron microscopists that cryotechniques too are neither simple nor necessarily more life-like in their outcome.

Not too long ago, experts in cryotechniques strictly denied that a cell could truly be vitrified, i.e. that all the solutes and macromolecules could be fixed within non-crystalline, glass-like solid water without the dramatic shifts and segregation effects caused by crystallization. We now know that vitrification is indeed possible. Growing insight into the fundamentals of the physics of water and ice, as well as increasing experience of how to cool cells rapidly enough have enlivened the interest in cryofixation and produced a wealth of successful applications.

The abundance of methods and the proliferation of papers using cryotechniques make the choice difficult for the newcomer and also an old hand may run into problems on starting out in a new direction. Therefore, we thought the time was ripe to sift the grain from the chaff and to evaluate the present state of the art. With a field so diverse, this cannot be the task for a single authority. We succeeded in getting leading experts to write selective and critical reviews on aspects of major interest. The selection of topics is subjective as we do not intend to present a comprehensive textbook. For example, we refrained from including a chapter on freeze-fracturing and -etching, since these well-established methods are exhaustively treated in earlier books and review articles.

The baseline for the whole book is given by three chapters on the physics of water and ice, the physics of specimen preparation and the effects of radiation damage. From these contributions, it should become clear what is known and what has to be borne in mind when planning cryopreparation. This part is followed by four chapters on the general methodology: cryofixation, cryoelectron microscopy of vitrified specimens, cryoultramicrotomy, freeze-substitution and freeze-drying. None of these is a mere description of equipment or a collection of recipes, because we believe that, at present, the prescription of methods without critical advice would impede rather than promote the "art" of specimen preparation. Six shorter contributions treat various, more specialized techniques, each of them with its own particular potential for the future: high pressure freezing, freeze-etching of macromolecules, high resolution metal replication, immunolabelling, autoradiography and scanning electron microscopy of cryospecimens. The last two chapters treat the cryofixation of dynamic events and defined physiological states, two fields where the advantages of rapid freezing techniques are especially evident. The appendix contains a short recapitulation of safety hazards and how to prevent them.

A novel feature of this multi-author book was an editorial conference of the authors *after* completion and precirculation of their contributions. This meeting was held at Ringberg Castle in Bavaria, November 19–21, 1986. We are greatful to the Max-Planck-Society for their hospitality at this convention centre and to the German Society for Electron Microscopy for generous sponsorship. The conference not only helped to avoid overlap, but also to speak the same language (see Glossary). Due to open-minded and constructive criticism from all participants, every single contribution has been substantially improved. We hope that the spirit and the enthusiasm of this meeting have been transferred to this book and will stimulate further research.

Seewiesen and Dortmund R. A. STEINBRECHT
July 1987 K. ZIEROLD

Contents

Part I Fundamentals

Chapter 1 **Physics of Water and Ice: Implications for Cryofixation**

L. BACHMANN and E. MAYER (With 18 Figures) 3

1 Introduction 3
2 Functions of Water in Cryospecimens 4
3 Water Below Room Temperature 9
4 Aqueous Solutions Below Room Temperature 17
5 Specimen Cooling 24
References 31

Chapter 2 **The Response of Biological Macromolecules and Supramolecular Structures to the Physics of Specimen Cryopreparation**

E. KELLENBERGER (With 9 Figures) 35

1 Introduction 35
2 The Hydration Shells of Biological Macromolecules and Supramolecular Structures 35
3 Phases and Interface Phenomena 47
4 The Particular Case of the Biological Membrane 56
5 A Potpourri of Collapses 57
References 60

Chapter 3 **Electron Beam Radiation Damage to Organic and Biological Cryospecimens**

Y. TALMON (With 7 Figures) 64

1 Introduction 64
2 Electron Beam/Specimen Interaction 65
3 Radiation Damage to Organic Materials at Low Temperature 67
4 Radiation Damage to Ice 69

5 Radiation Damage to Frozen-Hydrated and Vitrified-Hydrated Specimens 74
6 Conclusions 81
References 82

Part II General Methodology

Chapter 4 **Cryofixation Without Pretreatment at Ambient Pressure**
H. SITTE, L. EDELMANN, and K. NEUMANN (With 9 Figures) 87

1 Introduction 87
2 Cryofixation with Liquid Cryogen 88
3 Impact Cryofixation (Slamming) 95
4 Discussion 105
References 110

Chapter 5 **Cryoelectron Microscopy of Vitrified Specimens**
J. DUBOCHET, M. ADRIAN, J.-J. CHANG, J. LEPAULT, and A.W. MCDOWALL (With 3 Figures) 114

1 Introduction 114
2 Vitrification 116
3 Preparation of Thin Specimens 120
4 Image Formation 124
5 Beam Damage 126
References 128

Chapter 6 **Cryoultramicrotomy**
K. ZIEROLD (With 19 Figures) 132

1 Introduction 132
2 Technical Aspects of Cryoultramicrotomy 133
3 Physical Aspects of Cryoultramicrotomy 140
4 Conclusions 145
References 146

Chapter 7 **Freeze-Substitution and Freeze-Drying**
R.A. STEINBRECHT and M. MÜLLER (With 6 Figures) 149

1 Introduction 149
2 Methodology: Theoretical and Experimental Data 150
3 Procedures and Instrumentation 155
4 Critical Evaluation 160
References 168

Part III Special Aspects

Chapter 8 **Theory and Practice of High Pressure Freezing**
H. MOOR (With 6 Figures) 175

1 Introduction 175
2 Freezing Under Atmospheric Pressure 176
3 Freezing Under High Pressure 179
4 The Main Practical Problems of Pressure-Freezing 181
5 The High Pressure Freezing Machine (Balzers HPM 010) 182
6 Some Practical Advice 187
7 Discussion of Results 189
References 190

Chapter 9 **Freeze-Etching of Dispersions, Emulsions and Macromolecular Solutions of Biological Interest**
L. BACHMANN (With 6 Figures) 192

1 Introduction 192
2 Specific Problems of Specimen Preparation 192
3 Determination of Particle Concentrations and Molecular Weights 196
4 Measurements of Size and Shape 198
5 Structure of Dispersions in Bulk and at Interfaces 201
References 202

Chapter 10 **High Resolution Metal Replication of Freeze-Dried Specimens**
H. GROSS (With 5 Figures) 205

1 Introduction 205
2 Electron Microscopy and Image Processing 205
3 Characterization of the Test Specimens 206
4 Controlled Freeze-Drying 207
5 High Resolution Shadowing 210
References 215

Chapter 11 **Immunogold Labelling of Cryosections and Cryofractures**
J. BOONSTRA, P. VAN MAURIK, and A. J. VERKLEIJ (With 6 Figures) 216

1 Introduction 216
2 Cryoultramicrotomy 217
3 Cryofractures 220

4 Label Efficiency 225
5 Conclusions 228
References 229

Chapter 12 **Cryoultramicrotomy for Autoradiography and Enzyme Cytochemistry**
P. M. FREDERIK (With 1 Figure) 231

1 Introduction 231
2 Cryoultramicrotomy for the Autoradiography of Diffusible Substances 232
3 Thin Cryosections in Histochemistry 235
4 Conclusions 237
References 238

Chapter 13 **Scanning Electron Microscopy and X-Ray Microanalysis of Frozen-Hydrated Bulk Samples**
A. T. MARSHALL (With 8 Figures) 240

1 Introduction 240
2 Morphology 240
3 Analysis 243
4 Freezing 243
5 Instrumentation 244
6 Electron Interactions 245
7 X-Ray Emission 250
8 The Fracture Surface 250
9 Quantitative Analysis 251
10 Water Content or Dry Weight Fraction 252
11 Beam Damage and Mass Loss 253
12 Detection Limits 254
13 Conclusions 254
References 254

Chapter 14 **Cryofixation of Dynamic Processes in Cells and Organelles**
G. KNOLL, A. J. VERKLEIJ, and H. PLATTNER (With 3 Figures) 258

1 Introduction 258
2 Cooling Rate, Freezing Time, Time Resolution 259
3 Cellular Structures: Rapid Freezing Versus Chemical Fixation 260
4 Membrane Dynamics 263
5 Conclusions and Outlook 267
References 268

Chapter 15 **Cryofixation of Diffusible Elements in Cells and Tissues for Electron Probe Microanalysis**
K. ZIEROLD and R.A. STEINBRECHT (With 1 Figure) ... 272

1 Introduction ... 272
2 Specimen Preparation ... 273
3 Some Biological Applications and Results ... 278
4 Conclusion and Outlook ... 279
References ... 279

Part IV **Appendix**

Chapter 16 **Safety Rules for Cryopreparation**
H. SITTE, K. NEUMANN, and L. EDELMANN (With 1 Figure) ... 285

1 Introduction ... 285
2 Asphyxiation with Gaseous Nitrogen ... 285
3 Gaseous Propane Explosions ... 286
4 Burns Caused by Secondary Cryogen Splashing ... 286
5 Burns Caused by Primary Cryogen Splashing ... 287
6 Ignition of Combustible Secondary Cryogens ... 288
7 Bursting of Cryogen Containers ... 288
8 Transport and Disposal of Cryogens ... 288
9 Concluding Remarks ... 288
References ... 289

Glossary ... 291

Subject Index ... 293

Contributors

You will find the addresses at the beginning of the respective contribution

ADRIAN, M. 114
BACHMANN, L. 3, 192
BOONSTRA, J. 216
CHANG, J.-J. 114
DUBOCHET, J. 114
EDELMANN, L. 87, 285
FREDERIK, P.M. 231
GROSS, H. 205
KELLENBERGER, E. 35
KNOLL, G. 258
LEPAULT, J. 114
MARSHALL, A.T. 240
MAURIK, P. VAN 216
MAYER, E. 3
McDOWALL, A.W. 114
MOOR, H. 175
MÜLLER, M. 149
NEUMANN, K. 87, 285
PLATTNER, H. 258
SITTE, H. 87, 285
STEINBRECHT, R.A. 149, 272
TALMON, Y. 64
VERKLEIJ, A.J. 216, 258
ZIEROLD, K. 132, 272

Abbreviations

BE	backscattered electrons
BSA	bovine serum albumin
CPD	critical point drying
CURL	compartment of uncoupling receptor and ligand
DMEM	Dulbecco's modified Eagle's medium
DMSO	dimethylsulfoxide
DNA	desoxyribonucleic acid
DSC	differential scanning calorimetry
EELS	electron energy loss spectroscopy (-metry)
EF-face	exoplasmic fracture face
EGF	epidermal growth factor
EM	electron microscopy (microscope)
EPXMA	electron probe X-ray microanalysis
ESR	electron spin resonance
FD	freeze-drying
FS	freeze-substitution
$G_{(cryst, liquid)}$	free energy in the crystalline, liquid state
GN_2	gaseous nitrogen
HPI-layer	hexagonally packed intermediate layer
$I_{h, c, v}$	hexagonal, cubic, vitreous forms of ice
$I_{vl, vh}$	low density, high density form of vitreous ice
IMP	intramembrane particle
LC-display	liquid crystal display
LHC	light-harvesting chlorophyll complex
LHe(HeII)	liquid helium (helium II)
LN_2	liquid nitrogen
NMR	nuclear magnetic resonance
Nu	Nusselt number
PBG	PBS containing BSA and gelatine
PBS	phosphate buffered saline
PF-face	protoplasmic fracture face
PLT-technique	dehydration at progressively lowered temperature

PMMA	poly(methylmethacrylate)
PS	polystyrene
PVC	poly(vinylchloride)
RNA	ribonucleic acid
SDS	sodiumdodecylsulfate
SE	secondary electrons
SEM	scanning electron microscopy (microscope)
STEM	scanning transmission electron microscopy (microscope)
TEM	transmission electron microscopy (microscope)
T_b	temperature of boiling
T_d	temperature of devitrification
T_g	glass transition temperature
T_h	homogeneous nucleation temperature
T_m	temperature of melting
T_r	temperature region of recrystallization
TMV	tobacco mosaic virus
UHV	ultra high vacuum
ZAF	atomic number/absorption/fluorescence
Z-contrast	contrast depending on atomic number

Part I Fundamentals

Chapter 1 Physics of Water and Ice: Implications for Cryofixation

LUIS BACHMANN[1] and ERWIN MAYER[2]

1 Introduction

Cryotechniques in biological electron microscopy have one aspect in common: the use of low temperatures to stabilize or "fix" ultrastructure as it exists under physiological conditions. Ideal fixation and specimen preparation require that the constituents of the system keep their position within a range which is smaller than the resolution of the observation technique. The molecules and ions within a biological specimen interact in very complex ways. The kind of interaction depends strongly on temperature. What type of arrangement or structure is stable at a given temperature is determined by the laws of thermodynamics which state that any system tends towards a state of minimal free energy. The rate at which a system follows this tendency is determined by kinetics. The rearrangement or movement of molecules which is necessary to adapt a system to a change in temperature is usually an activated process, meaning that the molecules involved need excess energy (activation energy) in order to be able to change their positions. This "activation energy" is released after the event has taken place. Molecules which are not at the activated energy level will not react, although the reaction would result in a gain of stability. The pool out of which the activation energy is taken is the thermal energy of the system. Thus, the probability that a molecule is activated in a given time increases with temperature. As a result, a change of temperature has a dual influence on a specimen: it determines the equilibrium or stable structure at the new temperature and also the speed or rate at which the system adjusts to the new equilibrium state.

In the specific case of cryofixation, cooling of the specimen gives rise to ultrastructural changes by destabilizing the original structure in the sense of thermodynamics. At the same time, cooling tends to slow down any reaction rates which might lead to artefication, and increases the lifetime of the original structure by depriving the specimen of the activation energy. Once water is cooled to 100 K we believe that within practical time scales structural changes have come to a stop. For the goals of practical cryoelectron microscopy we consider 130 K as a "safe" temperature for specimen handling. For extended storage the use of

1 Institut für Technische Chemie, Technische Universität München, D-8000 München, FRG
2 Institut für Anorganische und Analytische Chemie, Universität Innsbruck, A-6020 Innsbruck, Austria

Cryotechniques in Biological Electron Microscopy
Edited by R. A. Steinbrecht and K. Zierold

liquid nitrogen is common practice. Thus, the "frozen-in state" can have practically an infinite lifetime unless activation energy is provided by further specimen preparation, e.g. by fracturing, vacuum coating or irradiation with the electron beam. The effect of these non-thermal activation energies cannot be excluded even at the lowest temperatures. The formation of cryoartefacts is limited to the time in which the specimen is between the original, e.g. the physiological, and the "safe" temperature. The extent of artefact formation is, therefore, the result of two kinds of processes which take place simultaneously during cooling: the extraction of thermal energy from the specimen and the formation of artefacts, i.e. the adjustment of specimen structure to the changing temperature. Since none of these processes occur "instantaneously", the degree of ultrastructural changes is simply the result of which of these two competing processes is faster: immobilization or reconformation. Heat dissipation and artefact formation are both complex reactions consisting of a series of steps, their thermodynamic driving force as well as their speed depending on the local composition and temperature.

In the following it is shown why water plays a unique part in cryofixation and some basic properties of water and aqueous solutions as a function of temperature are discussed. Preference is given to aspects of cryoelectron microscopy over phenomena which are important in cryobiology. Cryoelectron microscopy is much more connected to chemical kinetics than cryobiology; the latter aims at the preservation of life at low temperature. Cryofixation takes place within a fraction of seconds or even milliseconds. As in any reaction which is composed of several steps the overall speed of both, the favourable process (immobilization) and the unfavourable process (structural adaptation to lower temperature) is determined by their slowest elementary reaction. A discussion of the individual reaction steps is necessary for a basic understanding of what can be expected from cryofixation, and for establishing criteria by which cryofixation can be judged. A discussion of the mechanism should also be helpful in selecting the proper technique and in finding the most promising ways of improving it for specific applications.

2 Functions of Water in Cryospecimens

All constituents of biological systems change their interrelation when the temperature is changed, they are therefore, in principle, all subject to cryoartefact formation. There are, however, reasons why water justifies special attention when biological cryotechniques are discussed. For one, water is the major component in biological specimens, usually by more than 50%. In addition, water has some unusual properties which have made it one of the most investigated liquids whose structure is still not fully understood. By means of hydrogen bonding the water molecules interact with each other and with almost all molecules of biological interest. The hydrogen bond energy contributes substantially to solvation enthalpies. Liquid water is the most mobile component in biological systems. The main task of cryofixation is, therefore, to keep the water at the same place it occupied at room temperature. This is impossible for the individual water molecule. However, there is a fair chance to keep the water in a volume element smaller than the resolution of the EM. Water belongs to the compounds which are most difficult to transfer from the liquid into a non-crystalline solid state. The formation of ice is the most important single cause which leads to cryoartefacts: the solvent/solute segregation by the exclusion of dissolved and/or dispersed substances from the forming ice crystals. Therefore, the susceptibility to cryoartefact formation runs parallel to the water content of the specimens and their structural elements respectively.

2.1 Water – The Major Reservoir of Thermal Energy

Water is the main reservoir of thermal energy within a biological specimen since it is its main constituent and also because of its large heat capacity and heat of fusion. The difference between the enthalpy of water at 293 K and ice at 130 K is 685 J g^{-1}. The heat of fusion (334 J g^{-1} at 273 K) contributes about one-half of this value. The temperature range between 240 K and 190 K is probably the most decisive one for the formation of ice during fast cryofixation because of the temperature dependence of crystal growth. The crystallization enthalpy represents 70% of the total difference in heat content in this temperature range.

Frozen biological specimens contain a considerable amount of non-frozen water as will be shown later on. The amount and distribution of this non-crystalline water within a cell or tissue depends on the local chemical composition and on the cooling rate. From this it follows that one cannot predict the amount of thermal energy which has to be dissipated during the most critical phase of cryofixation. Only thermoanalysis of various biological samples, frozen at well-defined cooling conditions, could give us this value. The uncertainty regarding the enthalpy difference is enhanced by the irregular specific heat/temperature function of supercooled water and dilute solutions (see Sect. 3.3.1). Altogether the uncertainties regarding the enthalpy difference, i.e. the amount of thermal energy which has to be removed, and the anomalous relaxation behaviour of supercooled water and solutions make precise calculations of cooling rates and phase transitions under the conditions of fast cryofixation impossible. At present, such predictions cannot even be made for pure water let alone for biological specimens.

2.2 Water as Medium for Heat Transfer

Heat transport within a specimen occurs by conductivity only, convection is of no importance even in droplets of dilute dispersions or solutions. The thermal conductivity of ice decreases with increasing temperature, while water in the liquid, supercooled or vitrified state shows the opposite temperature dependence (Fig. 1). Porosity in quenched samples might also be of importance by reducing the heat conductance.

The heat conductance of unfrozen muscle tissue is approximately 80% of that of water, thus essentially corresponding to its water content. In frozen muscle tissue a mean heat conductance of 0.014 W cm^{-1} K^{-1} was observed at 248–263 K which is about 60% of the ice value (Bowman et al. 1975; Jones 1984).

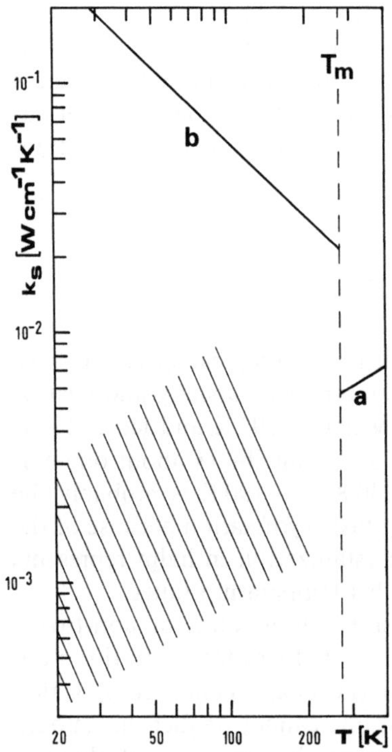

Fig. 1. The thermal conductivity, k_s, as a function of temperature: *a* liquid water, *b* hexagonal ice (according to Klinger 1980). The *shaded area* covers k_s values of amorphous dielectric substances. According to the theory, the values for vitrified water and solutions are also in this regime (Pohl 1975)

2.3 Water as the Dispersing and Embedding Medium

Ice formation causes a redistribution of dispersed material. Even 100-nm-sized particles can be moved hundreds and thousands of nm by the growing ice front. Since the viscosity of the water increases rapidly at lower temperature (Fig. 2a), the rate of displacement of the dispersed material in the water is greatly reduced with the degree of supercooling. This is discussed in more detail in Sections 4 and 5.

In cryosectioning and to a lesser degree in freeze-fracturing the hardness and brittleness of solidified water can be an important property. It has frequently been suggested that "vitrified" specimens cut easier than ice-containing samples. In our laboratories we found no different behaviour when 5% glycerol, splat-frozen in an X-ray-amorphous state, was fractured at 130 K, i.e. below the crystallization temperature, or at 170 K, i.e. after crystallization to cubic ice. This observation, together with our experience on freeze-fractured, spray-frozen (crystalline) solutions, indicates that the size of the pure ice crystals, the amount of dispersed material and the dimensions of the quenched droplets influence the fracture behaviour more than the mere presence or absence of crystalline ice (see also Zierold, Chap. 6, this Vol.). The observation of McDowall et al. (1983) on the ease of sectioning vapour-deposited, amorphous solid water may also be interpreted by the high porosity of this material (Ghormley 1967; Mayer and Pletzer 1986).

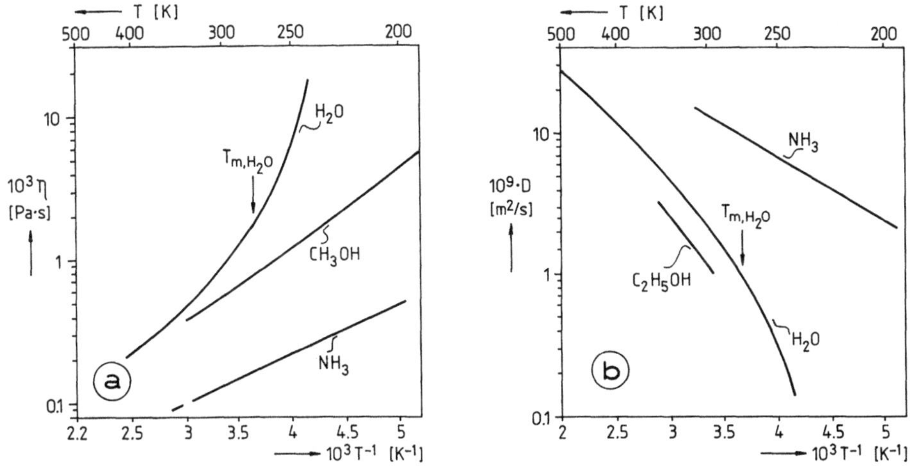

Fig. 2a, b. Temperature dependence of **a** the shear viscosity (η) and **b** the self-diffusion coefficient (D) at saturation pressure; unlike most liquids, water shows a non-linear behaviour which is even more pronounced in the supercooled state. (After Lang and Lüdemann 1982)

Vapour pressure and evaporation rate of the solidified water are the decisive properties for the question whether or not vitrified dilute solutions can be freeze-etched without devitrification. Dilute solutions when frozen in an amorphous state, e.g. by splat-quenching, start to crystallize in the same temperature region needed for a "limited" sublimation of ice. At 153 K the sublimation rate of pure ice is approximately 2×10^{-2} nm s^{-1} (Fig. 3). In conventional freeze-etching, this would be about the lowest feasible temperature for removing a 5- to 10-nm-thick matrix layer, since lowering the specimen temperature by 10 K results in a ten times lower sublimation rate. The vapour pressure of vitreous water must be

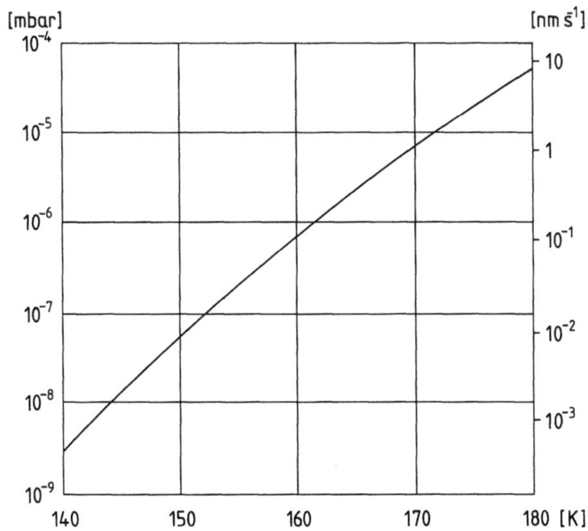

Fig. 3. Temperature dependence of the vapour pressure of hexagonal ice and the corresponding maximal sublimation rates calculated with the Knudsen formula. The reported experimental sublimation rates are generally in fair agreement with an evaporation coefficient close to 1. (After Umrath 1983)

higher than that of ice which is the more stable configuration, but we do not know the absolute difference at this temperature (see Sect. 3.3.1). It is, therefore, presently undecided whether there is a chance of relief formation by etching vitreous dilute solutions in conventional freeze-etching units at a temperature and within a time where crystallization is still negligible.

2.4 Water as an Intrinsic Part of the Ultrastructure

In biological specimens water not only serves as the dispersion medium for the various components. Specific structures of macromolecules and membranes frequently can exist only by interaction with water molecules which therefore cannot be replaced by other substances or the vacuum without artefact formation (see Kellenberger, Chap. 2, this Vol.). One of the main advantages of some cryotechniques is the possibility of observing hydrated structures in the electron microscope.

The water molecules in the hydrated structures have different physical and chemical properties than bulk water, and respond differently to changes in temperature. In the literature, these water molecules are frequently referred to as bound, chemisorbed or surface-modified water to distinguish them from free or bulk water. As in chemisorption there are no distinct classes of binding; there is a gradual transition from very specific interactions to bulk properties. Because of the diversity of hydrated structures, the amount of bound water depends on the nature of the physical or chemical probe used. Recently, it has been argued not to use the expression "bound"water (Franks 1986). Indeed, this expression has frequently been used indiscriminately for non-frozen, unfreezable or vitrified water and also for water showing reduced evaporation rates. It is also true that in many of the discussed cases the reason for not forming ice crystals or for not evaporating is of kinetic rather than thermodynamic nature. Nevertheless, there exists unfreezable water in the sense that at no temperature a true ice-like structure is stable.

In the praxis of cryofixation it is not of great importance whether water has a reduced availability or mobility in the vicinity of macromolecules or membranes, in other words, whether the reason is of thermodynamic or kinetic nature. Phase transitions of water from the fluid to the vapour or ice phase in dilute, gelled, colloidal solution are, for instance, restrained both by chemisorption, i.e. by binding, and by slower transition rates due to geometrical hindrance. The changed freezing behaviour of bound water is discussed in Section 3.5. A consequence of reduced sublimation rates of surface-modified water, caused either by a reduced vapour pressure or by steric hindrance, must be mentioned here: when measurements of deep-etched structures (membranes, macromolecules) are attempted in freeze-etch replicas, they might appear "too large" because of the surface-bound water.

EM artefacts on hydrated ultrastructure, which are caused by the changed physical and chemical parameters of liquid water at low temperatures, have not been reported although they are expected to be of importance in cryobiology. Possible implications of the changed solvent properties of water in the supercooled state are discussed in Section 4.2.

3 Water Below Room Temperature

The ability of the water molecule to form four hydrogen bonds gives rise to a three-dimensional H-bonded network. In hexagonal ice (Fig. 4a), every oxygen atom is at the centre of a tetrahedron formed by four oxygen atoms, each about 0.276 nm away. The O−H···O hydrogen bond is linear, though asymmetrical. In the lattice, the smallest polygons are non-planar hexagonal rings of water molecules having either "chair" or "boat" form (Eisenberg and Kauzmann 1969; Kuhs and Lehmann 1986).

The structure of *liquid* water is still not fully understood. Diffraction evidence supports the persistence of a tetrahedral hydrogen-bond order above the melting point, but with substantial disorder (Narten and Levy 1972; Dore 1985). Computer simulations suggest that in addition to 6-membered rings, as in ice, other polygons, such as 5- and 7-membered rings, are present. In the latter, the H-bonds are bent, i.e. strained (Rahman and Stillinger 1973; Stillinger 1980). Figure 4b shows a picosecond snapshot of a structural element containing polygons of various sizes. The view of liquid water that emerges from these studies is that of a highly dynamic, random, three-dimensional network of hydrogen bonds. This network has a local preference for tetrahedral geometry, but it contains a large proportion of strained and broken bonds. These strained bonds appear to play a fundamental role in kinetic properties. The calculations give no indication of large, low-density clusters or "iceberg" structures.

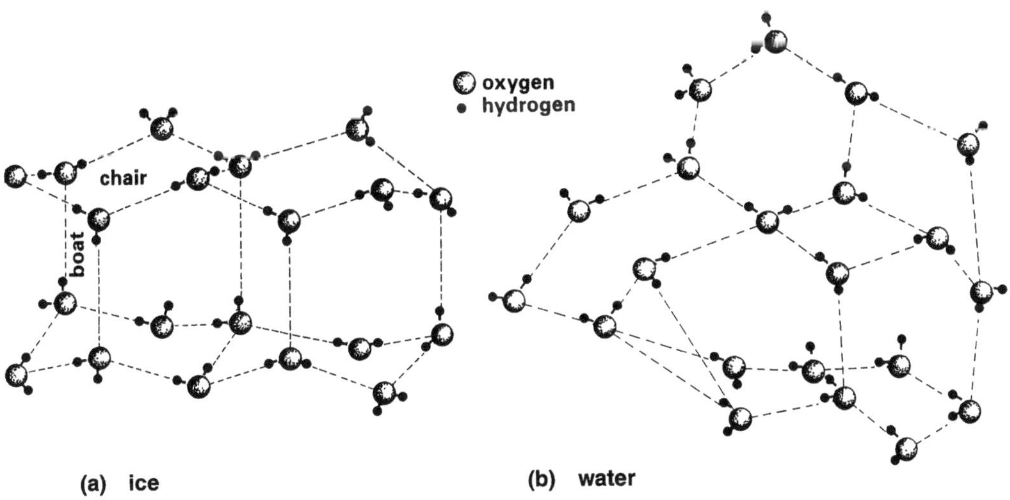

Fig. 4. a Lattice structure of hexagonal ice, showing the 6-membered rings of water molecules in "chair" or "boat" form. **b** A structural element in liquid water according to computer simulations: this picosecond snapshot shows in addition other polygons, e.g. pentagons, with non-linear hydrogen bonds. (After Angell 1982b)

3.1 Definitions

For an understanding of the complexities accompanying quenching of biological specimens, it is essential to discuss first the cooling of *pure* liquid water. The relevant region of the phase diagram is depicted in Fig. 5. The important temperatures a sample has to pass during a quench at normal pressure (1 bar) are: (1) T_m, the temperature of melting (273 K); (2) T_h (~232 K), the homogeneous nucleation temperature at which structural fluctuations *within* the liquid phase cause formation of "embryo" nuclei and subsequent crystallization. T_h is essentially kinetic in nature and has been found to be a function of cooling rate, higher rates yielding lower T_h values (Wood and Walton 1970; MacKenzie 1977; Katz and Donohue 1979; MacFarlane et al. 1983a). With increasing pressures both T_m and T_h are lowered, the overall appearance of the T_m/p and T_h/p curves being surprisingly similar. The application of high pressure in cryofixation is covered by Moor (Chap. 8, this Vol.). (3) A third important temperature is the glass transition temperature, T_g. It is obtained by risky extrapolations from concentrated binary solutions and therefore somewhat doubtful (Angell and Tucker 1980, Angell 1982a: approx. 140 K). T_g represents the temperature *region* at which the supercooled, albeit very viscous, liquid transforms into a rigid non-crystalline solid. It also depends on the rate of cooling and is therefore kinetic in origin (Kauzmann 1948; Angell 1970; Johari 1974). This can be explained best by considering the time scales of the relaxation processes (τ_{in}, see also Sect. 3.2) that occur as a liquid supercools, and their relation to the time scale of the experiment used to observe the glass transition. The configurational changes connected with relaxation of the supercooled liquid become increasingly slow with decreasing temperature until, at a given temperature (i.e. T_g), the material behaves as a solid. When the times of observation, t_{obs}, are long compared with the structural relaxation time, τ_{in} (i.e. $t_{obs} > \tau_{in}$) the material appears "liquid-like", whereas for $t_{obs} < \tau_{in}$ the

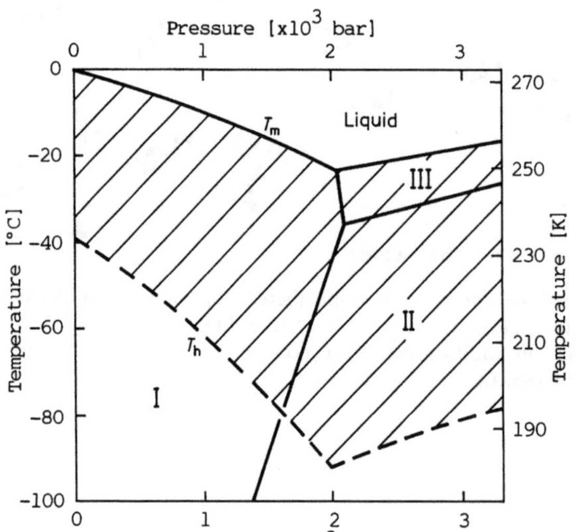

Fig. 5. Part of the pressure/temperature phase diagram for water, showing melting (T_m) and homogeneous nucleation (T_h) temperatures as a function of pressure. *Roman numerals* indicate the different stable ice polymorphs. The *cross-hatched* area shows the region within which supercooled water can exist. (According to Robards and Sleytr 1985; based on Kanno et al. 1975)

material behaves "solid-like". A "transition" will appear to have taken place if values of liquid-like parameters differ significantly from solid-like ones, as is the case for the heat capacity. The glass transition is observed therefore at T_g when $t_{obs} \simeq \tau_{in}$, viewed in terms of relaxation (Elliott 1983; p. 31). For cooling or heating rates of $10\,K\,min^{-1}$, often used in differential scanning calorimetry (DSC) and in other techniques, τ_{in} reaches about 10^3 s at the observed T_g (Angell et al. 1985). The corresponding viscosity at T_g is about 10^{12} Pa s (10^{13} poise). This implies that only at temperatures considerably below T_g the molecular mobility is slowed down sufficiently such that even over longer periods of time (years) structural changes are unlikely to occur.

Liquid water below 273 K (1 bar) is said to be supercooled (or undercooled); it is metastable with respect to hexagonal ice which at normal pressure is the stable crystalline modification. Cubic ice is also metastable with respect to hexagonal ice; it is often the product of devitrification of water-containing systems.

Freezing of liquids generally takes place by a two-step mechanism of nucleation and consecutive growth. If crystal growth is triggered by foreign particles or surfaces, it is called *heterogeneous* nucleation. If foreign particles do not play a role in the nucleation, the ice phase must be initiated by water molecules combining together to form an "embryo" which can grow spontaneously; the ice formation is then said to start by *homogeneous*, or *spontaneous* nucleation (for review, see Hobbs 1974). At every temperature crystallization induced by heterogeneous nucleation is faster than by homogeneous nucleation; in other words, only the absence of heterogeneous nucleation makes homogeneous nucleation effective. Due to heterogeneous nucleation it is not possible to cool *large* samples of liquid water far below T_m without initiating freezing. Even the most elaborate cleaning procedures do not eliminate all the foreign particles and therefore the onset of freezing is lowered only by a few more degrees. However, it is quite easy to supercool liquid water close to T_h and to avoid heterogeneous nucleation if it is in the form of small droplets, at most several µm in diameter, because the foreign particles are then confined to a few droplets only. Droplets of this size can be prepared either in water/oil emulsions (Rasmussen and MacKenzie 1973) or in the form of aerosols. Especially water/oil emulsions have been used to investigate the anomalous properties of supercooled water at leisure (Angell 1982a, 1983). However, further cooling below T_h requires extreme rates of cooling to avoid the growth of nuclei and the formation of crystalline ice.

3.2 Kinetics of Crystallization

Whether or not a liquid vitrifies and thus forms a glass is determined by crystallization kinetics (i.e. the nucleation frequency and the crystal growth rate) and the rate at which the liquid is cooled (Cohen and Turnbull 1959; Turnbull 1969). It is found that molecular liquids for which the reduced temperature $T_{boiling}/T_{melting}$ (T_b/T_m) has the value $\geq$ 2.0 will yield glasses under easily achieved quenching conditions (Cohen and Turnbull 1959; Angell et al. 1985; Angell and Choi 1986). For liquid water, the value is unusually low (1.37), and correspondingly, it belongs to the most difficult molecular liquids to vitrify.

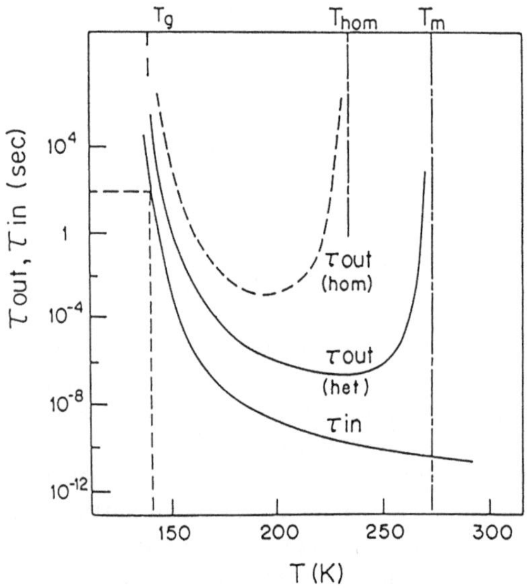

Fig. 6. Relationship between escape time, τ_{out}, and internal relaxation time, τ_{in}, for heterogeneously and homogeneously nucleating liquid systems; the high-temperature branch of τ_{out} is more nucleation controlled, the low-temperature branch reflects more the aspects of crystal growth. (After Angell and Choi 1986)

A better understanding and perception of the dynamics of the quenching process can be obtained by considering two "time scales" as shown in Fig. 6 (Uhlmann 1972, 1982; Angell et al. 1981; Angell and Choi 1986): (1) τ_{in} is the time scale for relaxation of the molecules *within* the supercooled liquid. To first approximation, τ_{in} scales with the viscosity, η (Turnbull 1969; Angell et al. 1985). The viscosities of most common liquids (e.g. water, ethanol) at room temperature are in the order of 10^{-3} Pa s (10^{-2} poise), corresponding to a τ_{in} in the order of 10^{-12} s. η and τ_{in} increase continuously with decreasing temperature. At T_g observed by DSC with 10 K min^{-1} cooling rate, η is about 10^{12} Pa s (10^{13} poise) and τ_{in} about 10^3 s. (2) The second time scale, called escape time, is τ_{out}; it is the time scale for crystallization of a given volume fraction of the liquid, e.g. 50%. τ_{out} reflects in a complicated combination the rates for nucleation and crystal growth. Starting from T_m, τ_{out} initially decreases with decreasing temperature as the thermodynamic driving force ($G_{cryst} - G_{liquid}$) builds up. The minimum or "nose" in the τ_{out} curve results from competition between this driving force and the molecular mobility, which decreases with decreasing temperature. The minimum ("τ_{nose}") corresponds to the *least time* for the given volume fraction to crystallize.

To avoid crystallization during quenching it is necessary to "bypass the nose", i.e. to use a cooling process which is fast with respect to the minimum of crystallization time (τ_{nose}). Other strategies to bypass τ_{nose} without crystallization, for example by addition of a cryoprotectant, are discussed in Section 4.3. In Fig. 6 τ_{out} is shown separately for heterogeneous and homogeneous nucleation, illustrating that heterogeneous nucleation must be avoided when aiming for vitrification. The simple strategy to avoid heterogeneous nuclei by dividing the sample into small compartments cannot be followed, however, in many biological specimens. The actual value of τ_{nose} for the crystallization of pure water is not known, although several estimates of the value for τ_{out} (homogeneous) have been made. They range from 10^{-5} to 10^{-10} s (Sargeant and Roy 1968; Turnbull 1969; Fletcher 1971; Uhlmann 1972). These values are the basis for estimates of cooling rates necessary for vitrification (see Sect. 3.4).

3.3 Supercooled Water

A complication in the discussion of nucleation and crystal growth for pure water is raised by anomalous variations in the physical properties of the supercooled liquid which enter the theoretical expressions. In careful studies of its properties, mainly by Angell and collaborators, it has been observed that virtually all properties measured at normal pressure appear headed for an anomaly at approximately 228 K. This temperature is just below the limit of attainable supercooling (about 232 K; for reviews see Angell 1982a, 1983; Lang and Lüdemann 1982). The strong implication is that these anomalies are caused by structural changes in the hydrogen-bond network of supercooled water within a small temperature range (Stillinger 1980). The general view seems to be that the anomalies of supercooled water have a similar origin as the well-known anomalies of the stable liquid above 273 K, being only more pronounced in the supercooled regime.

3.3.1 Thermodynamic Properties

Figure 7 shows the temperature dependence of the constant-pressure heat capacity, C_p, and of the density of supercooled water. Water is the only known liquid in which C_p raises with supercooling (Angell et al. 1973; Rasmussen and

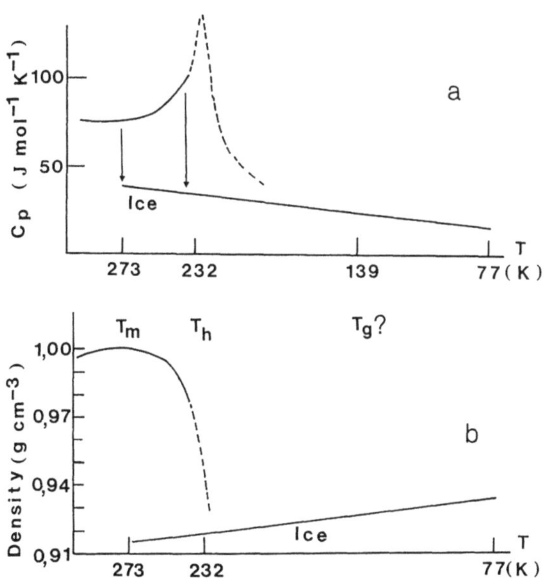

Fig. 7. a Heat capacity and **b** density of liquid water and ice as a function of temperature. *Solid lines* are experimental data, *broken lines* extrapolations, with the postulated λ-type transition. The *arrow at 273 K* indicates the phase transition at equilibrium. The *second arrow* indicates crystallization from the supercooled liquid. The *broken line* suggests one of the possible transitions to a vitrified state. (According to Angell and Tucker 1980; based on data from Angell et al. 1982; Hare and Sorensen 1986)

MacKenzie 1973). A λ-type transition in the C_p vs. T curve has been proposed as one of the interpretations of the anomalies of supercooled water (Speedy and Angell 1973). The density drops from its maximum value at 277 K (1 000 g cm^{-3}) to 0.978 g cm^{-3} at 239 K (Hare and Sorensen 1986). An extrapolation of the low-temperature density values suggests that it would fall to that of stable ice somewhat below T_h, the plunging decrease of density implying a very open structure of the deeply supercooled liquid. The vapour pressure of supercooled water was recently redetermined between 273 and 260 K. At 260 K it is only 0.25 mbar, i.e. 13% higher than that of stable ice (Kraus and Greer 1984). For possible consequences in freeze-etching, see Section 2.3.

3.3.2 Transport and Relaxation Properties

In Fig. 2. a, b the temperature dependencies of the viscosity, η, and the self-diffusion coefficient, D, are shown and compared with those of other hydrogen-bonded liquids. These transport properties govern the dynamics of quenching and the rate of displacements of molecules from their temporary positions at equilibrium. The strong temperature dependence of the viscosity in cold, stable water is even further amplified in the supercooled liquid. The activation energy of diffusion is 19 kJ mol^{-1} at 298 K, and approx. 46 kJ mol^{-1} at 242 K. The latter value already approaches the 60.7 kJ mol^{-1} reported for ice (Gillen et al. 1972). In liquid water near T_m, the molecules experience roughly 10^{11} or 10^{12} reorientational and translational movements per second, in ice at the same temperature only about 10^5 or 10^6. This is one of the chief differences between ice and water (Eisenberg and Kauzmann 1969, p. 205).

Deepest penetration of the supercooled regime down to 181 K was possible, using pressures up to 3 kbar. The observational technique was nuclear magnetic resonance, NMR (Lang and Lüdemann 1982; Lang et al. 1984b). The experiment gives the macroscopic spin-lattice relaxation time T_1, i.e. the time needed to achieve equilibrium along the magnetic field axis after a sudden increase of the magnetic field strength. From T_1 the microscopic orientational correlation time τ_c, which is roughly speaking the average period of time for which its orientation in space persists with little or no change, can be calculated. A decrease of T_1 indicates a decreasing mobility of the species investigated (Conway 1981, p. 191). It has been observed that for lower pressures ($p \leq 1.5$ kbar) τ_c behaves anomalously like the transport properties η and D, decreasing with falling temperature much faster than "normal" liquids and heading for a singularity at approx. 228 K. In contrast, the anomalies disappear at high pressures ($p \geq 2$ kbar) and liquid water behaves like a normal, viscous liquid. In addition, it was concluded from these NMR measurements that water near 193 K at 2 kbar is highly viscous, comparable with glycerol at room temperature (i.e. approx. 1500 times that of pure water; Lang and Lüdemann 1980). This high viscosity must have implications for cryofixation at high pressures (see Moor, Chap. 8, this Vol.).

The effect of solutes is similar to that of pressure: with increasing concentration the anomalies of supercooled water gradually disappear. However, at physiological concentrations the anomalies are on the whole still present (see Sect. 4).

3.4 Vitrification

In order to vitrify pure liquid water, the supercooled temperature region below T_h with its observed anomalies and the postulated structural transition has to be passed. In the C_p vs. T diagram (Fig. 7a) the broken line suggests one of the possible transitions to a vitrified state (Angell and Tucker 1980). Until now, four methods have been developed to vitrify μm-sized samples of pure liquid water: three of these need a liquid cryomedium such as liquid propane or ethane; the fourth method – rapid cooling of aqueous aerosol droplets on a solid cryoplate – is the only one which works without a liquid cryomedium for heat transfer (reviewed by Mayer 1985). The claims pertaining to the vitrification of pure water have been questioned with the argument that the quenched samples might contain some of the liquid cryomedium incorporated, forming something like an amorphous clathrate (Franks 1981; Angell 1983, p. 625). The vitrification of water droplets without liquid cryomedium should settle this point.

Direct EM observations on vitrified specimens are described by Dubochet et al. (Chap. 5, this Vol.). Aerosol droplets vitrified on a solid cryoplate have been investigated by freeze-etching (L. Bachmann and E. Mayer, unpubl.).

So far, the cooling rates necessary for vitrification have not been measured. They can only be extrapolated from existing techniques or calculated on a theoretical basis. Measured cooling rates of $10^4\,K\,s^{-1}$ on 200 μm-sized droplets resulted in segregation considerably larger than in spray-frozen, 20 μm-sized droplets. The rates in spray-freezing must therefore be at least in the order of $10^5\,K\,s^{-1}$ (Plattner and Bachmann 1982). From this we assume that vitrification may occur at cooling rates of 10^6 or $10^7\,K\,s^{-1}$. Estimates based on the theory of crystallization range from 10^7 to $10^{10}\,K\,s^{-1}$ for 1 μm layers or droplets (Fletcher 1971; Uhlmann 1972). For 10 μm layers or droplets the estimates are higher by two to three orders of magnitude. These discrepancies are not surprising considering the anomalies of supercooled water. For example, Angell and Choi (1986) have pointed out that the *rate of increase* of the thermodynamic driving force $G_{cryst} - G_{liquid}$ is abnormally small for water due to the exponential increase of heat capacity in the supercooled state.

The temperature of devitrification, i.e. of conversion to cubic ice, depends on the sample and on the measuring technique. DSC of optimally quenched aerosol droplets shows for a heating rate of $10\,K\,min^{-1}$ this process taking place with a peak maximum temperature of 163 K; the enthalpy of the phase transition is $-1.33 \pm 0.02\,kJ\,mol^{-1}$ (Hallbrucker and Mayer 1987). Vitrified specimens have been reported to devitrify in the cryoelectron microscope at approx. 135–140 K for longer observation times (Dubochet and MacDowell 1981; Dubochet and Lepault 1984), or at 150.8 ± 1.7 K for a heating rate of $12\,K\,min^{-1}$ (Dubochet et al. 1983). X-ray diffraction showed that the transition of cubic to hexagonal ice occurs at a very slow rate, beginning at approx. 200 K (Mayer and Hallbrucker 1987).

3.5 Non-Freezing Water at Membranes and Colloidal Interfaces

It is a matter of opinion whether to discuss water at interfaces in dispersed systems under the section of pure water or solutions. We treat it here, neglecting thereby low molecular weight components which in reality are of course also present at colloidal interfaces.

In colloidal solutions of proteins and polysaccharides and in biological tissues, a fraction of water amounting from about 30% to 60% of the dry weight remains unfrozen and mobile to temperatures below 200 K as indicated by the relatively sharp ^{1}H-NMR spectrum (Packer 1977; Derbyshire 1982). The general view is that on surfaces a mono- or bilayer of water is surface-modified in a thermodynamic sense and behaves substantially different from bulk water (Cooke and Kuntz 1974; Berendsen 1975; Conway 1981, p. 644; Finney 1986). This is in analogy to the Brunauer, Emmett and Teller multilayer adsorption isotherm (BET). Corresponding values were obtained using porous glasses as model surfaces (Belfort 1974; Brüggeller 1983). We deliberately avoid the discussion whether this fraction of water does not freeze due to thermodynamics or kinetics because there is obviously a gradual transition from "bound" water to bulk water (see Sect. 2.4).

The observation of non-freezing water at interfaces contributed to the more general discussion whether or not long-range effects on the dynamic and thermodynamic properties of water due to the presence of interfaces exist. However, the bulk of modern experimental evidence available is more consistent with short-range effects involving not more than one or two layers of water molecules. For recent discussions on hydration forces see Israelachvili (1985) and Rand et al. (1985).

Water near an interface with which it interacts will find itself in a state of dynamic anisotropy due to the influence of the interface on its tumbling motions,

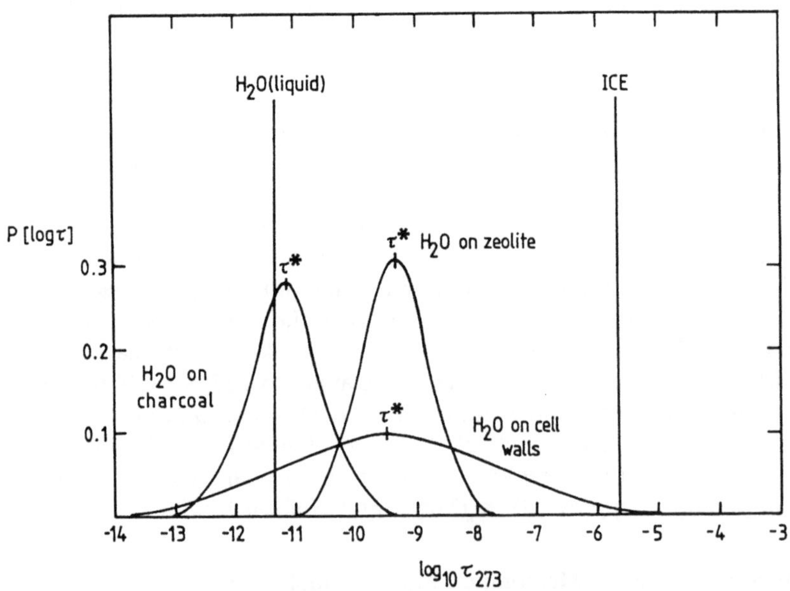

Fig. 8. Distribution functions of the jump time (τ) for bulk water, ice and for water adsorbed on various substrates at 273 K. The median jump time is marked as τ^*. (According to Resing 1972)

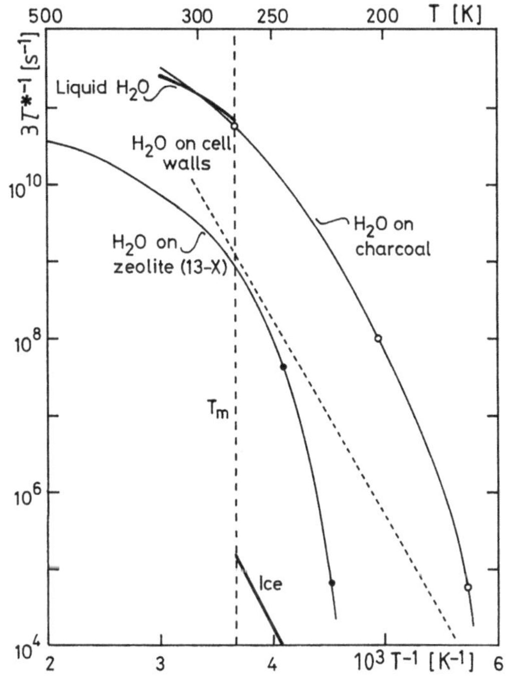

Fig. 9. Jump frequency, an inverse function of the medium jump time τ^*, vs inverse temperature for water, ice and water adsorbed on various substrates. Note the drastic difference between the values of ice and bulk water, and of ice and non-freezing, surface-modified water on zeolite. Measurements on several different cell wall-containing materials can be represented by the *dashed line*. (According to Resing 1972; the extrapolated curve for bulk water below T_m was omitted because the supercooled water anomaly has not been considered)

which will not be directionally random as in a bulk liquid phase. This is reflected in the NMR relaxation behaviour. To give some idea of the range of mobilities for water absorbed on various substrates, the distributions in the rates of motion, or jump time, and their dependence upon temperature are shown in Figs. 8 and 9 (Resing 1972). By these criteria the mobility of water on cell walls and on zeolite 13-X is more than 20 times lower than that of bulk water at room temperature. As Fig. 9 shows, the jump times of surface-modified water drop continuously from 300 to about 200 K without any evidence for a phase transition, i.e. for freezing. The mobility of water on cell walls varies over a wide range, reaching from the jump time of liquid water to that of ice (Fig. 8). Nevertheless, the median is more than an order of magnitude closer to the jump time of liquid water than that of ice. In conclusion, water on cell walls is not "ice-like" in terms of mobility. For a somewhat contrary view based on thermodynamic data, see Kellenberger (Chap. 2, this Vol.).

4 Aqueous Solutions Below Room Temperature

The complexity of quenching water is greatly enhanced by the presence of solutes. Artefact formation can be caused either by freeze-concentration of the solute due to crystallization of ice, or by changes in the dispersed phase structure during quenching, or by a combination of both. In the following, the properties of solutions in equilibrium are discussed first. Second, the properties of metastable, supercooled solutions and of vitrified solutions are discussed. Finally, the most complex and also the most general case will be treated which is the quenching of samples with concomitant non-equilibrium segregation and ice formation.

4.1 Equilibrium Properties of Solutions

Figure 10 shows a phase diagram of a simple eutectic binary system. Starting from the left side of the diagram, the melting point of the dilute solution decreases with increasing solute concentration until the eutectic temperature is reached: further cooling leads to segregation, i.e. formation of ice and the solid phase of the solute. The reason for a lowering of the melting temperature in solutions is that the free energy of liquid water is depressed when it is diluted with other molecules (Fig. 11a). The free energy of the ice is not changed because the solute is not incorporated in the ice lattice; therefore, in solutions the temperature at which the curves for the free energies of G_{ice} and G_{water} intersect, i.e. the melting point, is decreased. In general, the solubility limit of solutes in ice is less than 10^{-4} mol l^{-1}; only NH_4F, HF and HCl, the most soluble compounds, have higher solubility limits of 1×10^{-1}, 2×10^{-3} and 1×10^{-4} mol l^{-1} (Gross et al. 1974a, b). It is this property of the ice lattice, not to incorporate solute molecules, which is the basic cause of freezing artefacts. If the ice lattice would tend to form "solid solutions", one would not have to worry about segregation artefacts.

Attainment of equilibrium during cooling may not be easy, especially for the solute component. In concentrated solutions, e.g. of glycerol/water, it is practically impossible. Even very slow cooling of dilute solutions can lead to substantial supercooling. For rapidly quenched biological material, phase diagrams represent the conditions most remote from reality.

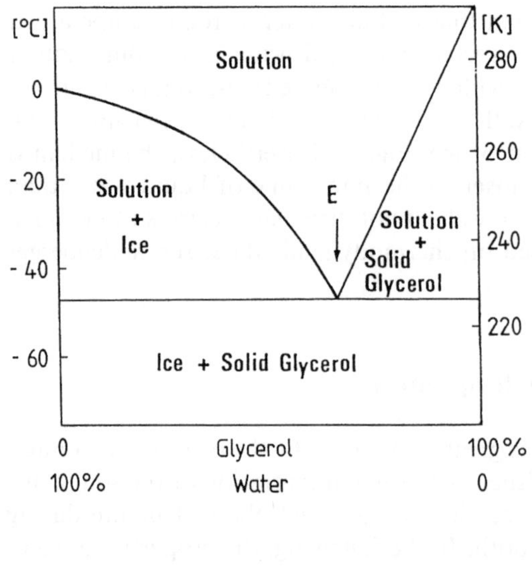

Fig. 10. Equilibrium phase diagram of glycerol-water; *E* eutectic point at approx. 67% glycerol and 227 K

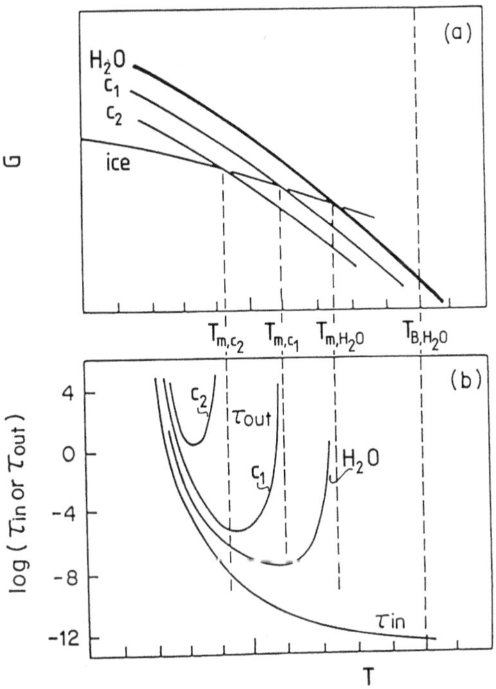

Fig. 11

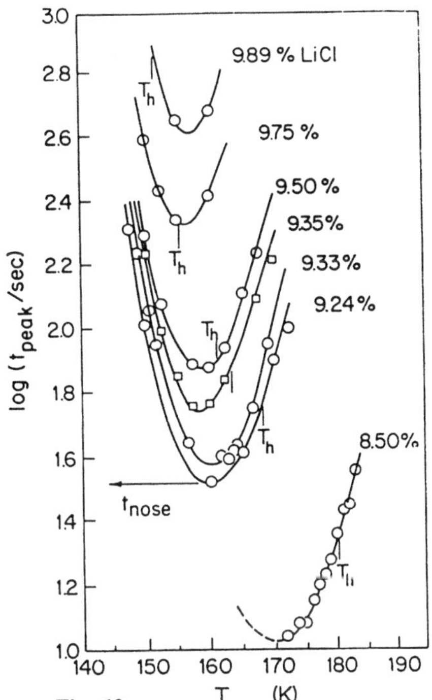

Fig. 12

Fig. 11. a Relations of free energy vs temperature, showing the effect of solutes (concentration $c_2 > c_1$). **b** Effect of solutes on τ_{out}. τ_{in} is assumed to be little influenced by concentration. The diagram explains the principle of enhanced glass-forming of solutions over the pure liquid. The *dashed lines* relate the thermodynamic data in **a** with kinetics in **b**. (According to Angell and Choi 1986)

Fig. 12. Measured τ_{out} curves for a series of LiCl+H_2O solutions showing the sensitivity of behaviour like that of Fig. 11b to increasing concentration of the second component (MacFarlane et al. 1983b). The plotted t_{peak} is a measure of τ_{out} as explained in the reference

4.2 Metastable Solutions

Concentrated aqueous solutions can be supercooled and vitrified rather easily with conventional cooling techniques. This is the basic idea for using cryoprotectants. Unfortunately, the concentrations generally needed for an effective cryoprotection are far from being physiological. Therefore, artefact formation can occur even without cooling. The emphasis here is on "dilute" aqueous solutions. A practical definition for "dilute" might be physiological concentrations or, that the usual standard methods of cryofixation will give solute/solvent segregation.

4.2.1 Thermodynamics of Supercooled Solutions

The major emphasis of recent work on supercooled solutions was to use solutes as probes to find out more about the anomalies of water. Figure 13 shows the heat

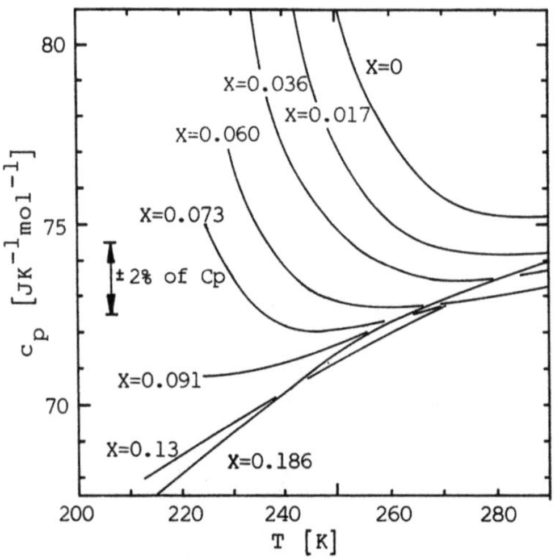

Fig. 13. Heat capacities of water and $H_2O+H_2O_2$ solutions as functions of temperature (X = mol fraction). (After Oguni and Angell 1980). We do not know of any comparable measurements on cryoprotective substances

capacities of H_2O/H_2O_2 solutions as a function of temperature (Oguni and Angell 1980). The anomalous, large heat capacity of supercooled pure liquid water decreases gradually with increasing H_2O_2 concentration until at approximately 30% "normal" behaviour is observed. The temperature of maximum density, one of the most characteristic anomalies of water, is also changed by solutes, this indicating structural changes (Sorensen 1983). A comparable behaviour is expected for physiologically important solutes.

The ionization constant of water, K_w, and its temperature dependence is a further property which might lead to artefacts during quenching of a specimen (Franks 1985, p. 38). K_w is known to decrease with decreasing temperature. An extrapolation of the values of the stable liquid leads to a pK_w of ~17 at 238 K. Therefore, it is to be expected that the dissociation constants of acids and bases used in buffer mixtures will also exhibit temperature dependencies which result in changes of pH. Although these pH changes cannot be avoided even at high cooling rates, they might lead to artefacts more in the field of cryobiology than in ultrastructural research.

Temperature effects on the solubility of non-polar solutes in water are to be expected: non-polar solutes dissolve in water with a *negative* entropy of dissolution, characterizing "hydrophobic bonding" or "hydrophobic interaction" (reviewed by Franks 1975; Conway 1981, p. 507). This negative entropy term leads to increasing solubility of the non-polar solute with decreasing temperatures. It has further been suggested that deeply supercooled water might be a good solvent for non-polar solutes due to the formation and clumping tendency of relatively strain-free, clathrate-like, convex cages and polyhedra, with a special preference for pentagons. The drastic increase in the number of such polyhedra near 228 K (i.e. the previously discussed temperature of the singularity) would provide many additional sites for non-polar solutes (Stillinger 1980; Stillinger and Weber 1983; Speedy 1984).

4.2.2 *Relaxation Properties*

The dynamics of water in supercooled salt solutions has also been investigated, mostly by NMR spectroscopy (Lang and Lüdemann 1982, 1985; Lang et al. 1984a). In one attempt to understand ionic hydration various ions have been classified either as "structure formers" or "structure breakers". A definition of these terms is that the mobility of water close to a structure-forming ion is decreased and vice versa (Hertz 1973; Conway 1981, p. 193). An indication for the mobility of water is obtained from T_1 measurements by NMR (see Sect. 3.3.2). The experiments show that at 283 K the mobility of water is lowered by structure formers (LiCl, NaCl, $MgCl_2$) at all pressures up to 2 kbar, whereas the mobility of water in 3 molal CsBr (a structure-breaking salt) is higher than in neat water at all pressures. The same ("normal") behaviour is observed in the supercooled state, at 238 K, at *high* pressures.

However, at *low* pressures at 238 K (e. g. 1 bar), which are the conditions of most interest, the ("anomalous") low mobility of neat supercooled water is still observable in dilute solution, i. e. in a 0.3 molal LiCl solution. The anomalies only disappear completely at high salt concentrations; the mobility of neat water is then lower than even in concentrated solutions of structure-forming salts (11 molal LiCl, 3 molal NaCl and $MgCl_2$).

In conclusion, these experiments show that in dilute solutions, e. g. a physiological NaCl solution, the anomalies of supercooled pure water are on the whole still preserved. They disappear only at high concentrations and/or high pressures. Therefore, these anomalies might contribute to the final immobilized state of a quenched sample. Even more important, they cannot be avoided simply by the most rapid cooling techniques because the mobility is still too high.

4.3 Complete Vitrification of Solutions

Complete vitrification means the absence of crystalline ice in a quenched specimen and has to be judged, e. g. by X-ray or electron diffraction. This is an essential point because even a dilute solution, cooled "slowly" by immersion in liquid nitrogen, normally contains some vitrified, highly concentrated solution. In this section we discuss only complete vitrification, while freeze-concentration is treated in Section 4.4.

The strategies for the vitrification of pure liquid water in terms of two competing time scales, τ_{in} and τ_{out}, have been discussed in Section 3.2 and are here extended to solutions, following thereby Angell and Choi (1986). Figure 11b shows how the addition of solute shifts the minimum (nose) of the τ_{out} curve towards lower temperatures and longer times. This slows down nucleation and growth of ice and thereby favours vitrification. This effect is a general one: probably most solutes could act as cryoprotectants. The enhanced chance to vitrify concentrated aqueous solutions is in accordance with the T_b/T_m rule (Sect. 3.2). A specific example for the dependence of the nose on concentration is shown in Figure 12 for concentrated LiCl solutions (MacFarlane et al. 1983b). Small changes in solute concentration have a strong influence on the position of the minimum. Comparable studies on "dilute" solutions have not been made.

Complete vitrification of dilute aqueous solutions has only recently been possible, using techniques as described in Section 3.4. In dilute aqueous solutions of alkali metal perchlorates, no interaction between cation and anion can be observed. However, in *vitrified*, dilute ($\geqslant$0.5 M), lithium and sodium perchlorate solutions strong perturbation of the anion by the cation was observed by infrared spectroscopy, and interpreted as formation of contact ion pairs, $M^+ \cdot ClO_4^-$, solvated by the water (Mayer 1986). In other words, ion association increases – at least in these systems – with decreasing temperature. This is contrary to Raman spectroscopic work where contact ion-pair formation was found to be favoured by high temperatures (Irish and Jarv 1977, 1983). The latter measurements, however, were made using equilibrium conditions, whereas for the vitrification of dilute solutions extreme non-equilibrium conditions are necessary. It is possible that rapid quenching of electrolyte solutions generally favours ion association in the vitrified state. This might be important for the cryofixation of biological polyions. It also shows that even complete vitrification does not preserve the room temperature structure of small, highly mobile solutes because ion-association processes generally are very fast (James 1985). However, the solute remains on the whole distributed randomly in the vitrified solution. For cryofixation this is the essential issue.

4.4 Freeze-Concentration

A concentration of the dissolved and dispersed material must occur in biological specimens during freezing whenever ice is formed. A "freeze-concentration factor", which is the ratio of the concentrations after and before freezing, is frequently used to quantify artefactive changes during cryopreservation (Franks 1985, p. 58). If the final concentration of the vitrified solution is assumed to be independent of its initial concentration, freeze-concentration factors will be much larger for dilute than for concentrated solutions. They are accompanied by drastic local changes, e. g. in osmolality, pH or possibly of enzyme activities.

The freeze-concentration factor does not take into account the size of the segregated zones ("segregation compartments"), but is only a measure for the total volume of ice. In other words, a high-pressure or spray-frozen sample with approximately 5.0 -nm ice compartments may be completely phase-separated, i. e. has the maximum freeze-concentration factor, but the size of the ice crystals does not interfere with EM observation. Freeze-concentration factors may therefore be indicative in connection with cryobiology, but hardly for EM cryofixation. The critical factor for cryofixation is the size and the distribution of the segregation compartments. The size of these segregation compartments may be very different from the crystal size determined by diffraction if these compartments are or have been interconnected and show identical orientation. The mode of growth, e. g. whether it is dendritic or more compact, is not determined by the thermodynamic driving force, but by the very complex kinetics of crystal growth. Thus, it is impossible to calculate the size and shape of the ice compartments, the rate of crystal growth or the local variations of the freeze-concentration factor. Only on specimens of exactly the same composition, e. g. solutions or well-defined biological specimens, can the size of the segregation compartments in the partially segregated state be used for judging relative cooling rates.

Near T_m the growth rate of ice in solutions was observed to be unaffected for solute concentrations $\leqslant 10^{-2}$ mol l^{-1}, and progessively retarded for higher concentrations (Pruppacher and Klett 1980, p. 558). In a simplified way it has been further assumed that the growth of ice crystals stops at a given concentration of the solution which is in contact with the growing ice front.

For evaluating the final quenched state, a so-called supplemented phase diagram which shows transition temperatures based on kinetics can be of help (Fig. 14): in addition to the melting temperature, the diagram contains the temperatures of glass transition, devitrification, recrystallization and of homogeneous

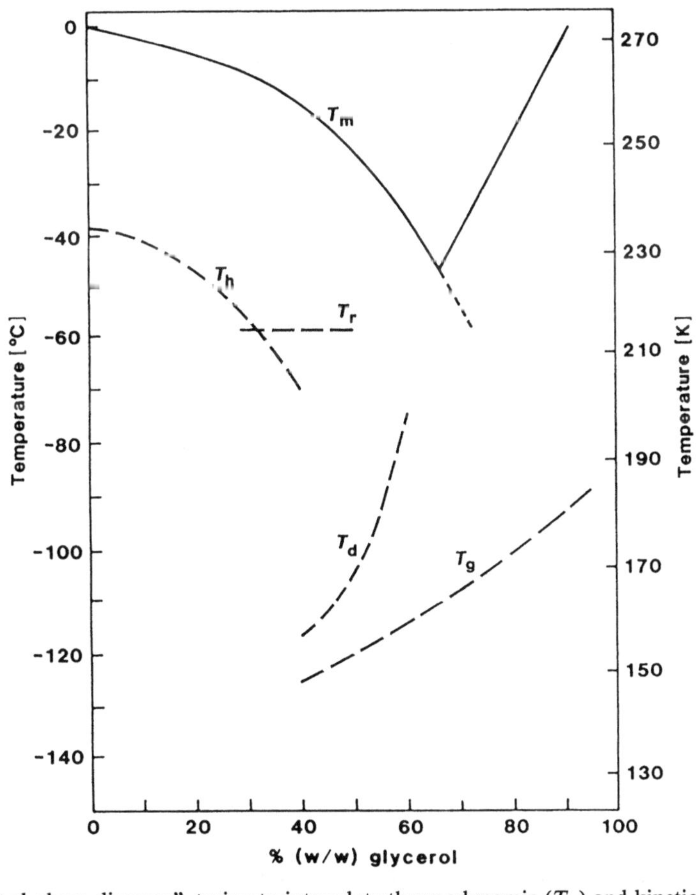

Fig. 14. "Supplemented phase diagram", trying to interrelate thermodynamic (T_m) and kinetic quantities, such as T_g, glass transition temperature, T_h, homogeneous nucleation temperature, T_d, devitrification temperature, and T_r, the temperature region of recrystallization, as a function of concentration. *Solid lines* are equilibrium data, *broken lines* are based on kinetics. (According to Robards and Sleytr 1985; based on data from Luyet and Rasmussen 1968)

nucleation as a function of concentration (Luyet and Rasmussen 1968; MacKenzie 1977). For several of the popular cryoprotectants T_g and T_d are concentration-dependent; therefore, it is possible to determine the amount of freeze-concentration by thermal analysis (Boutron 1984). Unfortunately, solutions with low initial concentration, which for biological work are the most interesting ones, are least sensitive to this procedure. It should be noted that there is not necessarily a change in crystal structure during recrystallization: the driving force may arise from the difference in free energy between the strain-free and the distorted states.

5 Specimen Cooling

Cooling rates within the specimen have been subject to measurements, estimates and speculations ever since cryofixation was introduced to electron microscopy. Recent reviews of the literature were given by Plattner and Bachmann (1982) and Robards and Sleytr (1985). In the following, only the possibilities and limitations of cooling speed, which are determined by the properties of aqueous specimens, shall be discussed.

5.1 Mechanism of Heat Dissipation

The thermal energy of the specimen is extracted into the quenching medium through their area of contact. Within the specimen and within *solid* heat sinks, heat transport occurs by conductance only while forced convection is an additional, in most cases, the crucial way of energy transport within *liquid* cooling media.

For convenience, the heat flow can be subdivided into the transport within the specimen, within the heat sink and the transfer between the two. The intention of the different cooling techniques is to ensure optimal or "ideal" cooling of a specimen of given size and geometry. This optimum is obtained when there is "perfect thermal contact" between specimen and substrate; in other words, when the surface of the specimen attains essentially the same temperature as the bulk of the heat sink. This can only be obtained when the conduction of heat within the quenching medium and the transfer through the contact area occur much faster than the heat conduction within the specimen and are thus not rate-limiting.

This relation between factors inside and outside the specimen can be mathematically expressed by the quantity $Nu = \frac{h \cdot d}{k_s}$ (Nusselt or Biot number), where h is the effective heat transfer coefficient, d is the effective dimension (thickness) and k_s is the heat conductance of the specimen. If $Nu \geqslant 30$, cooling is essentially "ideal", that means that the cooling rate cannot be increased any more, i. e. further technical improvements, for example by using heat sinks with still higher thermal heat conductance, can in this case have no consequence for the specimen. In other words, the cooling rates are determined by the object. It must be emphasized that ideal cooling as defined above does not necessarily imply high cooling rates within the specimen, because Nu also contains the dimension of the specimen. If Nu $\leqslant 0.1$, the cooling rate is limited by factors outside of the specimen, which means that the cooling rate can still be increased using better techniques of heat extraction. Between these extremes is the regime of intermediate cooling.

Freezing tissue by pushing it against a cold metal (van Harreveld and Crowell 1964) and spraying of suspensions onto a "cold block" (Williams 1954) were among the earliest attempts of EM cryofixation. Recently, quenching of aerosol droplets impinging on cold metal at supersonic speed has become a reliable method of vitrifying pure liquid water (Mayer 1985). In metallurgy and material science non-equilibrium and amorphous phases of metals and alloys have been obtained by "splat-cooling", a process by which liquid metals are quenched from the melt by ejecting them onto a cooled metallic substrate.

Cooling rates have frequently been calculated and experimentally tested using solid heat sinks where it is easier to calculate the heat transport than with fluid cryomedia. Ruhl (1967) has done this for liquid metals splat-cooled on solid heat sinks, and has analyzed the effect of the process variables h, d and k_s. These calculations have been compared with measured cooling rates up to $5 \cdot 10^8$ K s^{-1}. Since no corresponding measurements exist on water, we will discuss here Ruhl's analysis. Unidimensional heat flow is assumed from a sample whose diameter is much larger than its thickness, to a much thicker metal substrate. For a given splat thickness and thermal conductivity, three types of cooling have been characterized: (1) "ideal cooling", where the heat transfer coefficient approaches ∞ and the thermal gradients in the splat and the substrate are large. (2) "Newtonian cooling", where h is small. The thermal gradients in the splat are negligible and the cooling is therefore completely interface controlled. (3) "Intermediate cooling" for situations between (1) and (2) in which the thermal gradients are not negligible and the contact between the splat and the substrate is not ideal. Figure 15 shows temperature-time curves for various positions within a 10µm-thick iron splat, assuming ideal cooling, and for Newtonian cooling of the same specimen. For any given time the temperature gradients can be read from the diagram, the highest cooling rates occurring near the interface. The singularity at 8 µs signifies the time at which the solid/liquid interface has reached the outer surface of the splat and solidification is complete. For Newtonian cooling, solidification occurs in the horizontal region of constant temperature. Plotting the average cooling rates vs. time illustrates the difference between these two extremes more drastically (Fig. 16): during ideal cooling, solidification causes only an apparent discontinuity of the rate, while in the case of Newtonian cooling, the cooling rate is effectively zero during solidification. The latter is based on a model in which the release of melting enthalpy is slower than the rate of cooling and in which no supercooling occurs. The calculations start at 60 µs.

The cooling rate varies inversely as the square of the splat thickness in ideal cooling and inversely as the splat thickness in Newtonian cooling. As the specimen thickness d becomes smaller, the cooling rate will cease to be ideal unless the heat transfer coefficient is very large. For the case of infinite heat transfer, the square law holds for rates up to which atomic vibration frequencies become limiting.

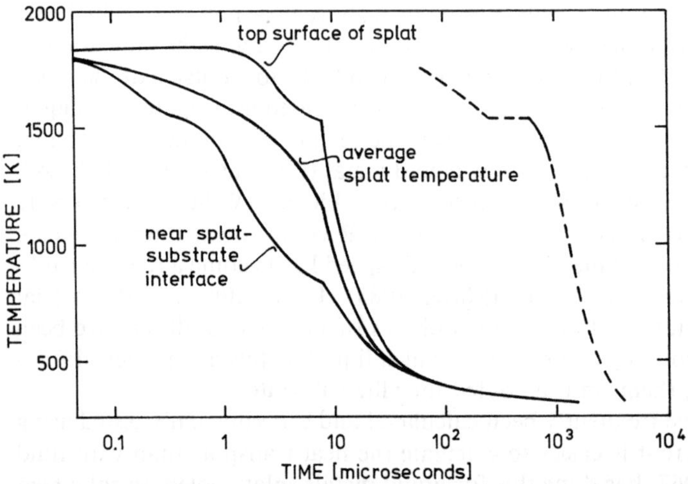

Fig. 15. Temperature vs. time calculated for Fe splat on Cu substrate: splat thickness 10 μm, initial splat temperature 1870 K, melting temperature 1570 K, substrate temperature 300 K. *Solid lines*: "Ideal" cooling, with heat transfer coefficient h = ∞. *Dashed lines*: "Newtonian" cooling, with $h = 4.2\ J\ cm^{-2}\ s^{-1}\ K^{-1}$, starting at 60 μs. Note the logarithmic time scale. (According to Ruhl 1967)

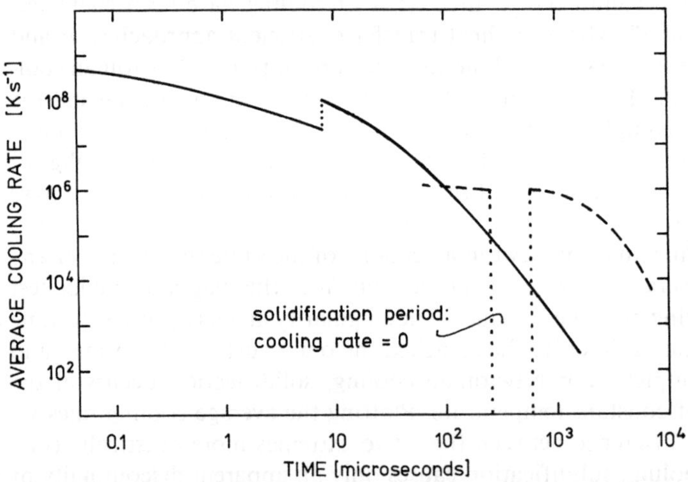

Fig. 16. Average cooling rate vs time for 10 μm Fe splat as in Fig. 15. *Solid lines*: Cooling rate for "ideal" cooling with h = ∞. *Dashed lines*: "Newtonian" cooling, with $h = 4.2\ J\ cm^{-2}\ s^{-1}\ K^{-1}$, starting at 60 μs. (According to Ruhl 1967)

Ruhl's somewhat simplified, but elucidating calculations for metal splat-cooling are, in principle, also applicable to aqueous specimens. The most important difference is that water conducts heat more than 100 times slower than iron. The ratio of heat conduction in heat sink and sample is larger by the same factor and therefore tends to increase the effective heat transfer coefficient. Nusselt's number implies that for water "ideal", i. e. specimen-limited cooling is obtainable either with much poorer heat transfer coefficients or down to a much smaller specimen size. In other words, compared to cooling rates within a 100-μm iron splat, the cooling rate in a biological specimen of the same thickness will be less sensitive to variations in the heat transfer coefficient, i. e. to the modes of quenching. The Newtonian behaviour will become dominant in specimens a hundred times smaller.

This conforms with the general experience in cryofixation: down to the 100 to 20 μm range the reduction in specimen size has a much more pronounced effect on the quality of cryofixation than modifications of established quenching modes, i. e. of the heat transfer coefficient. When tissue blocks are frozen, the depth of the well-frozen zone shows surprisingly little correlation with the quenching method (Elder et al. 1982). This indicates that heat transfer by routine freezing techniques (e. g. using copper or propane at liquid nitrogen temperature) is sufficient to take care of all the thermal energy which can possibly escape from a layer more than approximately 20 μm thick. For such samples, improved cryofixation cannot be expected from the use of optimized heat sinks (Bald 1983), because the heat transport within the coolant does not seem to be the rate-determining step. Only when the sample size (thickness) reaches the μm range, will improving heat transfer again become effective: a 10μm-thick jet of water injected into liquid propane at 80 K with a driving pressure of 400 bar gives mainly vitrified material (Mayer and Brüggeller 1982), while a driving pressure of 20 bar results in almost pure crystalline ice. In both cases, the fraction of vitrified water can be increased by vigorous stirring of the liquid propane. Thus, the cooling rate within the μm-sized samples clearly responds to these variations in the heat transfer coefficient.

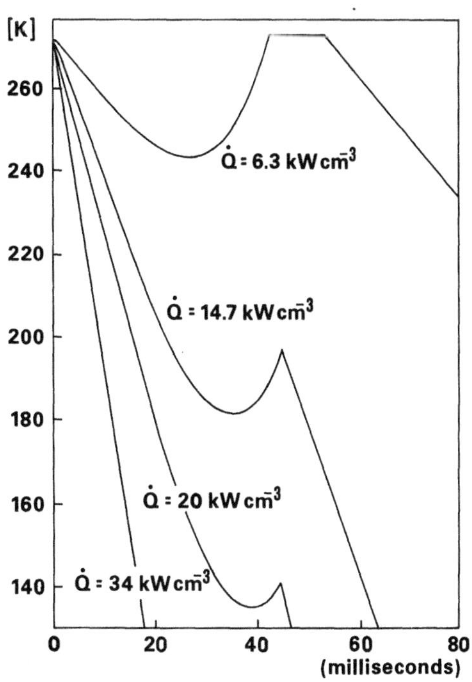

Fig. 17. Calculated cooling curves for water at different rates of heat removal; since the rate of heat release by the water is not known, this is only to demonstrate the principle. (According to Stephenson 1956)

5.2 The Temperature of the Growing Ice Front

It is frequently suggested that during the growth of the ice crystal, the temperature of the solid/liquid interface is raised by the heat of fusion to the equilibrium melting temperature until solidification is finished. This, however, is only the case when the rate of heat dissipation is lower than the rate of heat liberation by crystallization. Stephenson (1956) has computed theoretical curves for the temperature within freezing water for various rates of heat removal (Fig. 17). The calculations are based on some estimated rates of nucleation and growth, neglecting their temperature dependencies and those of the heat of fusion, of the heat capacities and the cooling rate. Although the absolute values should not be taken at face value, the plot seems rather instructive. It shows that "cryofixation times" in the order of milliseconds would prevent any significant temperature rise during crystallization. In biological specimens, the rate of crystallization compared to that in pure water is assumed to be lowered by a larger factor than heat conductance. Therefore, the temperature of the interfacial volume will not rise even at slower cooling rates.

5.3 Probes for Cooling Rates

Measurements with thermocouples have given valuable information on the approximate cooling rates within specimens of 100µm-size and larger (see Sitte et al., Chap. 4, this Vol.). These direct measurements are also important for a comparison of different quenching techniques (for reviews see Plattner and Bachmann 1982; Robards and Sleytr 1985). Especially for fixation techniques in which the sample is divided into small volumes (spraying, jetting), cooling rates can only be estimated by a risky scaling down of data obtained on larger samples or determined indirectly, by the presence or absence of temperature-dependent specimen alterations (=artefacts). The alteration in the quenched specimen can be observed by various means. The most pragmatic approach is the EM observation of the "quality" of cryofixation. In particular, the preservation of chromatin structure may serve as a relative cooling-rate indicator (see Sitte et al., Chap. 4, this Vol.). In this connection, it is important that preparation techniques following quenching do not enhance or diminish freezing artefacts. Freeze-fracturing might be a relatively safe procedure in this respect. Up to now, these quality criteria are mainly changes in connection with the phase transitions of water.

The presence of crystalline ice can be observed in the cryoelectron microscope, particularly on specimens frozen in very thin layers. The instability of the ice layer in the electron beam has to be considered when the observation temperatures are below 70 K (Lepault et al. 1983; Heide and Zeitler 1985). Spray-, jet- and splat-frozen samples have been investigated by electron and X-ray diffraction (Dubochet and MacDowell 1981; Dubochet et al. 1982; Mayer and Brüggeller 1982; Mayer 1985).

The presence and relative amount of vitreous water can be determined using DSC, by measuring the amount of heat released when the sample is warmed above its crystallization temperature (MacFarlane and Angell 1984; Hallbrucker and Mayer 1987). So far, the most sensitive way to measure segregation in dilute salt solutions down to 10^{-3} M is the determination of the freeze-concentration factor by electron spin resonance measurements. This technique can be used to discriminate cooling rates from about $1000\,K\,s^{-1}$ up to vitrification conditions (Brüggeller and Mayer 1980, 1981; Brüggeller 1983). We know, however, that even in pure water these transition rates depend heavily, e. g. on droplet size and pressure. All these techniques can "only" give information about the relative merits of quenching techniques with respect to ice damage.

Cryoartefacts can also be induced in non-aqueous components of the sample, e. g. by phase transformation and segregation processes in lipids and in micellar systems. These transitions could give us information on the thermal history of the sample even above the freezing temperature of water. The properties of model sensors should be that their rate of transition depends only on the temperature and that this rate can be determined on an absolute scale, e. g. by temperature-jump techniques. A novel method by Aurich and Förster (1984) in which the temperature-dependent fluorescence in a dye, umbelliferone, was used for determining temperature during rapid cooling, might have these properties. Possible other models are molecules with different rates, e. g. of cis-trans isomerization; the cooling rates could be calculated by the relative amount of the configurations in the quenched state. Knowing the thermal history of specimens obtained under varying conditions would have an interesting feedback for a more quantitative understanding of the processes occurring when water is quenched to low temperatures.

For studies of membrane dynamics, time resolution is an essential factor (see Knoll et al., Chap. 14; Kellenberger, Chap. 2, this Vol.). If the absolute rate at which such a process occurs could be determined by other techniques, it could then serve as a probe for cooling rates.

5.4 Cooling Rates and Limits of Cryofixation

As already stated, true cooling rate measurements in aqueous specimens end at approx. $10^4\,K\,s^{-1}$. There is reason to believe that the highest cooling rates which have so far been obtained are at most in the $10^7\,K\,s^{-1}$ range. Based on this assumption, we want to estimate what kind of molecular motions can be arrested in a structural state close to that at room temperature.

In a review, Packer (1977) compiled data concerning the time scale of dynamic changes in water molecules which might influence biological systems, e. g. a protein molecule in water. From these data an estimate is possible on the response of various motions to cooling rates. Figure 18a shows the type of motions in bulk water which are of importance: the correlation time for translational diffusion (τ_D) and the correlation time for reorientation (τ_r) are of the order of 10^{-12}–10^{-11} s at 290 K; the exchange of protons between water molecules (τ_e) occurs with a lifetime of 10^{-3} s. Figure 18b shows the additional influences of an interface, e. g. of a macromolecule, on water molecules. Those water molecules which are free of the influence of the interface (shaded area) diffuse and rotate and exchange protons with times τ_D, τ_r and τ_e characteristic for bulk water. Water

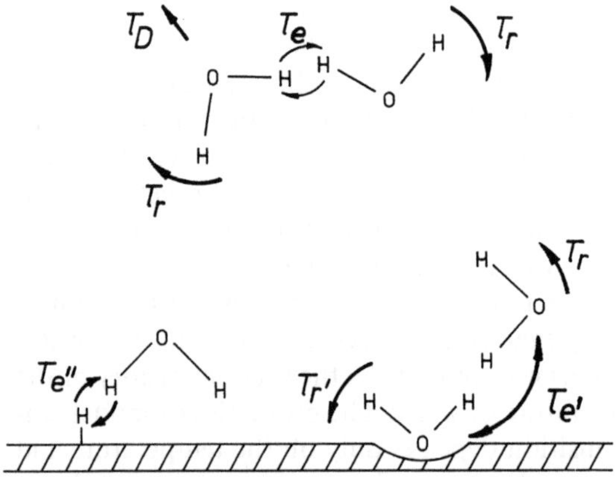

Fig. 18 a, b. Dynamic processes **a** in bulk water, and **b** in water near an interface, e.g. of a macromolecule or a membrane. (According to Packer 1977)

molecules interacting with the macromolecule tumble anisotropically, this process being represented by a collective correlation time, $\tau_{r'}$, of approx. 10^{-9} s at 260 K; these "bound" or surface-modified molecules exchange with bulk water with a lifetime, $\tau_{e'}$, of the order of $10^{-5}-10^{-6}$ s at 300 K. Proton exchange between water protons and protons on the interface is characterized by $\tau_{e''}$. Time scales of internal motions in macromolecules, e. g. a small-scale unfolding of a protein chain, vary over an enormous range, some being 10^{-3} s or slower.

The question is whether or not experimentally accessible cooling rates can immobilize, "freeze-in", these various motions close to room temperature. We discuss the effect of two cooling rates: (1) 10^7 K s^{-1}, which might be the highest rate for μm-sized samples and might have been obtained by quenching water droplets on a cryoplate; and (2) 10^3 K s^{-1}, the so-called standard freezing method of μl droplets. We further assume that a temperature drop of 100° is necessary to immobilize the various motions, and we neglect the temperature dependence of the relaxation times (Fig. 9). For a rate of 10^7 K s^{-1}, this gives a "fixation" time of 10^{-5} s: that means that of the various motions, only τ_e, to some extent $\tau_{e'}$ and the hypothetical internal protein motion of 10^{-3} s are arrested close to room temperature. All the other much faster motions are arrested at much lower temperatures. For a cooling rate of 10^3 K s^{-1}, the fixation time is 10^{-1} s. Therefore, *all* the various processes with time scales $< 10^{-1}$ s continue, possibly leading to drastic structural rearrangements even in the total absence of freezable water. In conclusion, complete vitrification of two identical specimens does not imply that the final immobilized states are identical. If one of them has been quenched with a higher cooling rate, the original structure may be better preserved. This by definition means better cryofixation.

References

(References marked by an asterisk are review papers or books respectively)

*Angell CA (1970) The data gap in solution chemistry. J Chem Educ 47:583–587
*Angell CA (1982a) Supercooled water. In: Franks F (ed) Water, a comprehensive treatise, vol 7. Plenum, New York London, pp 1–81
*Angell CA (1982b) Les anomalies de l'eau. Recherche 133:584–593
*Angell CA (1983) Supercooled water. Annu Rev Phys Chem 34:593–630
*Angell CA, Choi Y (1986) Crystallization and vitrification in aqueous systems. J Microsc (Oxford) 141:251–261
Angell CA, Tucker JC (1980) Heat capacity changes in glass-forming aqueous solutions and the glass transition in vitreous water. J Phys Chem 84:268–272
Angell CA, Shuppert J, Tucker JC (1973) Anomalous properties of supercooled water. Heat capacity, expansivity and proton magnetic resonance chemical shift from 0 to −38 °. J Phys Chem 77:3092–3099
*Angell CA, Clarke JHR, Woodcock LV (1981) Interaction potentials and glass formation: a survey of computer experiments. Adv Chem Phys 48:397–453
Angell CA, Oguni M, Sichina WJ (1982) Heat capacity of water at extremes of supercooling and superheating. J Phys Chem 86:998–1002
*Angell CA, Busse LE, Cooper EI, Kadiyala RK, Dworkin A, Ghelfenstein M, Szwarc H, Vassal A (1985) Glasses and glassy crystals from molecular and molecular ionic systems. J Chim Phys 82:267–274
Aurich F, Förster T (1984) Temperature measurement during rapid cooling in µl-volumina using a fluorescence label. Cryoletters 5:231–238
Bald WB (1983) Optimizing the cooling block for the quick freeze method. J Microsc (Oxford) 131:11–33
Belfort G (1974) Structure of water on porous glass. Nature (London) 249:593–594
*Berendsen HJC (1975) Specific interactions of water with biopolymers. In: Franks F (ed) Water, a comprehensive treatise, vol 5. Plenum, New York London pp 293–330
Boutron P (1984) More accurate determination of the quantity of ice crystallized at low cooling rates in the glycerol and 1,2-propanediol aqueous solutions: Comparison with equilibrium. Cryobiology 21:183–191
*Bowman HF, Cravalko EG, Woods M (1975) Theory, measurement and application of the thermal properties of biomaterials. Annu Rev Biophys Bioeng 4:43–80
Brüggeller P (1983) Evaluation of modified water by electron paramagnetic resonance: no evidence for long-range order of water on pore glasses. J Colloid Interface Sci 94:524–530
Brüggeller P, Mayer E (1980) Complete vitrification in pure liquid water and dilute aqueous solutions. Nature (London) 288:569–571
Brüggeller P, Mayer E (1981) EPR Investigation of the nonfreezing water/free water ratio in disperse systems. J Phys Chem 85:4135–4139
Cohen MH, Turnbull D (1959) Molecular transport in liquids and glasses. J Chem Phys 31:1164–1169
*Conway BE (1981) Ionic hydration in chemistry and biophysics. Elsevier, Amsterdam
*Cooke R, Kuntz ID (1974) The properties of water in biological systems. Annu Rev Biophys Bioeng 3:95–126
*Derbyshire W (1982) The dynamics of water in heterogeneous systems with emphasis on subzero temperatures. In: Franks F (ed) Water, a comprehensive treatise, vol 7. Plenum, New York London, pp 339–430
*Dore JC (1985) Neutron diffraction studies of water structure. In: Franks F (ed) Water science reviews 1. Univ Press, Cambridge, pp 3–92
Dubochet J, Lepault J (1984) Cryo-electron microscopy of vitrified water. J Phys C7:85–94
Dubochet J, McDowall AW (1981) Vitrification of pure liquid water for electron microscopy. J Microsc (Oxford) 124:RP3–RP4
Dubochet J, Lepault J, Freeman R, Berriman JA, Homo JC (1982) Electron microscopy of frozen water and aqueous solutions. J Microsc (Oxford) 128:219–237
Dubochet J, Adrian M, Vogel RH (1983) Amorphous solid water obtained by vapour condensation or by liquid cooling: a comparison. Cryoletters 4:233–240

*Eisenberg D, Kauzmann W (1969) The structure and properties of water. Univ Press, Oxford
Elder HY, Gray CC, Jardine AG, Chapman JN, Biddlecombe WH (1982) Optimum conditions for the cryoquenching of small tissue blocks in liquid coolants. J Microsc (Oxford) 126:45–61
*Elliott SR (1983) Physics of amorphous materials. Longman, London New York
*Finney JL (1986) The role of water perturbations in biological processes. In: Neilson GW, Enderby JE (eds) Water and aqueous solutions. Hilger, Bristol Boston, pp 227–244
*Fletcher NH (1971) Structural aspects of the ice-water system. Rep Prog Phys 34:913–994
*Franks F (1975) The hydrophobic interaction. In: Franks F (ed) Water, a comprehensive treatise, vol 4. Plenum, New York London, pp 1–94
Franks F (1981) Glassy water-recent developments. Cryoletters 2:69–71
*Franks F (1985) Biophysics and biochemistry at low temperatures. Univ Press, Cambridge
*Franks F (1986) Metastable water at subzero temperatures. J Microsc (Oxford) 141:243–249
Ghormley JA (1967) Adsorption and occlusion of gases by the low-temperature forms of ice. J Chem Phys 46:1321–1325
Gillen KT, Douglass DC, Hoch JR (1972) Self-diffusion in liquid water to −31 °C. J Chem Phys 57:5117–5119
Gross GW, McKee C, Wu C (1974a) Concentration dependent solute redistribution at the ice/water phase boundary. I. Analysis. J Chem Phys 62:3080–3084
Gross GW, Wu C, Bryant L, McKee C (1974b) Concentration dependent solute redistribution at the ice/water phase boundary. II. Experimental investigation. J Chem Phys 62:3085–3092
Hallbrucker A, Mayer E (1987) Calorimetric study of the vitrified liquid water to cubic ice phase transition. J Phys Chem 91:503–505
Hare DE, Sorensen CM (1986) Densities of supercooled H_2O and D_2O in 25 μm glass capillaries. J Chem Phys 84:5085–5089
Harreveld A van, Crowell J (1964) Electron microscopy after rapid freezing on a metal surface and substitution fixation. Anat Rec 149:381–385
Heide HG, Zeitler E (1985) The physical behavior of solid water at low temperatures and the embedding of electron microscopical specimens. Ultramicroscopy 16:151–160
*Hertz HG (1973) Nuclear magnetic relaxation spectroscopy. In: Franks F (ed) Water, a comprehensive treatise, vol 3. Plenum, New York London, pp 301–399
*Hobbs P (1974) Ice physics. Clarendon, Oxford
Irish DE, Jarv T (1977) Temperature dependence of Raman band parameters of aquated cations. Discuss Faraday Soc 64:95–101
Irish DE, Jarv T (1983) Vibrational spectral studies of solutions at elevated temperatures and pressures. IV. Raman spectra of aqueous zinc nitrate solutions. Appl Spectrosc 37:50–55
*Israelachvili JN (1985) Measurements of hydration forces between macroscopic surfaces. Chem Scr 25:7–14
*James DW (1985) Spectroscopic studies of ion-ion-solvent interaction in solutions containing oxyanions. Progr Inorg Chem 33:353–391
*Johari GP (1974) Introduction to the glassy state in the undergraduate curriculum. J Chem Educ 51:23–27
Jones GJ (1984) On estimating freezing times during tissue rapid freezing. J Microsc (Oxford) 136:349–360
Kanno H, Speedy RJ, Angell CA (1975) Supercooling of water to −92 °C under pressure. Science 189:880–881
Katz JL, Donohue MD (1979) A kinetic approach to homogeneous nucleation theory. Adv Chem Phys 40:137–155
* Kauzmann AW (1948) The nature of the glassy state and the behaviour of liquids at low temperatures. Chem Rev 43:219–256
Klinger J (1980) Influence of a phase transition of ice on the heat and mass balance of comets. Science 209:271–272
Kraus GF, Greer SC (1984) Vapor pressures of supercooled H_2O and D_2O. J Phys Chem 88:4781–4785
*Kuhs WF, Lehmann MS (1986) Oxygen disorder and the geometry of the water molecule in ice Ih. In: Neilson GW, Enderby JE (eds) Water and aqueous solutions. Colston Pap 37. Hilger, Bristol Boston, pp 75–82

Lang EW, Lüdemann HD (1980) Pressure and temperature dependence of the longitudinal deuterium relaxation times in supercooled heavy water to 300 MPa and 188 K. Ber Bunsenges Phys Chem 84:462–470

*Lang EW, Lüdemann HD (1982) Anomalies of liquid water. Angew Chem Int Edn Engl 21:315–329

Lang EW, Lüdemann HD (1985) p, T, c-Dependence of ^{2}H Spin-Lattice Relaxation Rates in Supercooled LiCl-D_2O Solutions. Ber Bunsenges Phys Chem 89:508–516

Lang EW, Fink W, Lüdemann HD (1984a) The p, T, c-dependence of deuterium spin-lattice relaxation rates in supercooled LiCl-, NaCl- and $MgCl_2$-D_2O solutions. J Phys C7:173–177

Lang EW, Lüdemann HD, Piculell L (1984b) Nuclear magnetic relaxation rate dispersion in supercooled heavy water under high pressure. J Chem Phys 81:3820–3827

Lepault J, Freeman R, Dubochet J (1983) Electron beam induced "vitrified ice". J Microsc (Oxford) 132:RP3–RP4

Luyet B, Rasmussen D (1968) Study by differential thermal analysis of the temperatures of instability of rapidly cooled solutions of glycerol, ethylene glycol, sucrose and glucose. Biodynamica 10:167–191

MacFarlane DR, Angell CA (1984) Nonexistent glass transition for amorphous solid water. J Phys Chem 88:759–762

MacFarlane DR, Kadiyala RK, Angell CA (1983a) Cooling rate dependence of the ice I nucleation temperature in aqueous LiCl solutions. J Phys Chem 87:235–238

MacFarlane DR, Kadiyala RK, Angell CA (1983b) Homogeneous nucleation and growth of ice from solutions. TTT curves, the nucleation rate, and the stable glass criterion. J Chem Phys 79:3921–3927

*MacKenzie AP (1977) Non-equilibrium freezing behaviour of aqueous systems. Philos Trans R Soc London Ser B 278:167–189

*Mayer E (1985) Vitrification of pure liquid water. J Microsc (Oxford) 140:3–15

Mayer E (1986) Vitrified dilute aqueous solutions 1. Infrared spectra of alkali metal nitrates and perchlorates as solutes. J Phys Chem 90:4455–4461

Mayer E, Brüggeller P (1982) Vitrification of pure liquid water by high pressure jet freezing. Nature (London) 298:715–718

Mayer E, Hallbrucker A (1987) Cubic ice from liquid water. Nature (London) 325:601–602

Mayer E, Pletzer R (1986) Astrophysical implications of amorphous ice-a microporous solid. Nature (London) 319:298–301

McDowall AW, Chang JJ, Freeman R, Lepault J, Walter CA, Dubochet J (1983) Electron microscopy of frozen hydrated sections of vitreous ice and vitrified biological samples. J Microsc (Oxford) 131:1–9

*Narten AH, Levy HA (1972) Liquid water: Scattering of X-rays. In: Franks F (ed) Water, a comprehensive treatise, vol 1. Plenum, New York London, pp 311–331

Oguni M, Angell CA (1980) Heat capacities of $H_2O+H_2O_2$, and $H_2O+N_2H_4$, binary solutions: Isolation of a singular component for C_p of supercooled water. J Chem Phys 73:1948–1954

*Packer KJ (1977) The dynamics of water in heterogeneous systems. Philos Trans R Soc London Ser B 278:59–87

*Plattner H, Bachmann L (1982) Cryofixation: A tool in biological ultrastructural research. Int Rev Cytol 79:237–304

*Pohl RO (1975) Phonon scattering in amorphous solids. In: Challis LJ, Rampton VW, Wyatt AFG (eds) Phonon scattering in solids. Plenum, New York London, pp 107–114

*Pruppacher HR, Klett JD (1980) Microphysics of clouds and precipitation. Reidel, Dordrecht Boston London

Rahman A, Stillinger FH (1973) Hydrogen-bond patterns in liquid water. J Am Chem Soc 95:7943–7948

*Rand RP, Das S, Parsegian VA (1985) The hydration force, its character, universality and application: some current issues. Chem Scr 25:15–21

Rasmussen DH, MacKenzie AP (1973) Clustering in supercooled water. J Chem Phys 59:5003–5013

*Resing HA (1972) NMR relaxation of adsorbed molecules with emphasis on adsorbed water. Adv Mol Relaxat Processes 3:199–226

*Robards AW, Sleytr UB (1985) Low temperature methods in biological electron microscopy. In: Glauert AM (ed) Practical methods in electron microscopy, vol 10. Elsevier, Amsterdam
Ruhl RC (1967) Cooling rates in splat cooling. Mater Sci Eng 1:313–320
Sargeant PT, Roy R (1968) A new approach of prediction of glass formation. Mater Res Bull 3:265–280
Sorensen CM (1983) Densities and partial molar volumes of supercooled aqueous solutions. J Chem Phys 79:1455–1461
Speedy RJ (1984) Self-Replicating Structures in Water. J Phys Chem 88:3364–3373
Speedy RJ, Angell CA (1973) Isothermal compressibility of supercooled water and evidence for a thermodynamic singularity at −45 °C. J Chem Phys 65:851–858
Stephenson JL (1956) Ice crystal growth during the rapid freezing of tissues. J Biophys Biochem Cytol 2:45–51
*Stillinger FH (1980) Water revisited. Science 209:451–457
Stillinger FH, Weber TA (1983) Inherent structure in water. J Phys Chem 87:2833–2840
*Turnbull D (1969) Under what conditions can a glass be formed? Contemp Phys 10:473–488
*Uhlmann DR (1972) A kinetic treatment of glass formation. J Non-Crystalline Solids 7:337–348
*Uhlmann DR (1982) Kinetics of glass formation and devitrification behavior. J Phys C9:175–190
*Umrath W (1983) Berechnung von Gefriertrocknungszeiten für die elektronenmikroskopische Präparation. Mikroskopie (Wien) 40:9–37
*Williams RC (1954) In: Harris RJC (ed) Biological applications of freezing and drying to electron microscopy. Academic Press, London New York, p 303
Wood GR, Walton AG (1970) Homogeneous nucleation kinetics of ice from water. J Appl Phys 41:3027–3036

Chapter 2 The Response of Biological Macromolecules and Supramolecular Structures to the Physics of Specimen Cryopreparation

EDWARD KELLENBERGER[1]

1 Introduction

In the preceding chapter we have learned about the problems of freezing and have seen that one of the major problems in cryotechniques is the formation of ice crystals (see Bachmann and Mayer, Chap. 1, this Vol.). Due to their size, they disturb the fine structure of cells (see also Sitte et al., Chap. 4, this Vol.; Moor, Chap. 8, this Vol.). In the present chapter, we consider the problems which we face when we are able to work in vitreous ice, i.e. what happens with biological macromolecules and supramolecular structures. In these considerations we distinguish two levels of resolution: at the highest resolution (1 – 10 nm), we consider conformational changes of macromolecules, particularly of proteins. This resolution is, in general, only achieved with procedures that allow for averaging from identical subunits. At the other level, we consider rearrangements and small aggregations of macromolecules due to the specimen preparation. This is the resolution of 10 – 20 nm achieved normally with the current methods of preparation of biological material.

We will base our considerations on the modern knowledge of "biological colloids" without always providing references for basic phenomena. These are found in any modern treatise of biophysics and protein chemistry (e.g. Hoppe et al. 1977; Cantor and Schimmel 1980). We will attempt to develop very simple, fundamental concepts, which should be useful for interpreting the results obtained by the manifold cryopreparation techniques in electronmicroscopy.

2 The Hydration Shells of Biological Macromolecules and Supramolecular Structures

2.1 The Nature of Hydration Shells

When "biological colloids" were studied and defined at the turn of this century, it was already well known that living organisms always contained large proportions of water. Biomolecules have evolved genetically in and with water. Extracts

1 Abteilung für Mikrobiologie, Biozentrum, Universität Basel, CH-4056 Basel, Switzerland

Cryotechniques in Biological Electron Microscopy
Edited by R. A. Steinbrecht and K. Zierold

of organisms, organs or cells are always solutions of biomolecules or suspensions of larger, assembled structures with hydrophilic properties. When such suspensions, e.g. of cell walls, are tested for their preference of aqueous media versus organic liquids, they choose the former. This is tested by phase separation: the aqueous particle suspension is overlaid by a non-water miscible organic liquid, e.g. an aliphatic hydrocarbon and then slightly shaken. After the two liquids have separated again the particles will still be in the aqueous medium.

Nearly all biological substances turn out to be hydrophilic: polysaccharides, nucleic acids, most proteins and even lipid bilayers. For the last two classes of substances, however, the above described phase-separation test is not feasible, because – as we will see below – they are of a particular, composite nature involving hydrophobic parts inside and hydrophilic parts at their surfaces.

Biological macromolecules, macromolecular assemblies and biological membranes are hydrophilic, because of the nature of the chemical groups at their surfaces. Water molecules are bound to them in a more ordered manner than they are in liquid water. They form a hydration shell, the properties and the behaviour of which are different from that of the surrounding aqueous medium. For macromolecules or biological structures, these hydration shells appear to be essential for solubility or monodispersive suspension in aqueous media. The hydration shells prevent them from aggregating. After random encounters, two particles do not stick together, but separate again.

The water molecules on the surface of biological material are oriented according to the local charges of the surface. Water molecules are elongated, polar molecules. The surface charge will therefore be transferred into and through the hydration shell (Fig. 1). One still disputes the number of water layers constituting this shell; in fact, there is no real limit, but a continuous decrease of the order, so as to reach the (relatively) disordered state of the water molecules in the aqueous medium. It is important to remember that the innermost layer of the

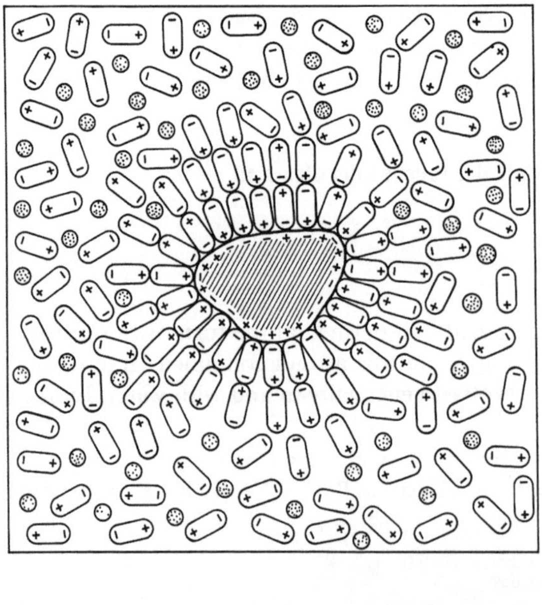

Fig. 1. A protein molecule in aqueous solution. The polar water molecules form a hydration shell of pure water on the surface of the hydrophilic protein. With increasing distance from the surface, the order of the shell steadily decreases to reach that of the surrounding medium. The *circles* symbolize solutes (i.e. salts, metabolites) of the medium

shell is pure water; the solutes of the medium are not penetrating down to the surface of the biological matter. The more distant we are from the surface, the more the concentration of solutes approaches that of the suspension medium and the more disordered are the water molecules. Also, the charge pattern becomes blurred so that further outside only the average charge predominates.

Very basic thermodynamic experiments (reviewed by Lauffer 1975) have shown that the ordered water of hydration shells is, in a first approximation, comparable to that of ice (but see also Bachmann and Mayer, Chap. 1, this Vol.). The same heat of fusion is needed to open the weak bonds between the water molecules. The water molecules of the hydration shell are only in a slow dynamic exchange equilibrium with the free water, exactly as with ice. It is therefore correct to define melting or fusion as the weakening of bonds needed for passing from a solid to a liquid. It is also correct to speak of melting of a protein when the thermal movements have reached energies which are above the weak interactions that hold the polypeptide chain in a given, specific conformation. Although commonly used with the thermal denaturation of DNA, the term melting is here less well chosen, because it describes only the separation of the two strands by opening of the hydrogen bonds. Before being "melted", a DNA molecule is not really a solid, as is for instance the hydration shell.

The melting temperature of the hydration shell is not 273 K, but depends very strongly on the chemical nature of the surface. A sort of "degree of hydrophilicity" of any part or area of a surface can be defined by the melting temperature of this part of its hydration shell. The lower it is, the more hydrophobic is this surface area.

The above mentioned studies were made in an attempt to understand the so-called hydrophobic interaction (Lauffer 1975; Tanford 1980), which dominates in the biological world. Subunits of tobacco mosaic virus (TMV) are brought to assemble by *raising* the temperatures. As parts of the hydration shell are removed, "melted away", the sticky surface of the protein underneath is uncovered. When two such parts encounter each other, they will stick. The interaction is the stronger, the more the two surfaces are complementary in charge and hydrogen-bonding properties and the larger the interacting areas are. This is how most supramolecular structures made of proteins assemble by virtue of specific interaction sites. Rise of temperature, but also an increase of the salt concentration of the surrounding medium, promote assembly: by the addition of salt we lower the melting point again in exact analogy to the melting of ice in brine.

Proteins, which are not designed for assembly, aggregate randomly and precipitate when the temperature is sufficiently increased or after increasing the salt concentration. They are "salted out". In the case of temperature-induced aggregations, as a rule a denaturation of the protein is also involved; lowering the temperature is not sufficient for resuspending the molecules of a heat-precipitated, random aggregate. Salted-out proteins, in general, can be redissolved.

Ordered protein assemblies dissociate upon lowering the temperature and/or when decreasing the concentration of solutes. This phenomenon, discovered with TMV, was later found to be true in nearly all cases of protein assemblies that were investigated. It apparently contradicted the laws of entropy which states that the order decreases with increasing temperature. Thermodynamic measurements (see Lauffer 1975) explained this paradox: the decrease of order that accompanies the removal of the hydration shells has more weight than the increase of order by the polymerization of protein subunits into regular arrays.

We have mentioned that the "melting temperature" of the hydration shell is a sort of measure for the degree of hydrophilicity. The lower this temperature, the less hydrophilic a protein is. Thus, even hydrophobic molecules might eventually acquire a hydration shell at low temperatures and become soluble in water. This has indeed been shown for many organic molecules, like the very hydrophobic methane the solubility of which in water at room temperature is virtually zero, but strongly increases with lower temperatures (see Franks 1973, pp. 7 and 376). The increased hydrophilicity at low temperatures is particularly important when considering freeze-substitution and low temperature embedding (see Steinbrecht and Müller, Chap. 7, this Vol.; see also Sect. 2.2.4).

The hydration shells of nucleic acids have been intensively studied (Hearst and Vinograd 1961). Operationally, a number of water molecules bound per nucleotide was calculated from the experimentally determined buoyant density (in heavy metal salts like CsCl) and put in relation to a parameter that describes the "amount"of water in the solution, the so-called water activity. In pure water 20–50 water molecules are bound per nucleotide. In physiological buffers this amount decreases and is less than ten.

For polysaccharides and proteins the available data are not as precise and ordered as those for nucleic acids. Polysaccharides are supposed to have rather substantial hydration shells equal in size or even superior to those of nucleic acids. Proteins have hydration shells (Kuntz and Kauzmann 1974) constituted on the average of some two layers. We have confirmed such a value with M. Moncany on virus proteins (unpubl. data).

The composition of the polar heads of lipids predicts hydration layers for lipid bilayers to be very similar to those of proteins. Modern biophysical methods, like nuclear magnetic resonance confirm these predictions (Killian and de Kruijff 1985; van Duijn et al. 1985). The surfaces of biological membranes (see Sect. 4) show, in addition, parts of the emerging proteins and many short oligosaccharide chains; both are also hydrophilic. Some ionophores and oligomeres of proteins form hydrophilic channels through the lipid bilayer (for references see Killian and de Kruijff 1985; Rosenbusch et al. 1982).

2.2 The Fate of Biological Macromolecules in Organic Liquids and After Fixation

2.2.1 The Constitution of Cellular Plasms

Electron microscopy of thin sections has enabled the cytologists to distinguish in the cell between membranes, plasms and a few species of particles like the ribosomes. By other procedures cell biologists have learned very early to distinguish between sols and gels. Sols are able to flow and allow the movement of enclosed organelles and particles, gels do not. Electron microscopy has revealed that the shape of biological macromolecules is either fibrous or average globular. It has eliminated the intermediate model of elongated, elliptic structures, proposed to explain their hydrodynamic behaviour which is now understood to be the consequence of intrinsic hydration.

Sols are solutions of biological macromolecules that do not interact with each other and therefore are freely mobile. The viscosity of sols increases with the concentration. At the same concentration, it is much higher with fibrous particles than with globular ones. Most biological gels known are based on fibres; they form gels because they are able to interact locally. By these contact sites the relative movement of the fibres is abolished and the gel, viewed macroscopically, behaves as a solid. Sol-gel transitions can occur, for example with solutions of gelatine at 310 K. When gelatine or agar melts, the weak bonds at the contact points are opened. With fibres gels can eventually be formed that have solid contents below 1% (agar). The meshwork is thus wide open (like a sponge) and can be filled with a sol of globular molecules.

In order to produce a gel with globular molecules much higher concentrations are needed (20 – 50% solid content). The interactions needed for gelling globular molecules are still very poorly understood. The artificially produced protein crystals used for crystallographic investigations are gels formed by ordered arrays of identical particles. Crystalline gels are found only very rarely in nature (for instance, the toxin in *Bacillus thuringiensis*), while increasing evidence favours the idea that many ground plasms, for instance of the mitochondria, are much less ordered gels of various species of densely packed globular protein molecules (Srere 1981). Such gels obviously have very small "pores" through which only small molecules (substrates and products) can diffuse.

The ground plasm of the cytoplasma, i.e. the "background" solution in the absence of ribosomes, organelles and membranes (endoplasmatic reticulum, etc.), is thought to be formed of both globular and fibrous elements. As soon as the fibrous elements are made visible in the electron microscope (microfilaments, intermediate filaments and microtubules) they are, in general, no longer considered as part of the ground plasm; they are then part of the cytoskeleton. In the DNA-containing plasms (e.g. interphase nucleus) the DNA is still an integral part of the ground plasm as long as we cannot resolve individual DNA molecules in thin sections (see Sect. 2.2.3).

Cellulosic cell walls are gels that contain very large amounts of water. The walls of bacteria (polysaccharide in gram-positive or peptidoglycan in gram-negative), for instance contain more than 85% aqueous medium (Gerhardt and Judge 1964; Hobot et al. 1984).

All constituents of the cell, except the biological membranes (see Sect. 4) and lipid vesicles, are in an aqueous phase (see Sect. 3). All macromolecules are surrounded by a hydration shell and therefore soluble, even when gelled by local interactions.

What happens now with aqueous sols and gels if the aqueous medium is removed by freeze-drying or replaced by organic liquids or ice?

2.2.2 *The Fate of Biological Plasms*

In the preceding sections we have seen that nucleic acids and proteins are not soluble in organic liquids. As an exception among hydrophilic biological material, some polysaccharides are soluble in a few organic solvents. When we replace water by an organic liquid, a biological ground plasm cannot remain as it was. Figure 2a shows a solution of a protein in water. The molecules are in random motion, but they repulse each other so that they are not connected together. In the organic liquid they are not soluble; each encounter leads to an irreversible contact. The best possible case is given in a′, where they form a relatively fine network. This situation will rarely occur; rather the formation of larger aggregates in a much coarser network (Fig. 2a″) is the rule. It seems fairly obvious that high concentrations of proteins favour finer networks.

Figure 2c illustrates the response of a fibrous sol (e.g. DNA). Again aggregation occurs. If the concentration is high enough, a network might be produced with minor (c′) or major (c″) aggregations into bundles.

Figure 2b shows a typical ground plasm that is composed of a fibrous network and a solution of globular elements. The fibres are supposed to form a relatively rigid gel (e.g. microtubules). When transferred to an organic liquid, the globular macromolecules will have a tendency to aggregate onto the nearest fibres rather than making aggregates among themselves. We see here how microtrabeculae might be formed.

In summary Figure 2 describes schematically the aggreations that occur obligatorily when both substituting water by organic liquids and/or by chemical crosslinking (chemical fixation). The whole art of cytological electron microscopy

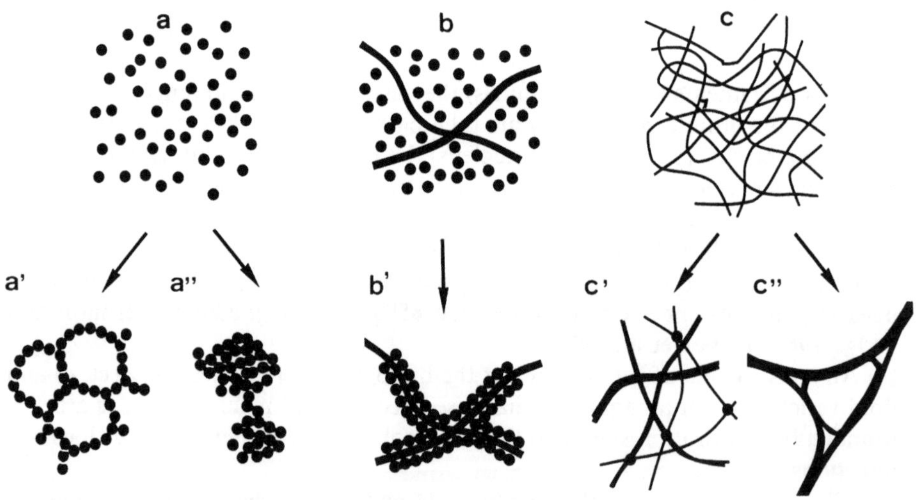

Fig. 2a–c. Typical biological ground plasms before and after transfer into organic liquids or after fixation. **a** A plasm formed of globular and **c** of fibrous macromolecules. **b** A mixture of fibrous and globular elements. The solvent- or fixation-induced aggregations are shown below with different degrees of coarseness. The number of globular macromolecules is the same before and after aggregation. The *dots* in **c′** show crosslinks necessary for forming a gel

is to reduce the size of aggregates which would occur necessarily in any organic solvent by producing first a network of relatively fine aggregates through chemical crosslinks which then is stable enough to successfully withstand the strong tendency of aggregation in the solvent; this tendency is also decreased by progressive dehydration.

Systematic studies on the fixation and dehydration of biological plasms have only been made for those containing DNA in addition to protein. We will describe in the next section that non-eukaryotic DNA-containing plasms, despite the presence of proteins, can become crosslinked only by heavy metal fixatives, but not by aldehydes. The necessity of also observing unstained material (to exclude so-called staining artefacts) led us to the discovery that aggregates become smaller, the lower the temperature is at which the water is substituted (Kellenberger et al. 1986a). It is most likely that this is also true for proteins alone, although not enough data are available to prove it.

With these observations in mind we are now able to understand the astonishingly good results obtained by cryofixation followed by cryosubstitution at low temperature into solvents that do not contain chemical fixatives. The coarseness of the subsequent aggregation is strongly reduced as a consequence of the low temperature of the solvent. Cryofixation itself is not crosslinking, not really "fixing" in the sense of the effects of chemical fixation.

2.2.3 Dehydration Effects on DNA and Non-Eukaryotic Chromatin

The notion of "DNA-containing plasm" was coined for cellular compartments which contain not only DNA, but presumably also proteins and polyamines in addition to ions that are present throughout the cell. The proteins could be either structural partners of DNA, for instance proteins that are particularly involved in compaction by promoting supercoiling, or freely soluble enzymes. Examples of partners are the nucleosomic proteins, on the one hand, and enzymes involved in metabolic functions such as gyrases and topoisomerases, on the other. In the extreme, a DNA plasm might turn out to be "naked DNA". We therefore include as a "DNA plasm" also solutions of chemically purified DNA, which is usually in the form of a sodium salt.

When studying thin sections of non-eukaryotic, DNA-containing plasms, such as the bacterial nucleoid, the pool of replicating and transcribing bacteriophage DNA and the chromosomes of dinoflagellates and mitochondria, it was found that they aggregate during dehydration when the usual fixatives are used (for references see Kellenberger 1962), suggesting insufficient or lack of crosslinking. A fine-stranded aspect was produced only by particular fixations (Kellenberger and Ryter 1964; Kellenberger et al. 1981). We now have a fair number of data that provide a coherent explanation of these facts.

Aqueous solutions of DNA, RNA, nucleoproteins of all types, as well as of proteins, are precipitated when a certain amount of a water-miscible organic solvent (usually ethanol, methanol, acetone) is added. Biochemists use this reaction currently for purification purposes. The size and shape of the aggregates depend strongly on the initial concentration of the solution, and, as was found recently

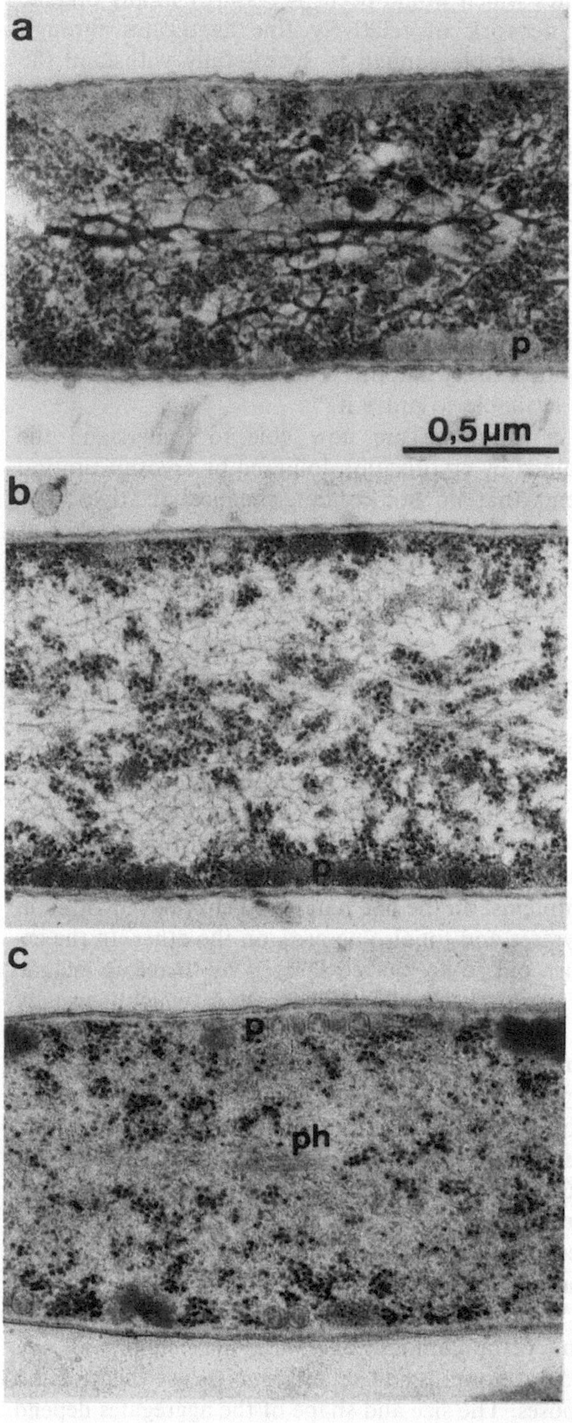

Fig. 3a–c

in our laboratory, fortunately also on the temperature (Kellenberger et al. 1986a). It was discovered indeed that the aggregation of the DNA plasm, constituted by the pool of replicating and transcribing T4-phage DNA, was coarse when dehydrated at room temperature, already less coarse when submitted to a progressive lowering of temperature (PLT) and least when cryosubstituted at 193 K in a fixative-containing solvent (see Fig. 3).

None of the non-eukaryotic DNA plasms studied as yet could be made resistant towards solvent-induced aggregation at room temperature by a prior aldehyde treatment (Kellenberger et al. 1981). Only some special fixatives containing heavy metals (U, Os, In) reduced aggregation. Interestingly, all the fixing conditions leading to reduced aggregations were those under which DNA plasms were gelled in vitro. This is reasonable since a gel is more resistant to aggregation than a sol. Unfortunately, all fixatives, even those that promote gelling, have a property in common: they very rapidly permeabilize plasma membranes (Malhotra 1962; Morel et al. 1971; Moncany 1982). The ensuing leakage of potassium, magnesium and possibly also of other ions has proven to affect the shape of DNA-containing compartments, for instance the nucleoids of bacteria (Hobot et al. 1985). Other visible side effects are possibly due to OsO_4-induced destruction of proteins (Emerman and Behrman 1982; Baschong et al. 1984) and maybe even of DNA (Lilley and Palecek 1984).

The aggregations described above are not observed to any comparable degree with the DNA plasm contained in eukaryotic interphase nuclei, as ascertained from the vast literature available. In the test tube, however, solutions of desoxyribonucleoprotein or chromatin show the same behaviour as described above for the non-eukaryotic DNA plasms (Kellenberger et al. 1981). We propose two putative explanations for this difference: first, the concentration of the material in the test tube is, at the best, about ten times less than in a nucleus and second, the amount of protein bound to DNA is also one to two orders of magnitude higher in eukaryotic chromatin than in non-eukaryotic. Both causes would greatly facilitate chemical crosslinking and also reduce the coarseness of aggregation as discussed above (Sect. 2.2.2).

These very severe limitations encountered when processing non-eukaryotic DNA plasms for thin sections by conventional methods seem now to be overcome by cryofixation and observation in the frozen-hydrated state (see Dubochet et al.,

Fig. 3a–c. Aggregation of the DNA-containing plasms represented by the replicating and transcribing bacteriophage T4-DNA within an infected cell. By T4 infection, the bacterial nucleoid has been destroyed and its DNA digested and the resulting nucleotides reused. **a** Preparation by glutaraldehyde fixation and dehydration in ethanol at room temperature. The DNA is aggregated into coarse threads. **b** Preparation by glutaraldehyde fixation and dehydration at progressively lowered temperatures (PLT) in ethanol (Carlemalm et al. 1985). The aggregation figures of DNA are much finer than in **a**. **c** Preparation by cryofixation and cryosubstitution in a solution of 2.5% OsO_4 in acetone at about 190 K. The DNA plasm is now extremely fine; the fibrous character is, however, not very striking. As discussed in the text, aqueous solutions of OsO_4 are able to fix DNA plasms at room temperature only when the R-K conditions (pH 6, presence of amino acids and divalent cations) are fulfilled (see Kellenberger and Ryter 1964). *p* denotes phage-T4 proheads and *ph* polyheads; they do not contain DNA and accumulate in the mutant of gene 21 that is considered here, because the pool is in this case exaggeratedly large in comparison to wild-type phage (Kellenberger et al. 1986a)

Chap. 5, this Vol.) or by cryosubstitution (Hobot et al. 1985; Kellenberger et al. 1986a).

2.2.4 Dehydration Effects on Polysaccharides

Polysaccharides do not respond uniformly. Some of them, like gels of agar-agar, can be dehydrated in ethanol and acetone without any important aggregation. The cellulosic cell walls of plants, as well as starch and glycogen granules never showed aggregations in thin-sectioned material, despite their high water contents.

Polysaccharidic capsules of some bacterial species showed typical aggregations upon dehydration. This was overcome by adding a dimethylformamide step in dehydration (Bayer et al. 1985). This procedure is based on the observation (E. Carlemalm, unpubl.), that agar-agar gels become dissolved in this solvent. The use of dimethylformamide was, however, successful only with the capsules of some bacterial strains.

2.2.5 Dehydration Effects on Fine Structural Details

The hydration shell (see Sect. 2.1) is essential in the establishment and stabilization of the folding of the polypeptide chain into a protein. In organic solvents it would fold completely differently. In contrast to what our students learn in textbooks, the folding of a polypeptide chain into a protein is not only dependent on the primary sequence of amino acids, but also on the environment in which this folding occurs. The hydration shell, at least the possibility to form one, is an integral part of the folding process. Removal of the hydration shell by organic solvents should therefore lead to a change of folding, i.e. a conformational change. In order to perform such a change, the bonds within the protein molecule acting between the different segments of the polypeptide chain have to be opened before they can be reformed in the new structure. At room temperature only single bonds are opened at any moment due to the thermal vibrations and are then rapidly reclosed. To unfold stretches of a polypeptide several bonds have to be opened simultaneously. In aqueous media this happens only at temperatures that are higher than some 320 K. In concentrated solutions of urea or guanidinium hydrochloride, complete unfolding occurs already at room temperature. No other organic liquids are as yet known to be equally good unfolding agents. In sodium dodecylsulphate (SDS) a refolding occurs, including now the detergent as part of the structure (Reynolds and Tanford 1970). Lowering the temperature counteracts conformational changes because the binding of the hydration shells is now strongly increased (Sect. 2.1). We would also predict that hydration shells are better preserved in non-polar organic liquids than in polar ones, which are able to compete, to a certain degree, with water.

A word of caution should be made here about the indiscriminate use of the word "denaturation" (Kauzmann 1959; Tanford 1968, 1970). We have to be aware that small conformational changes that affect positions in the range only of Ångström units might be sufficient for biological inactivation, although we will certainly not be able to see the change in the electron microscope, even with the best methods of image processing available. Other changes related to biological functions are supposed to be much larger, such that electron microscopy might image them in the future (discussed in Kellenberger 1978). It is a bad habit of electron microscopists to use "denaturation" for describing coarse aggregates in ground plasms, since we do not know at all if conformational changes of the macromolecules have really occurred.

2.3 The Fate of Hydration Shells During Cryofixation and After Subsequent Dehydration Processes

2.3.1 Hydration Shells in Ice

Organic molecules are predicted to become more hydrophilic with lower temperatures (see Sect. 2.1). Low temperatures would thus promote the formation of hydration shells. Moreover, hydration shells probably provide an excellent integration of biological material into surrounding ice [a type of good copolymerization, as investigated for resins by Acetarin et al. (1987)].

According to this theory, the protein molecule should not suffer from freezing as far as the hydration shell is concerned. The increase of these shells at low temperatures certainly cannot have as dramatic an effect as their removal by organic liquids at room temperature or above. By extrapolation of experimental data and thermodynamic considerations, however, a "cold" denaturation is predicted (Cantor and Schimmel 1980, pp. 1079–1081). It is intuitively understandable because of the decreased hydrophobicity of the amino acid residues of the interior of a protein. They could now interact with water and thus the protein would behave similarly as in an organic solvent: by turning the inside out. Fortunately, however, rearrangements of the peptide chain are extremely slow at low temperatures. Only temperatures above some 320 K allow for the required, simultaneous openings of weak interactions, so that conformational changes can rapidly occur. In the temperature range between 200 K and 270 K it is virtually impossible to predict conformational changes of proteins through rearrangement of the peptide chain. However, movements of flexible parts or domains (Colman et al. 1976) and those associated with other elastic deformations are obviously still foreseeable.

When not observing frozen-hydrated specimens, the ice is removed either by substitution with an organic liquid (freeze-substitution) or by sublimation (freeze-drying) (see Steinbrecht and Müller, Chap. 7, this Vol.). The fate of hydration shells in these procedures and the consequences of their removal is described below.

2.3.2 Effects of Freeze-Drying

Freeze-drying incurs many potential problems. The temperature must be above a certain threshold in order to allow sublimation within a reasonable time. Therefore, crystallization may occur with concomitant phase separation (MacKenzie 1977). Crystals of pure ice will form and the solutes will assemble in the grain boundaries. This freezing collapse is difficult to distinguish from aggregations and collapses that occur later with the already dried material (see Sect. 5).

In Chapter 10 (this Vol.), H. Gross reports studies about "residual" water that remains in biological material in increasing amounts with lowered temperatures of sublimation. It is most likely that this residual water is that of hydration shells. The strong binding of the hydration water reduces the vapour pressure of water as it reduces the exchangeability of water molecules between the shell and the surrounding medium. It is a very sound working hypothesis to assume that the protein molecules are protected by a persisting hydration shell against conformational changes regardless of whether these occur spontaneously or are induced by surface tension (see Sect. 3).

When considering dissolved molecules, it is obvious that upon freeze-drying they will adhere to the next "solid" of the neighbourhood as aggregates (Fig. 2). Even if freezing was perfectly vitreous and therefore the spatial distribution of solutes in ice maintained random as before, this state can obviously not persist once the ice is sublimated.

These types of aggregations and freezing collapses are of particular concern for low resolution work when freeze-dried or critical-point dried cells are to be observed in the high voltage electron microscope (Porter 1984) or when the freeze-dried material is infiltrated by an organic liquid in view of a resin embedment and sectioning (Sjöstrand and Kretzer 1975; Chiovetti et al. 1986; see also Sect. 5).

Experiments made 10 years ago in our laboratory (J. Dubochet, unpubl.; Kellenberger and Kistler 1979) are in agreement with the arguments given by H. Gross (Chap. 10, this Vol.) and support the view that protein conformation is better preserved when maintaining the hydration shell. They were produced on sensitive structures for which it is relatively easy to decide which structure is most faithful to the native state: bacterial flagella were shadowed immediately after freeze-drying and later observed together with their shadows. The shadows showed much smoother bends than the flagella themselves. Whether this distortion was due to the removal of the hydration shells or to subsequent, electron beam-induced destruction remained open and is answered by the following experiment. Flagella, dehydrated at room temperature in ethanol and then dried either directly or by passing through ether, showed similarly distorted features which are not visible on flagella prepared by air drying on a supporting film. Together, these observations are in favour of the hypothesis that the distortions are caused by the removal of the hydration shells.

2.3.3 Effects of Freeze-Substitution

Freeze-substitution involves two levels of events. The first are the aggregations of macromolecules and supramolecular structures in organic liquids as discussed in Section 2.2 and summarized in Fig. 2. The second comprises much less defined events happening in the transition zone between ice-containing material and the material in pure substitution liquid. We know nothing about the depth of this zone, nor about the precise quantitative relations of the mixture of water and solvent (for the solubility of water in substitution liquids, see Steinbrecht and Müller,

Chap. 7, this Vol.). The substitution liquid frequently contains a chemical crosslinker, but we know nearly nothing about its reactivity at these temperatures and in the water-solvent mixtures of the transition zone. The only thing we fortunately know is that the lower the temperature, the finer are the aggregates of macromolecular solutes that are produced in organic liquids (see Sects. 2.2.2 and 2.2.3). This explains why freeze-substitution is apparently effective even without added fixatives.

The role and fate of hydration shells in freeze-substitution has not yet been studied in as precise a manner as has been done for freeze-drying (see previous section and Gross, Chap. 10, this Vol.). Since hydration shells are bound the tighter the lower the temperature is, it is reasonable to assume that some temperature limits exist, below which hydration shells are preserved. These limits certainly depend also on the polarity of the organic liquids involved in substitution and embedding. Preliminary experiments revealed that it is not justified to maintain uncritically the widespread belief that the preservation of the hydration shells is advantageous also in the case of resin embedding. It is predictable that the hydration shells prevent an efficient copolymerization between biological material and resin and that as a consequence, the two might separate too easily (Kellenberger et al. 1986c). This might have an advantage, on the one hand, for purposes of immunocytochemistry by revealing antigenic sites in the cleavage plane of sectioning (Kellenberger et al. 1987), but, on the other hand, the disadvantage in producing distortions during sectioning (Acetarin et al. 1987).

2.3.4 Consequences of the Removal of Water: Summary

1. Proteins and nucleic acids are not soluble in organic liquids, whether the hydration shell is removed or not. Thus, aggregations cannot be avoided, only their size can be minimized.
2. Biological macromolecules cannot form a gas. During freeze-drying and freeze-substitution macromolecules of a sol will aggregate, but the size of the aggregates and the resulting network are finer with lower temperatures.
3. Proteins are compact solids which in some parts are softer than in others. Having nothing to lose, they cannot shrink, but only change conformation. The hydration shell plays an essential role in determining the conformation of a protein besides the amino acid sequence. Removing the hydration shell potentially induces a conformational shift. Whether or not it occurs within a reasonable lapse of time depends on temperature and on the dielectric properties and other properties of the surrounding liquid.

3 Phases and Interface Phenomena

3.1 Phases in Biological Material

By virtue of mutual attraction, counteracted by thermal movement, molecules will very strongly cohere (solids), moderately cohere (liquids) or not cohere at all (gases). Biological macromolecules, and even their chemical constitutents (sugars, nucleotides, amino acids) are characterized by a virtually absent gas pressure because their cohesion forces are extremely strong. These substances do not exist as gases; when raising the temperature, they decompose before reaching the gaseous state. Practically, this means that biological macromolecules will not be lost upon freeze-drying. Since biological macromolecules are also insoluble in organic liquids, they will also not be lost during freeze-substitution. This situation might be very different for small molecules (see Steinbrecht and Müller, Chapt. 7, this Vol., Zierold and Steinbrecht, Chapt. 15, this Vol.).

The phases are discussed in more detail by Bachmann and Mayer (Chapt. 1, this Vol.). We summarize here some particular features that are relevant to our discussion on interfaces.

When slowly lowering the temperature of an aqueous protein solution below the freezing point, we will observe the progressive formation of crystals of pure ice. The concentration of the remaining solution will increase until it reaches the solubility point of the protein, which then will precipitate. The precipitate can be, in particular, in the form of a protein crystal (i.e. a water-containing gel, see Sect. 2.2.2) or, more generally, in that of a random aggregate, a "lump". Such lumps are currently produced in expressing clones of genes in bacteria (Guo et al. 1986). Interestingly enough, they cannot be dissolved by simple dilution; one has to pass them through denaturation in unfolding agents (urea or guanidinium chloride) and renaturation to obtain a soluble, active product. Something irreversible must have happened to the protein, as is also observed with many denaturations (e.g. by heat or foaming, see below).

The phase separation as described above (into three phases ice, aggregated protein and saturated protein solution) accounts for that reached at equilibrium. As is well known (see Chaps. 1 and 5, this Vol.) by rapid freezing a non-equilibrium vitreous state can be reached in which the random distribution of the dissolved macromolecules is maintained. The stability of the vitreous state is strongly temperature-dependent. The higher the temperature, the faster the equilibrium will be reached by crystallization into pure ice and concomitant fallout of solutes in the form of aggregates. This "freezing collapse" (MacKenzie 1977) occurs prior to that occurring later, i.e. during freeze-drying or freeze-substitution.

3.2 Interfaces

The limit between two clearly defined phases is called interface. We distinguish interfaces between solid and liquid and liquid and gases (or vacuum). A protein solution on a carbon film has two interfaces: solution to carbon film and solution to atmosphere. The mere existence of a boundary (surface, interface) in a liquid or solid is also the consequence of the cohesive forces between molecules. When comparing the cohesive forces acting on a molecule within the liquid or on its surface, we immediately see (Fig. 4a′) that the first is attracted equally (on the average) in all directions, while the other is strongly attracted towards the inside. The forces exerted on the molecule at the surface by molecules of any gas above the liquid are negligible. [Except under the so-called critical conditions of the gas,

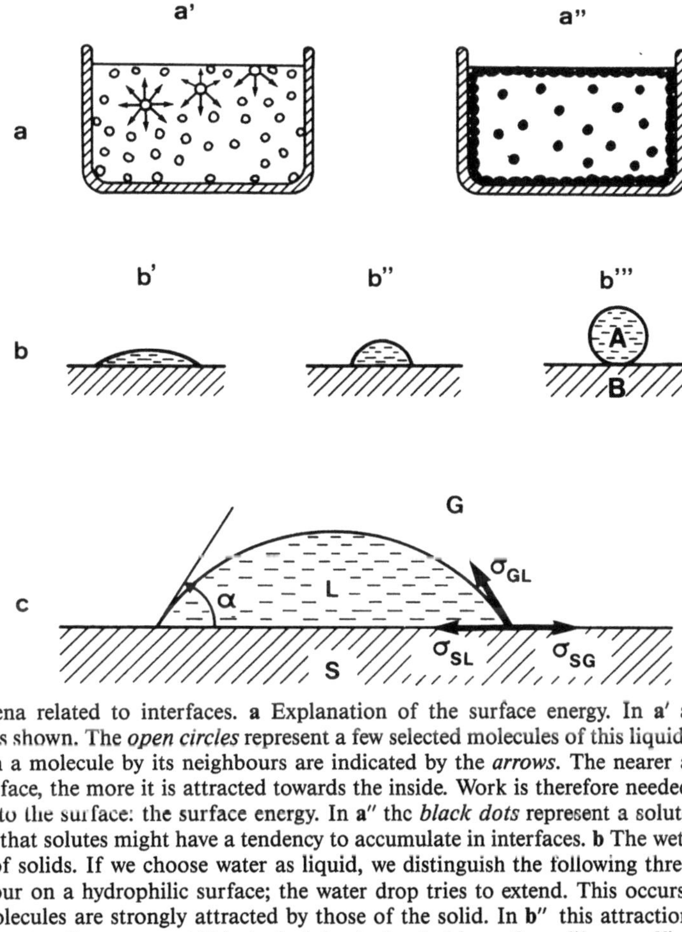

Fig. 4a–c. Phenomena related to interfaces. **a** Explanation of the surface energy. In **a'** a homogeneous liquid is shown. The *open circles* represent a few selected molecules of this liquid. The forces exerted on a molecule by its neighbours are indicated by the *arrows*. The nearer a molecule is to the surface, the more it is attracted towards the inside. Work is therefore needed to move molecules into the surface: the surface energy. In **a''** the *black dots* represent a solute in a solvent. It shows that solutes might have a tendency to accumulate in interfaces. **b** The wetting of the surfaces of solids. If we choose water as liquid, we distinguish the following three cases: **b'** The behaviour on a hydrophilic surface; the water drop tries to extend. This occurs, because the water molecules are strongly attracted by those of the solid. In **b''** this attraction is less, and in **b'''** no attraction occurs. **b'''** is typical for hydrophobic surfaces like paraffin, polyethylene, parafilm and silicone. Due only to gravity, the drop (*A*) is slightly flattened on its interface with the solid (*B*). **c** The surface tensions acting on the "edge" of a drop. The (σ's are the surface tensions between gas and liquid (σ_{GL}), liquid and solid (σ_{SL}) and solid and gas (σ_{SG}). The total surface energies are obtained by multiplying the σ's with their corresponding surfaces. The shape of the drop is determined by the sum of these energies being a minimum. The wetting angle a can be calculated from the three surface energies as $\sigma_{GL} \cos a + \sigma_{SG} = \sigma_{SL}$ and $\cos a = (\sigma_{SL} - \sigma_{SG})/\sigma_{GL}$; *G* gas, *L* liquid, *S* solid

where the interface disappears (Anderson 1951; Hayat and Zirkin 1973)]. When moving a molecule from the inside to the surface, work has to be done against the resulting force that pulls the molecule back to the interior. The work needed to form one unit area of surface is called the surface energy. According to minimal energy principles, it is clear that the surface wants to become as small as possible. The force which tries to contract the surface and acts in all directions parallel to it is called the surface tension; it is identical to the surface energy (an energy per unit area has the same dimension as a force per unit length).

In the interfaces between two liquids, two solids or between liquids and solids we also have surface energies. A molecule positioned in the interface will be attracted by the molecules of both phases (A) and (B) that are in contact. These forces will normally be unequal. Thus, we will observe different responses as described in Fig. 4b for a liquid drop (A) on a solid (B), depending on the wettability of the solid (B) by the liquid (A). These are situations that we commonly observe with water on hydrophilic or hydrophobic surfaces.

It is rather obvious that those substances that tend to form hydration shells are also those which are wettable by aqueous media. Hydrophobic substances are not wettable, do not form hydration shells (at room temperature) and are also not soluble or suspendable in aqueous media.

Figure 4c illustrates the principle of minimum energy: the drop of liquid will assume such a form that all the interface energies (liquid-gas, liquid-solid and solid-gas) and the potential energy of gravity will be minimal.

3.3 Surface Energy Effects in the Preparation of Thin Specimen Layers

Most macromolecules and biological particles in aqueous media will have a tendency to enrich in interfaces. The extent of enrichment depends on the respective nature of both the particle and the interface. In normal preparation procedures, one chooses supporting films of adequate charge and hydrophilicity to adsorb particles from an overlaid drop of suspension. This adsorption is nothing more than the enrichment in an interphase, here between a solid and an aqueous medium (see Sect. 3.4). Similar enrichments occur in liquid-gas interfaces, obviously again depending on the nature of the particle and on that of the interface (see Sect. 3.5). In all these interactions, accompanying molecules might either promote or prevent this enrichment. Although these predictions are only of a qualitative, conceptional nature, they should help in devising adequate enrichment procedures for frozen-hydrated specimens either with or without supporting film.

3.4 Adsorption to Supporting Films

Hollow cylinders of relatively large diameter and comparatively thin walls are suitable test specimens. Tubular protein assemblies occur frequently in nature, for instance as variants of many viruses or are produced in vitro from many enzymes and other proteins. They have walls that are between 4 and 10 nm thick and are relatively flexible. When a layer of an aqueous suspension of such tubes is deposited on a supporting film, and the tubes allowed to adsorb, it can easily happen that they flatten out completely (Fig. 5a). Whether this happens or not depends on the charges and wetting abilities of the two interacting surfaces. Flattening might thus already occur when the liquid is still present; it will then be visible after rapid freezing and observing in the frozen-hydrated state. This has been shown by Brisson and Unwin (1985) on tubular assemblies of the acetylcholine receptor.

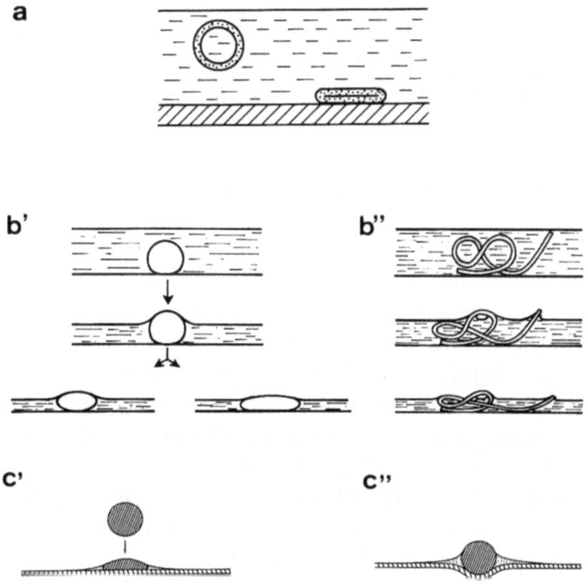

Fig. 5a–c. Different flattening collapses. **a** Flattening of a tubular structure by an interface liquid-solid. A tubular assembly of proteins suspended in water is brought onto a solid support. In many cases the tube will become flattened when adsorbing onto the solid. This depends again on the surface energies of the protein with water or with the solid. If the protein ist strongly attracted by the molecules of the solid, then it will adsorb and flatten. **b** The effects of surface tension during removal of a (wetting) liquid. **b′** The effect of surface tension on a deformable, wettable particle (e.g. a virus) adsorbed to a solid when the amount of liquid is steadily reduced. **b″** The same for a fibrous structure. **c** The effect of wrapping. **c″** The wrapping phenomenon that occurs when the supporting film is more deformable than the particle; it is compared with the other extreme of a rigid supporting film and flexible particle (**c′**). (**b** From Kellenberger and Kistler 1979; **c** after Kellenberger et al. 1982)

Flattening as a consequence of adsorption (interface solid-liquid) has to be distinguished from that produced when the surface of a liquid passes through a specimen during drying as will be discussed in Section 3.6. Examples for similar tubular structures as those discussed above, but prepared by drying in negative stain, have been reported by Yanagida et al. (1970) and Brisson and Unwin (1984).

For the adsorption to a supporting film, the complementarity of charges between the surfaces of the solid and the particle is a necessary condition. The better the adsorption, the stronger the tendency to flatten. A strong adsorption is a prerequisite in most of the current preparation methods for solutions in which the suspension medium is withdrawn. Such a strong adsorption is not needed in a preparation method in which the liquid is left, as is the case, for instance with frozen-hydrated layers (see Dubochet et al., Chap. 5, this Vol.), or in agar filtration (Kellenberger and Bitterli 1976). In this latter method particles are collected by depositing them by filtration through the supporting collodion membrane.

When prepared from suspensions, filaments, like DNA, will become partly attached to supporting films, provided the film is charged positively. Obviously, upon drying, they will be flattened down by the surface tension so as to adhere all along (Fig. 5b).

According to obvious geometrical and mechanical reasons, average globular particles are not able to spread-flatten as easily as a hollow cylinder. In these cases the supporting film frequently wraps itself around part of the particle (Fig. 5c″).

3.5 Events in the Interface Liquid-Gas

Many solutes are known which have a tendency to accumulate at the surface of aqueous media. These "surface-active" or "tensio-active" substances need less energy for movement to the surface than the solvent molecules. These substances form therefore a surface layer (Fig. 4a″). Well-known examples are the fatty acids and the lipids. Both have a hydrophobic part, the aliphatic chain, and a hydrophilic "head" that is polar or even ionizable. These molecules will arrange on a surface and form a closely packed monolayer with the aliphatic chains pointing into the gas. Other representatives of this class of substances are the detergents. It is rather obvious that such layers will also be formed in interphases between water and a hydrophobic solvent, or between water and hydrophobic solids (paraffin, parafilm, polyethylene).

Also, proteins frequently form surface layers, e.g. cytochrome C. Proteins of such layers are in most cases denatured and often are no longer able to return into solution. Foaming is known to irreversibly inactivate enzymes, viruses and other biological particles in suspension (Joly 1965). By foaming the surface is very much increased and continuously renewed, so that a large portion of the biological particles of the solution will eventually become positioned at the surface before they return into suspension. Denaturation in this case is defined by a loss of biological activity. As discussed already, such a loss can already be achieved by minute conformational changes that are far below the detectability of electron microscopy, although much stronger, visible changes are by no means excluded, e.g. flattening. Macromolecular solutes and/or suspended particles are thus likely to enrich on interfaces, whether these are formed with a solid (Sect. 3.4) or with a gas (or vacuum) (Sect. 3.5, Fig. 4a″). According to the nature of the solute, the reverse might also occur. Measurable parameters are not yet available for predicting the concentrations in the liquid and in the interfaces, which would be most useful for frozen-hydrated specimens. We have taken as an example a suspension of bacteriophage T4 prepared without supporting film to form a layer that is a little thicker than the width of this phage. If we would like to observe only one phage on every 10 μm^2 and assuming a layer thickness of 0.1 μm, then we would need a concentration of 10^{12} phages ml^{-1}. Experience shows that concentrations of one to two orders of magnitude less still produce good preparations. In fact, the required concentrations are similar to those needed for negative stain preparations. This can be explained by either or both of two events: (1) phages have enriched on the interface and/or (2) during preparation of the liquid layer a volume decrease by drying has occurred. J. Dubochet (pers. commun.; see also Chap. 5, this Vol.) has some evidence in favour of a preponderance of (1). The observed large differences between different specimens is in agreement with a particle-specific occupation of the interface as mentioned above. He found, in addition, that drying seems to become very slow, once a certain lower limit of thick-

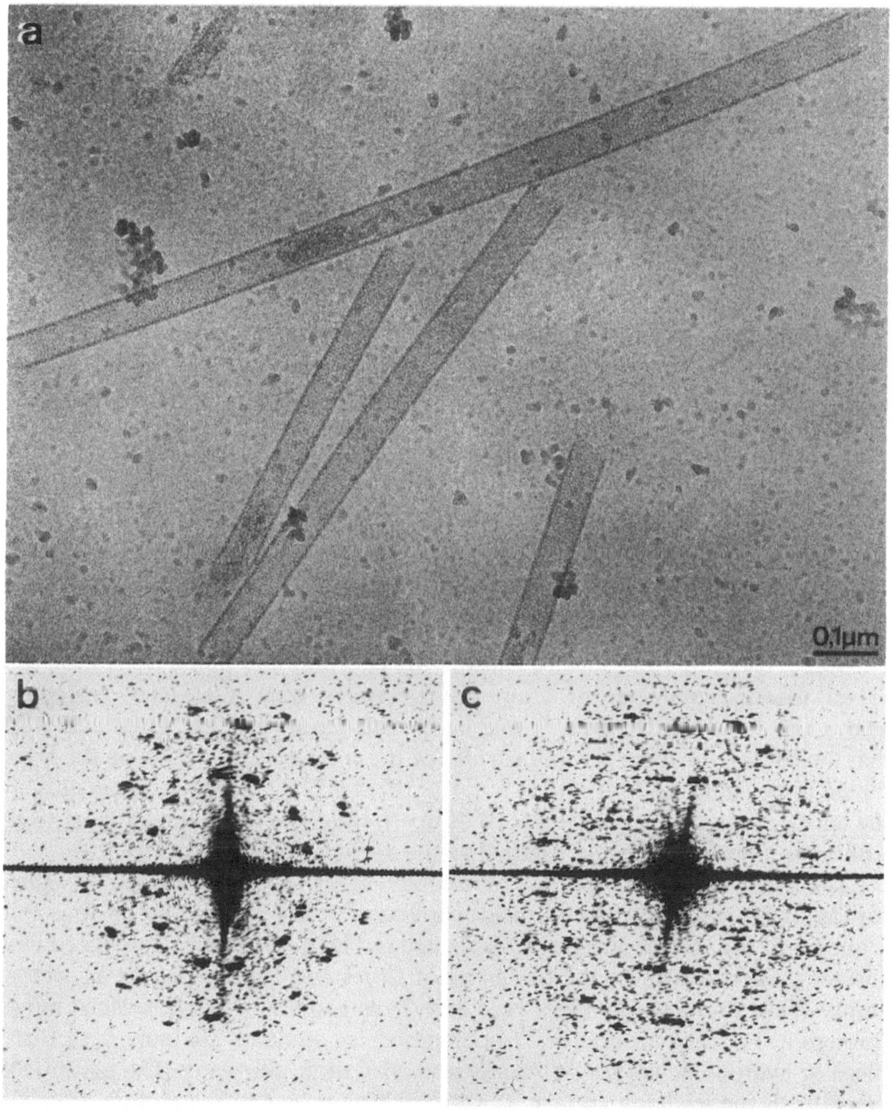

Fig. 6a–c. Tubular variant of the prohead of bacteriophage T4. **a** Micrograph of a frozen-hydrated specimen. A suspension of the so-called polyheads isolated by C. and W. Baschong in our laboratory was studied frozen-hydrated, without supporting film by J. Lepault, EMBL Heidelberg. He took the micrograph at an underfocus of about 4 µm, in order to produce sufficient phase-contrast. Preparations of this type were analyzed by diffraction, as shown in **b**, in order to judge flattening. **b, c** Diffractions from tubular prohead variants as shown in **a**. **b** The diffraction pattern of a relatively flattened particle with the sharp diffraction spots, as if produced by a completely flat lattice. **c** The diffraction pattern of non-flattened particles of elliptic or circular cross-section. They show the characteristic equatorial spreading of the spots according to Bessel functions. Most particles observed in this frozen-hydrated specimen prepared without supporting film were unflattened. This study of flattening on liquid-gas interfaces was made by J. Lepault, EMBL Heidelberg. (Reported in Kellenberger et al. 1986a)

ness is achieved. This could again be explained by the lowered water pressure of hydration shells: when the liquid layer has become reduced in thickness, such as to be formed essentially by particles with their hydration shells, the water pressure also becomes strongly decreased.

A bacteriophage T4, with mutations in the form-determining genes (but not in the gene producing the shell protein), produces tubular variants of the prohead, the so-called polyheads. J. Lepault at EMBL, Heidelberg, has used them to check whether they are flattened on the liquid-gas interface before freezing. He found only few completely flattened particles; most were still producing layer lines according to Bessel functions, typical for cylindrical or near cylindrical tubes (Fig. 6).

3.6 Surface Tension Effects Associated with Removing or Adding Liquids: The Flattening Effect of Capillarity

Very early T.F. Anderson (1952) had recognized the effects of surface tension on biological material when a liquid surface is moving "through" it. As we have seen above, isolated biological particles in aqueous suspension are all hydrophilic, i.e. wettable. Aqueous media will attempt to "sneak" along their surfaces. Hydrophilic fibres, like DNA, hanging into an aqueous medium will become pulled into the liquid, because it tends to "sneak" upwards along the fibres (Fig. 5b″). The forces resulting from surface tensions are enormous: to pull a DNA molecule of an assessed diameter of 2 nm out of water with a surface tension of $72\,\mathrm{dyn\,cm^{-1}}$ ($=72\times10^{3}\,\mathrm{N\,m^{-1}}$) one easily calculates the total force and from that the tension (= force per unit surface of the cross-section) acting on the DNA thread to be about $1.5\,\mathrm{tons\,cm^{-2}}$ ($=1.47\times10^{8}\,\mathrm{N\,m^{-2}}$). This value comes close to the tensile strength of steel ($5-7\times10^{8}\,\mathrm{N\,m^{-2}}$); thus, the DNA thread most likely will break under the tension exerted by the surface tension.

The phenomenon of liquids "sneaking" along wettable surfaces is called capillarity because it also occurs inside capillaries. It is a natural and obvious consequence of the surface energies as explained in Fig. 4. Capillarity associated with surface tension is seen by the flattening of fibres and fibrous plasms, as well as empty shells as shown in Figs. 5b, c and 7, as soon as a liquid surface moves through it. For the purposes of demonstration for students, we immerge a fluffy piece of hydrophilic cotton into water and then drain the water. The cotton collapses with the lowering surface. It will not regain its fluffy structure after air-drying nor when immerged into water again. Similar situations occur always when preparing a specimen by drying from a suspension. If the particle contains water, like a virus, it will flatten through capillarity, but after drying it will not regain its initial spatial extension, exactly as exemplified with cotton (Fig. 5b″). Most commonly, the suspension medium is water. It could, however, be any organic solvent or liquid. Some of these solvents have much lower surface tensions than water (Table 1). It has frequently been proposed to replace water by a sequential series of solvents, ending in one with low surface tension and thus in this way reducing the flattening effects. As we have briefly mentioned in Section 2.3.3, the results are disappointing, because the effects of the solvents per se are frequently more deleterious than those of surface tension.

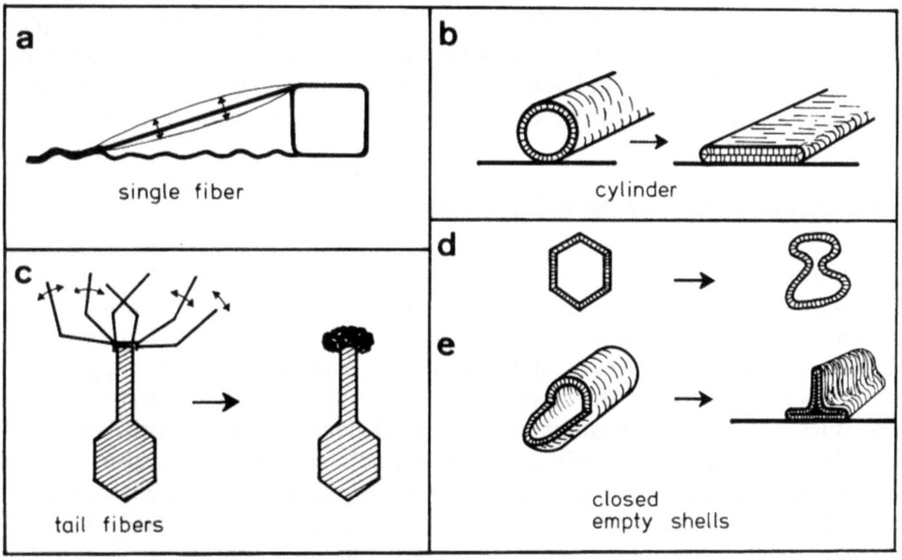

Fig. 7a–e. Collapses as consequences of thermal movements and as observed with freeze-dried material. **a** The thermal vibration of a suspended fibre will lead to an increased adhesion and thus to increased tension of the fibre. It will finally break (Anderson 1954). **b** A cylinder collapsing on the support. **c** The tail fibres of a bacteriophage will collapse onto themselves and form a knob. The actual micrographs are shown in Fig. 8. **d, e** Empty shells, e.g. empty capsids of viruses will have a tendency to collapse onto themselves. A micrograph of collapsed, empty bacterial envelopes is given in Fig. 9 (Kellenberger and Kistler 1979)

Table 1. Surface tension of various solvents (in contact with air or with their own vapour at 20 °C)

	$\text{dyn cm}^{-1} = 10^{-3}\text{N m}^{-1}$ $\text{erg cm}^{-2} = 10^{-3}\text{J m}^{-2}$
Water	72.5
Ethylene glycol	47.7
Acetone	23.7
Ethanol	22.7
Methanol	22.6
Diethylether	17.0

Critical-point drying (Hayat and Zirkin 1973) is obviously also an interesting method for avoiding surface-tension effects. It has given excellent results with relatively sturdy material, e.g. insects, in scanning electron microscopy. The use of the method for high resolution work with sensitive material (Porter 1984) is again limited by the aggregations associated with the necessary steps of substitution in organic solvents.

Capillarity effects, similar to those described above (Fig. 5b), but acting in a reversed manner, might occur when freeze-dried material (e.g. cells) will have to be infiltrated by a resin. Should the need become apparent, a reversed critical-point method can be designed for infiltration of freeze-dried material. It would first be penetrated by the gas (CO_2 or Freon) and then, by bringing the sample to the critical conditions of pressure and temperature, the highly compressed gas would transform into a liquid without forming an interphase. This liquid must then be substituted by adequate organic solvents and the uncured resin.

4 The Particular Case of the Biological Membrane

The biological membrane, composed of lipids and proteins, sometimes with polysaccharides added at the surface, is a very special structure. It forms a very narrow hydrophobic phase in which the two surfaces are hydrophilic in order to be able to exist in the aqueous medium of biological organisms. The argument of existence can be turned the other way around: without being "surrounded" on both sides by an aqueous medium, the biological membrane would lose its stability, as it does when it is brought into a hydrophobic environment (solvents). As we have mentioned previously, the membrane proteins have evolved such that hydrophobic residues are on that part of its surface which is "immersed" into the aliphatic phase. Ion pumps and pores are oligomeric proteins that form hydrophilic channels through the lipid phase (Rosenbusch et al. 1982). Membrane proteins can be brought into monodisperse aqueous solution with the help of adequate detergents that replace the lipids (Dorset and Rosenbusch 1983).

The fate of a biological membrane prepared by adsorption to a supporting film and drying with or without negative stain is virtually unpredictable. It depends on parameters like the ratio of proteins to lipids, which in most cases is rarely known (except average values!), to predict how the unfixed lipids will rearrange in or on a non-aqueous medium.

No wonder that also other preparation methods for the electron microscopy of biological membranes have been accompanied by accidents. Lipids are dissolved in all current dehydration liquids and monomeric resins (Weibull et al. 1983, 1984) unless they are fixed with OsO_4. Os uptake occurs only with unsaturated fatty acids, but the trilaminar aspect appears even with saturated ones (Rilfors and Weibull 1985). Aldehyde treatment does not prevent solubilization in the above mentioned liquids. When the temperature is lowered below about 200 K, most lipids become insoluble (Weibull et al. 1984). Some new resins allow one to work in this temperature range (Acetarin et al. 1986). For Nanoplast, a melamin-based resin (Bachhuber and Frösch 1983), some observations suggest the possibility, that with this resin lipids are preserved even when embedded at room temperature and without OsO_4 fixation (Westphal and Frösch 1984).

The rather queer structure and behaviour of biological membranes gave rise to the Davson-Danielli-Robertson model which was known to be incorrect from evidence of biochemistry and genetics long before the new model (Nicholson-Singer) was accepted as a replacement. The new model finally introduced the concept of transmembrane proteins which were predicted from cellular transports. Freeze-fracturing was the electron microscope method that gave support to this model. Unfortunately, this method mainly provides information on the arrangement of proteins in the plane of the membrane, while functionally more important is information concerning the functions across the biological membranes.

Theoretically, thin sections would be ideally suited for providing cross-sections of membranes. It is hoped that in this biologically so important field, modern cryomethods other than cryofractures will also become successful (Kellenberger et al. 1986b).

5 A Potpourri of Collapses

We will use "collapse" in the sense of "caving in", i.e. something getting distorted by aggregations on itself. With this definition, every collapse is therefore also accompanied by a local increase of compactness. Collapses with biological material are essentially constituted by aggregations of biological macromolecules. Protein molecules themselves cannot collapse, nor shrink, because they are by nature built most compactly. In their interior they contain no removable small molecules like water. The only change a protein can undergo is conformational. A fibre can for instance shrivel up to form a sphere, as we will see below. Assemblies of protein molecules (supramolecular structures) are always porous. These pores are filled with water or, in the biological membranes, with lipids. Nucleic acids are highly hydrated, even when closely packed. Such water-containing structures are susceptible to collapse when the water is removed, as was clearly illustrated by the collapse of purified glycogen particles (Arber et al. 1957).

Another typical collapse is illustrated in Fig. 8 which shows bacteriophage T4 tail fibres when collapsed to a knob (Fig. 8b) and when preserved in the extended, normal form after previous adsorption to a supporting film (Fig. 8a) (Kellenberger and Kistler 1979). Both preparations are by freeze-drying. In the first case it was made according to Williams (1953) by spraying aerosol droplets of a phage suspension onto a copper block cooled to the temperature of liquid nitrogen (77 K). The frozen microdroplets were then kept at the temperature of solid carbon-dioxide (194 K) and sublimated against a liquid nitrogen trap. In Fig. 8a, the phage suspension was applied to the film and phages allowed to adsorb to it, before freezing (Kistler and Kellenberger 1977). By the adsorption to the film, the fibres are prevented from collapse. With the first method they collapsed during sublimation before they reached the supporting film. This is confirmed by the occasional presence of phages with extended fibres which have reached the film before collapsing. Bacterial viruses had been prepared very successfully by freeze-drying (Williams and Fraser 1953). The "angular" aspect, even of small globular viruses, was interpreted as faithfully representing their polyhedral shape. The angularity was taken as a criterium of good preservation, although it was soon after well explained that the icosahedral symmetry of the arrangement of protein subunits is compatible with a perfect sphere (Crick and Watson 1956). With frozen-hydrated specimens (see Dubochet et al., Chap. 5, this Vol.) the "rounded" aspect of icosahedral viruses is now well demonstrated.

In the prevailing climate of that time of overestimating the virtues of freeze-drying explains why the distal knobs of the T2 bacteriophages (Fig. 8b) was also mistaken by many (but not by Williams himself!) as a definite proof for the non-existence of free tail fibers as were observed with other preparation methods. This incertitude lasted until the existence of tail fibers was proven definitively by sedi-

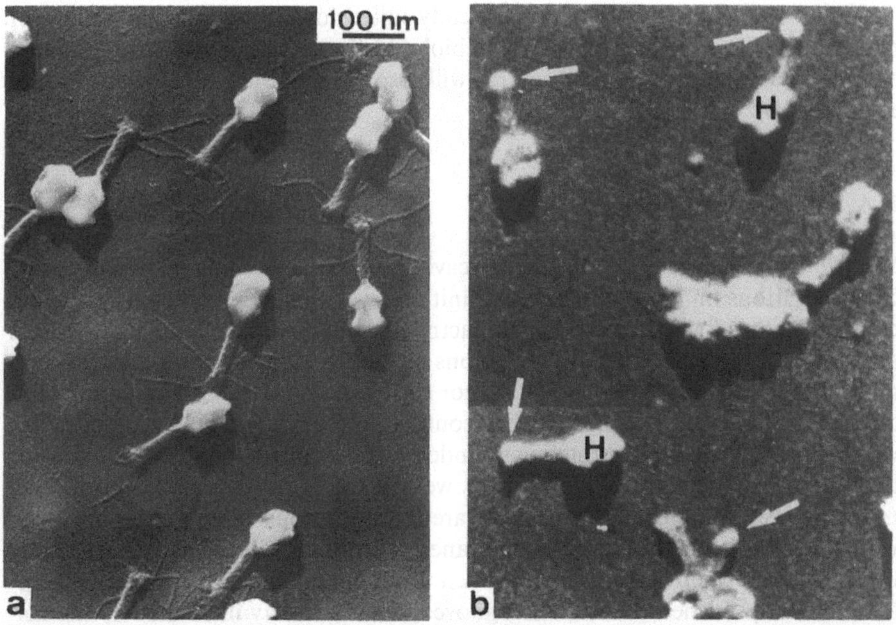

Fig. 8a, b. Bacteriophage T4 prepared with two different procedures of freeze-drying. **a** As discussed in the text, the particles were allowed to adsorb to the supporting film prior to rapid freezing and freeze-drying. The tail fibres are extended and well visible. **b** Preparation according to Williams (1953) by projection of aerosol droplets onto a copper block at the temperature of liquid N_2. The frozen droplets are then brought to the temperature of solid CO_2 in acetone and sublimated against a cooling trap with liquid N_2. The particles could attach to the supporting film on the copper block only when the ice had sublimated away. During this time, the fibres have collapsed into a knob at the distal part of the tail (*arrows*) (Williams and Fraser 1953). **a, b** were chosen from micrographs obtained from ghosts, in order to also show the typical collapses of the empty heads (*H*) (Kistler and Kellenberger 1977)

mentation studies with adequate mutants (Kellenberger et al. 1965). Now the experimental results reported above and in Fig. 8 demonstrate unambiguously the occurrence of a thermal collapse prior to shadowing and observation. The formation of the "knob" does not require any transfer of energy from the electron beam. The interpretation of this collapse as a consequence of the loss of the hydration shell associated with thermal movements is thus justified.

The role of hydration shells in stabilizing and possibly rigidifying flexible structures remains to be investigated. It is certain that the hydration shell stabilizes secondary and tertiary structures of proteins. Its removal would promote conformational changes. As we have seen, a rearrangement of the peptide chain of a globular protein at low temperatures is probably slow and thus conformational changes are not to be expected as long as higher temperatures are avoided (e.g. polymerization at 330–350 K). For fibrillar proteins conformational changes might be easier to induce as shown by the example of the phage tail fibres above.

Other collapses related to freezing have been studied by MacKenzie (1977). We have mentioned them before (Sect. 3.1); they are due to phase separations during ice-crystal formation. Collapses occur also after or during freeze-drying because

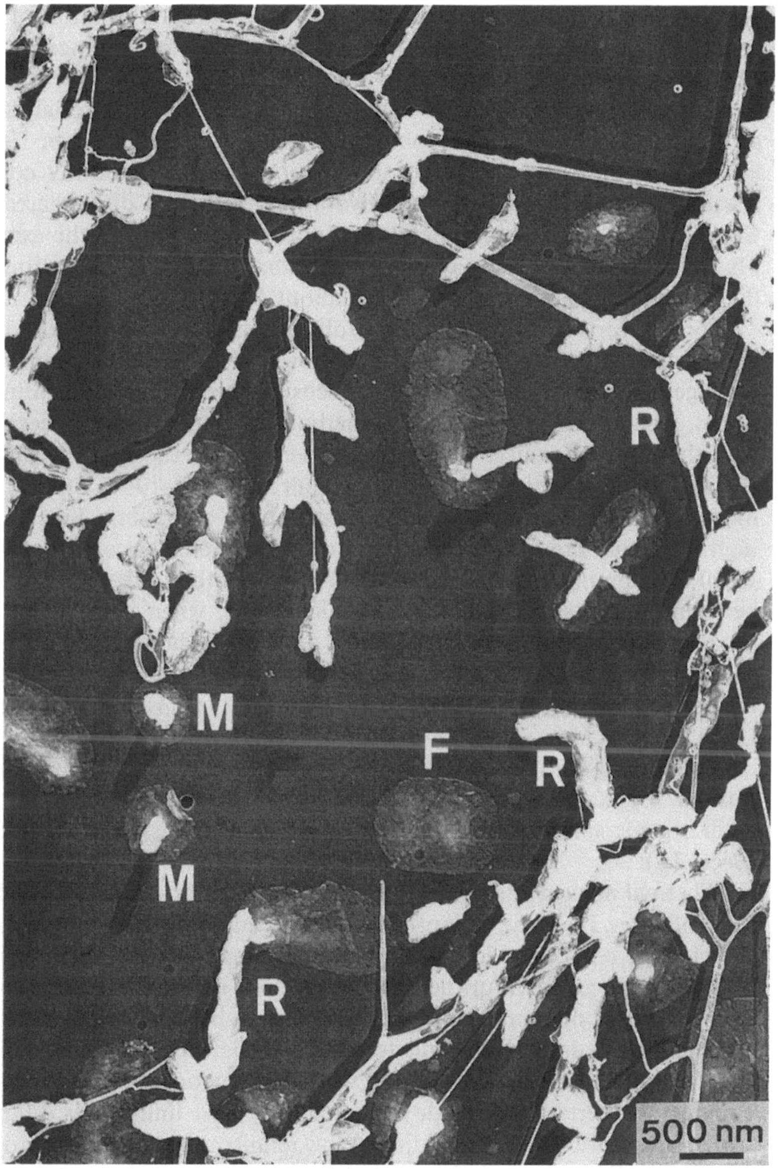

Fig. 9. Freeze-dried, empty bacterial (*E. coli*) envelopes. A suspension of empty bacterial envelopes was prepared in the same manner as the phages (Fig. 8a) by allowing for adsorption to the supporting film prior to freezing. Some envelopes have completely flattened down onto the film (*F*). Others, that have not yet adsorbed, have shrivelled up into dense rods (*R*). A few are only partially adsorbed and still show the remaining part as shrivelled rods (*M*) (Kellenberger and Kistler 1979)

at room temperature some structures vibrate as a manifestation of the thermal energy (Anderson 1954). These thermal vibrations, for instance of a flagellum, may be so strong that they lead to fractures. In Fig. 9 and schematically in Fig. 7, we show other examples of such collapses, of empty bacterial envelopes and of tubular variants of bacteriophage T4 (Kistler and Kellenberger 1977). In these latter cases the temperature of sublimation was not carefully controlled and possibly so high (238 K) that the hydration shells had disappeared (see Gross, Chap. 10, this Vol.); this has certainly not been the case in the example of Fig. 8b, where the phages had fallen onto the fibre-stabilizing supporting film before the temperature was increased above about 100 K.

For the further study of such problems, it is obviously most important to choose adequate test specimens which show structures that appear to be sensitive. Average globular proteins will not be sensitive even when studied in two-dimensional, regular arrays with modern methods of information processing. Proteins with thin "protuberances" have been observed, like those in the surface layer of *Bacillus brevis* (Aebi et al. 1973; Lepault and Pitt 1984) or of *Lampropedia hyalina* (R.G.E. Murray and J.W. Austin, pers. commun.). Single molecules of high sensitivity are for instance parts of the complement system (Tschopp et al. 1980) or laminin (Engel et al. 1981). There are other adequate test structures to be found and a search for them should always, when possible, also concern a broader biological interest than only a morphological description. The study of such problems has become more efficient through applying conditions in which the more or less good preservation of hydration shells is under control (see Gross, Chap. 10, this Vol.).

Other, well-defined collapses are those of many DNA-containing plasms or "naked" DNA. They occur with freeze-dried material (Richter et al. 1966) as well as in organic solvents. The latter cases we have discussed in some details above (Sect. 2.2).

Thermal movements of flexible parts are necessary to initiate collapses. Thermal collapses, i.e. those collapses which are caused by thermal movements, could therefore be reduced at lower temperatures. Observing a sample when still in ice is ideal in this respect. For the time being, these methods still suffer severe limitations. On the one hand, phase contrast has to be achieved by defocussing and, on the other, it is very difficult to obtain sufficiently thin, undistorted cryosections as required for high resolution work (see Dubochet et al., Chap. 5, this Vol.). Progress will certainly be made in overcoming these limitations. In the meantime, freeze-substitution and embedding at low temperatures has proven to be a promising avenue. Once the resin is cured at low temperature, the mobility of structural elements is now so much reduced that collapses are prevented even when the section is brought to room temperature. With freeze-dried material it is very likely that the observation at low temperatures (liquid nitrogen or even helium) might reduce collapses very substantially (Rachel et al. 1986).

Acknowledgements. I am grateful to the members of our research group for many fruitful discussions. This is particularly the case with Eric Carlemalm and Werner Villiger. I thank the editors and A. J. Verkleij for most useful comments on the first drafts of the manuscript. To Michel Wurtz and Margrit Jäggi I am most indepted for the preparation of the illustrations and to Marianne Schäfer and Regula Niederhauser for the reiterated typewriting.

References

(References marked by an asterisk are review papers or books respectively)

Acetarin JD, Carlemalm E, Villiger W (1986) Development of new Lowicryl resins for embedding biological specimens at even lower temperatures. J Microsc (Oxf) 143:81–88

Acetarin JD, Carlemalm E, Kellenberger E, Villiger W (1987) Correlation of some mechanical properties of embedding resins with their behaviour in microtomy. J Electron Microsc Techn 6:63–80

Aebi U, Smith PR, Dubochet J, Henry C, Kellenberger E (1973) A study of the structure of the T-layer of *Bacillus brevis*. J Supramol Struct 1:498–522

Anderson TF (1951) Techniques for the preservation of threedimensional structure in preparing specimens for the electron microscope. Trans N Y Acad Sci Ser II 13:130–134

Anderson TF (1952) A method for eliminating gross artifacts in drying specimens. In: C R 1er Congr Int Microsc Electron, Paris, pp 567–576

Anderson TF (1954) Some fundamental limitations to the preservation of 3-dimensional specimens for the electron microscope. Trans N Y Acad Sci 16:242–249

Arber W, Kellenberger E, Laszt L (1957) Morphologie von Leberglycogen. Kolloid Z 150:123–127

Bachhuber K, Frösch D (1983) Melamine resins, a new class of water-soluble embedding media for electron microscopy. J Microsc (Oxford) 130:1–9

Baschong W, Baschon-Prescianotto C, Wurtz M, Carlemalm E, Kellenberger C, Kellenberger E (1984) Preservation of protein structures for electron microscopy by fixation with aldehydes and/or OsO_4. Eur J Cell Biol 35:21–26

Bayer ME, Carlemalm E, Kellenberger E (1985) Capsule of *E. coli* K29: Ultrastructural preservation and immunoelectron microscopy. J Bacteriol 162:985–991

Brisson A, Unwin PNT (1984) Tubular crystals of acetylcholine receptor. J Cell Biol 99:1202 1211

Brisson A, Unwin PNT (1985) Quaternary structure of the acetylcholine receptor. Nature (London) 315:474–477

*Cantor Ch. R, Schimmel RP (1980) Biophysical chemistry. Freeman, San Francisco

Carlemalm E, Villiger W, Hobot JA, Acetarin JD, Kellenberger E (1985) Low temperature embedding with Lowicryl resins: two new formulations and some applications. J Mol Biol 140:55–63

Chiovetti R, Little SA, Brass-Dale J, McGuffee LJ (1986) A new approach to low temperature embedding: Quick freezing, freeze-drying and direct infiltration in Lowicryl K4M. In: Müller M, Becker R, Boyde A, Wolosewick J (eds) The science of biological specimen preparations 1985. SEM, AMF O'Hare, IL 60666, pp 155–164

Colman PM, Deisenhofer J, Huber R (1976) Structure of the human antibody molecule Kol (immunoglobulin G1): An electrondensity map at 5 Å resolution. J Mol Biol 100:257–282

Crick FHC, Watson JD (1956) The structure of small viruses. Nature 177:473–475

Dorset DL, Rosenbusch JP (1983) Properties of the anomeric octylglucosides used in the recrystallization of membrane proteins. In: Quagliariello E, Palmieri F (eds) Structure and function of membrane proteins. Elsevier, Amsterdam, pp 199–203

Duijn G van, Dekker J, Leunissen-Bijvelt J, Verkleij AJ, de Kruijff B (1985) Influence of trinitrophenylation on the structure and dynamics of phosphatidylethanolamine-containing model membranes. Biochemistry 24:7640–7650

Emerman M, Behrman EJ (1982) Cleavage and cross-linking of proteins with osmium (VIII) reagents. J Histochem Cytochem 30:395–397

Engel J, Odermatt E, Engel A, Madri JA, Furthmayr H, Rohde H, Timpl R (1981) Shapes, domain organizations and flexibility of laminin and fibronectin, two multifunctional proteins of the extracellular matrix. J Mol Biol 150:97–120

*Franks F (ed) (1973) Water, a comprehensive treatise, vol 2: Water in crystalline hydrates; aqueous solutions of simple nonelectrolytes. Plenum, New York London, pp 376, 399

Gerhardt P, Judge JA (1964) Porosity of isolated cell walls of *Saccharomyces cerevisiae* and *Bacillus megaterum*. J Bacteriol 87:945–951

Guo P, Grimes S, Anderson D (1986) A defined system for in vitro packaging of DNA-gp3 of the *Bacillus subtilis* bacteriophage 029. Proc Natl Acad Sci USA 83:3505–3509

Hayat MA, Zirkin BR (1973) Critical point drying method. In: Hayat MA (ed) Principles and techniques of electron microscopy, vol 3. Van Nostrand Reinhold, New York, pp 297–313

Hearst JE, Vinograd J (1961) The net hydration of T4 bacteriophage deoxyribonucleic acid and the effect of hydration on buoyant behaviour in a density gradient at equilibrium in the ultracentrifuge. Proc Natl Acad Sci USA 47:1005–1014

Hobot JA, Carlemalm E, Villiger W, Kellenberger E (1984) Periplasmic gel: New concept resulting from the reinvestigation of bacterial cell envelope ultrastructure by new methods. J Bacteriol 160:143–152

Hobot JA, Villiger W, Escaig J, Maeder M, Ryter A, Kellenberger E (1985) The shape and fine structure of the nucleoid observed on sections of ultrarapid frozen and cryosubstituted bacteria. J Bacteriol 162:960–971

*Hoppe W, Lohmann W, Markl H, Ziegler H (1977) Biophysik, ein Lehrbuch. Springer, Berlin Heidelberg New York

*Joly M (1965) A physico-chemical approach to the denaturation of proteins. Academic Press, London New York

Kauzmann W (1959) Some factors in the interpretation of protein denaturation. Adv Prot Chem 14:1–63

Kellenberger E (1962) The study of natural and artificial DNA-plasms by thin section. In: Harris RJC (ed) The interpretation of ultrastructure. Academic Press London New York, pp 233–247

Kellenberger E (1978) Possibility to detect reproducibly some 5–10 Å conformational differences by conventional techniques: physiologically defined lattice transformations in bacteriophage T4. In: Sturgess JM (ed) Electron microscopy 1978, vol 3. Proc 9th Int Congr Electron Microsc, Toronto, pp 441–449

Kellenberger E, Bitterli D (1976) Preparation and counts of particles in electron microscopy: application of negative stain in the agar filtration methods. Microsc Acta 78:131–148

Kellenberger E, Kistler J (1979) The physics of specimen preparation. In: Hoppe W, Mason R (eds) Unconventional electron microscopy for molecular structure determination. Vieweg, Braunschweig, pp 49–79

Kellenberger E, Ryter A (1964) Bacteriology. In: Siegel BM (ed) Modern developments in electron microscopy. Academic Press, London New York, pp 335–393

Kellenberger E, Bolle A, Boy de la Tour E, Epstein RH, Franklin NC, Jerne NK, Reali-Scafati A, Sechaud J, Bendet I, Goldstein D, Lauffer MA (1965) Functions and properties related to the tail fibers of bacteriophage T4. Virology 26:419–440

Kellenberger E, Carlemalm E, Stauffer E, Kellenberger C, Wunderli H (1981) In vitro studies of the fixation of DNA, nucleoprotamine, nucleohistone and proteins. Eur J Cell Biol 25:1–4

Kellenberger E, Häner M, Wurtz M (1982) The wrapping phenomenon in air dried and negatively stained preparations. Ultramicroscopy 9:139–150

Kellenberger E, Carlemalm E, Villiger W (1986a) Physics of the preparation and observation of specimens that involve cryoprocedures. In: Müller M, Becker R, Boyde A, Wolosewick J (eds) The science of biological specimen preparation, 1985. SEM, AMF O'Hare, IL 60666, pp 1–20

Kellenberger E, Carlemalm E, Villiger W, Wurtz M, Mory C, Colliex CH (1986b) Z-contrast in biology; a comparison with other imaging modes. Ann N Y Acad Sci 483:202–228

Kellenberger E, Villiger W, Carlemalm E (1986c) The influence of the surface relief of thin sections of embedded, unstained biological material on the image quality. Micron Microsc Acta 17:331–348

Killian JA, de Kruijff B (1985) Importance of hydration for gramicidin-induced hexagonal H_{11} phase formation in Dioleoylphosphatidylcholine model membranes. Biochemistry 24:7890–7898

Kistler J, Kellenberger E (1977) Collapse phenomena in freeze-drying. J Ultrastruct Res 59:70–75

*Kuntz ID, Kauzmann W (1974) Hydration of proteins and polypeptides. Adv Protein Chem 28:239–345

*Lauffer MA (1975) Entropy driven processes in biology. Springer, Berlin Heidelberg New York

Lepault J, Pitt T (1984) Projected structure of unstained, frozen-hydrated T-layer of Bacillus brevis. EMBO J 3:101–105

Lilley DMJ, Palecek E (1984) The supercoil-stabilised cruciform of Col E1 is hyper-reactive to OsO_4. EMBO J 3:1187–1192
MacKenzie AP (1977) The physico-chemical basis for the freeze-drying process. Int Symp Freeze-drying of biological products, Washington, DC. Dev Biol Standard 36:51–67
Malhotra SK (1962) Experiments on fixation for electron microscopy. Q J Microsc Sci 103:5–15
Moncany MLJ (1982) Determination des conditions intracellulaires chez *E. coli.* Conséquences biologiques de leur modification. Thèse Doct D'Etat, Univ Paris VII
Morel FMM, Baker RF, Weyland H (1971) Quantitation of human red blood cell fixation by glutaraldehyde. J Cell Biol 48:91–100
*Porter KR (1984) The cytomatrix: A short history of its study. J Cell Biol 99, 1, 2:3s–12s
Rachel R, Jakubowski U, Baumeister W (1986) Electron microscopy of unstained, freeze-dried macromolecular assemblies. J Microsc (Oxford) 141:179–191
Reynolds JA, Tanford C (1970) The gross conformation of proteinodium dodecyl sulfate complexes. J Biol Chem 245:5161–5165
Richter J, Koller R, Ramel A, Thürkauf M (1966) Observations on electron micrographs of DNA. Biochim Biophys Acta 123:286–297
Rilfors L, Weibull C (1985) The consumption of osmium tetroxide by components of the cytoplasmic membrane of *Acholeplasma laidlawii* and its morphological implications. Micron Microsc Acta 16:77–83
Rosenbusch JP, Garavito RM, Dorset DL, Engel A (1982) Structure and function of pore-forming transmembrane protein: high resolution studies of a bacterial porin. In: Peeters H (ed) Protides of the biological fluids. 20th Colloqu 1981. Pergamon, Oxford New York, pp 171–174
Sjöstrand FS, Kretzer F (1975) A new freeze-drying technique applied to the analysis of the molecular structure of mitochondrial and chloroplast membranes. J Ultrastruct Res 53:1–28
Srere PA (1981) Protein crystals as a model for mitochondrial matrix proteins. TIBS 6:4–6
*Tanford CH (1968) Protein denaturation, pt A and B. Adv Prot Chem 23:121–282
*Tanford CH (1970) Protein denaturation, pt C. Adv Prot Chem 24:1–95
*Tanford CH (1980) The hydrophobic effect: Formation of micelles and biological membranes, 2nd edn. Wiley Interscience, New York
Tschopp J, Villiger W, Lustig A, Jaton J-CL, Engel J (1980) Antigen-independent binding of JgG dimers to Clq as studied by sedimentation equilibrium, complement fixation and electron microscopy. Eur J Immunol 10:529–535
Weibull C, Christiansson A, Carlemalm E (1983) Extraction of membrane lipids during fixation, dehydration and embedding of *Acholeplasma laidlawii*-cells for electron microscopy. J Microsc (Oxford) 129:201–207
Weibull C, Villiger W, Carlemalm E (1984) Extraction of lipids during freeze-substitution of *Acholeplasma laidlawii*-cells for electron microscopy. J Microsc (Oxford) 134:213–216
Westphal CH, Frösch D (1984) Electron-phase-contrast imaging of unstained biological materials, embedded in a water-soluble melamine resin. J Ultrastruct Res 88:282–286
Williams RC (1953) A method of freeze-drying for electron microscopy. Exp Cell Res 4:188–201
Williams RC, Fraser D (1953) Morphology of the seven T bacteriophages. J Bacteriol 66:458–464
Yanagida M, Boy de la Tour E, Alff-Steinberger C, Kellenberger E (1970) Studies on the morphopoiesis of the head of phage T-seven: VIII. Multilayered polyheads. J Mol Biol 50:35–58

Chapter 3 Electron Beam Radiation Damage to Organic and Biological Cryospecimens

YESHAYAHU TALMON[1]

1 Introduction

Radiation damage, i.e. the introduction of structural (e.g. loss of crystallinity) or compositional (e.g. mass loss) changes by the electron beam, has been an inevitable and serious factor in limiting the amount of microstructural information that can be collected from an electron microscope specimen. Although radiation damage is always present because it is part of the electron beam/specimen interaction that produces the image, it may be reduced or minimized, if it and the physical factors that affect it are well understood.

The development of cryoelectron microscopy techniques has added a new dimension to the problem of radiation damage to organic and biological specimens. On the one hand, it was found that cryogenic temperatures slow down to some extent overall radiation damage, although not to the extent it was hoped and initially reported. On the other hand, many of these cryospecimens contain water as ice in one or more forms, either because of the nature of the specimen (e.g. "frozen-hydrated" specimens), or because of water condensation on the specimen in the microscope. The presence of water modifies, in some cases significantly, radiation damage to these specimens.

The main purpose of this chapter is to describe the progress that has been made in recent years in understanding radiation damage at low temperature, with emphasis on the effects of ice on these processes. This last point is emphasized because of the increasing interest in electron microscopy of frozen-hydrated specimens, especially vitrified ones. As the equipment and specimen preparation techniques become more readily available, more electron microscopists will have to tackle the unique problems of radiation damage in presence of ice, or use specimens cooled to liquid nitrogen or liquid helium temperatures to reduce radiation damage.

We begin with a brief description of electron beam/specimen interactions. More detailed treatments of the subjects can be found in earlier papers by Cosslett (1978), Glaeser (1979) and Hobbs (1979b). This is followed by another rather short section on the effects of low temperature on radiation damage with emphasis on work done since 1977. An excellent review paper by Glaeser and

1 Department of Chemical Engineering, Technion – Israel Institute of Technology, Haifa 32000, Israel

Cryotechniques in Biological Electron Microscopy
Edited by R. A. Steinbrecht and K. Zierold

Taylor (1978) details the work done prior to 1978. The next section deals with the peculiar behaviour of pure ice in its hexagonal, cubic and vitreous forms under the electron beam, and then we discuss the radiolytic process at the organic material/ice interface. The last section discusses the implications of the available knowledge to cryoelectron microscopy and the open questions that await further research.

2 Electron Beam/Specimen Interaction

To obtain a useful electron micrograph, one needs to produce detectable contrast on the image-recording film or electron detector. This requires a sufficiently strong interaction of the electrons, first with the specimen and then with the film or detector. Since both phenomena are statistical in nature, a minimum exposure of electrons is needed to produce acceptable contrast with the required resolution (Glaeser 1971). High resolution calls for high magnification, and the electron exposure required increases with the square of magnification. In addition to elastic scattering events that are the main contributors to useful image information, many inelastic events take place as electrons traverse the specimen. Some of the energy transferred from the electrons to the specimen causes radiation damage, which may be defined as changes in the microstructure or chemical composition of the specimen as a result of exposure to ionizing radiation, i.e. the electron beam in our case.

Beam/specimen interactions may be classified according to the nature of the physical processes involved. Primary events are: (1) excitation of an individual atom or a group of atoms (the latter effect is sometimes referred to as "plasmon", a term most applicable to the "electron gas" in metals); (2) ionization of an atom; and (3) displacement of an atom. Secondary effects are: (1) electron emission; (2) X-ray emission; (3) light emission; (4) temperature rise; (5) electrostatic charging; (6) bond scission; (7) crosslinking; (8) mass loss and (9) formation of carbonaceous coating ("contamination") (Cosslett 1978).

Most radiation damage to organic and biological specimens is caused by ionization in which the energy transferred to the specimen is 10 to 1000 eV per event, and plasmon excitations which give rise mostly to specimen heating. Less effective are X-ray emission and atom displacement also known as "knock-on" events.

A spectrum of energy losses in 100 kV electron beam transmitted through a thin film of carbon is shown in Fig. 1. The main loss peak is centered at about 24 eV. Part of this energy will cause ionization. Although the X-ray excitation peak is much smaller, the large energy transfer involved in the process makes it a non-negligible effect with a contribution of about 20% of the plasmon peak. For light elements, electron emission by the Auger process is more probable than X-ray emission. Thus, the tail end of the energy loss spectrum, especially K-shell excitation, is a more important factor in radiation damage than what is expected from its small intensity.

Electron exposures ("doses") are usually given in units of $e^- Å^{-2}$ or $e^- nm^{-2}$. Equivalently, electron exposure can be given as the electric charge per

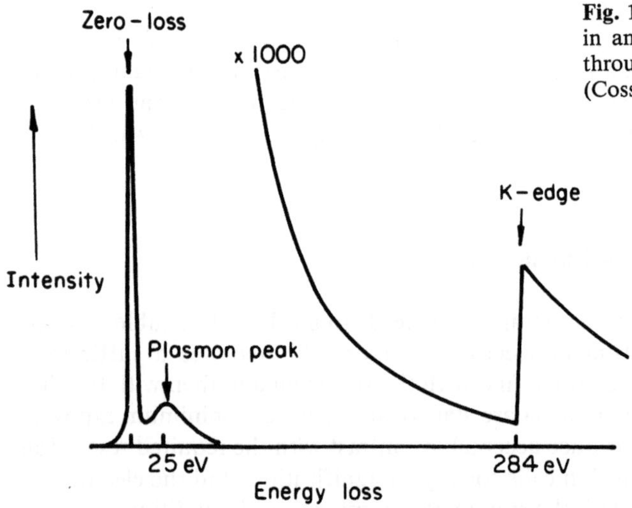

Fig. 1. Spectrum of energy losses in an electron beam transmitted through a thin film of carbon (Cosslett 1978)

unit area, $C\,m^{-2}$ (coulombs per m^2). Higher exposures are given in units of kiloelectrons per nm^2 ($ke^-\,nm^{-2}$) or kilocoulombs per m^2 ($kC\,m^{-2}$). The conversion between the two systems is given by: $1\,e^-\,nm^{-2} = 0.160\,C\,m^{-2}$. In the general radiation literature, however, the intensity of radiation is measured by the amount of energy desposited in the sample. Hence, conversion factors between the units used by electron microscopists and radiologists depend on the electron acceleration voltage, the thickness and elemental composition of the specimen. For thin specimens of light elements the conversion will be approximately 6.1×10^7 rad per $C\,m^{-2}$ at 60 kV and 4.0×10^7 rad per $C\,m^{-2}$ at 100 kV. One rad is equal to $6.25 \times 10^{13}\,eV\,g^{-1}$ absorbed energy. The effectiveness of a radiation-induced reaction may be measured by the number of molecules of a given chemical species produced per 100 eV energy deposited in the sample. This is called the "G value" for that particular molecule. To characterize the radiolysis of a given molecule, several G values, one for each product molecule, may be necessary.

The most extensive work on the secondary reactions following ionization has been done on long-chain polymers (Charlesby 1972, 1981), because of the technological and practical applications or consequences of radiation damage, i.e. modification of polymer properties either during their production or during their use. In many polymers the most obvious effect of radiation is the reduction of average molecular weight, termed "degradation" or "chain scission". Scission takes place at random along the chain. Only rarely are monomers produced, but small molecules may be produced, especially by a large electron exposure, and these smaller molecules may diffuse to the surface of the specimen, and leave it by sublimation. Polymers showing this type of behaviour are often characterized by the presence of two large sidegroups attached to the main chain. Poly(methylmethacrylate) (PMMA) is an obvious example. It is assumed that these bulky sidegroups pose a steric hindrance to the rejoining of the two ends of the broken chain. A similar situation may exist in many biological macromolecules such as proteins.

An alternative process to the above is crosslinking. Primary ionization may cause the formation of a sidechain free-radical and a hydrogen free-radical. The latter can easily diffuse through the polymer, and abstract another H· atom from another chain to give H_2, leaving another polymer radical behind. A distribution of various length polymer radicals builds up, and they combine in pairs to produce crosslinks. Since these radicals do not necessarily form in close proximity, this process involves complex problems of free-radical diffusivity in the specimen. The resulting crosslinked polymer is generally more radiation resistant than the original polymer. Although the amount of mass lost in crosslinking is relatively small, crystallinity is eventually totally lost.

In all polymers, as in other organic molecules, scission and crosslinking processes take place simultaneously. But in some polymers, e.g. PMMA as mentioned above, scission is the predominant reaction. These are called scission-type polymers. Polymers such as polystyrene (PS), in which the predominant reaction is crosslinking, are called crosslinking-type polymers.

The various methods for detecting radiation damage in electron microscopy were reviewed by Reimer (1984) not long ago. These include in situ methods such as evaluation of mass loss and concentration changes by electron energy loss spectroscopy (EELS), X-ray microanalysis and optical density changes, as well as the observation of crystallinity changes through changes in intensity of diffraction spots or rings, or changes in contrast in the imaging mode. Other methods involve analysis of the pre-irradiated sample outside the microscope, e.g. measurements of spectroscopical (absorbance) and optical properties, electrical conductivity, solubility or chemical analysis. They require a much larger sample than the methods of the first type.

The above discussion of radiolytic processes involves only organic matter. But even in the electron microscopy of biological specimens, inorganic materials such as various salts originally in the specimen, or stains that are added to it, are quite important. Some of these materials may be radiolysed by the beam and cause compositional and apparent structural changes in the specimen. The interested reader is referred to papers by Hobbs (1979a, b) covering this subject.

The last aspect of radiation damage to be considered in this short survey is that of beam heating. A substantial fraction of the energy absorbed by the specimen is eventually dissipated as heat. Under normal operating conditions (not necessarily "low-dose" conditions), beam-heating effects are limited to a few degrees (Talmon and Thomas 1977a, b). Only if the specimen is not mounted properly, and thermal contact to the substrate or specimen holder is poor, is local heating of tens of degrees possible.

3 Radiation Damage to Organic Materials at Low Temperature

Because the secondary reactions of radiation damage processes involve diffusion of free radicals and other reaction products, and since the diffusivities of these chemical species are reduced at lower temperatures, it seems obvious that a reasonable way to reduce radiation damage to electron microscope specimens is by cooling them. Of course, frozen-hydrated specimens are cooled anyway; these are kept at low temperature to reduce vapour pressures to levels negligible compared to the microscope column vacuum, and to arrest all supramolecular motion (e.g. Brownian motion) in an originally fluid or semifluid system.

To attain the theoretical advantage of using a low temperature specimen holder ("cold-stage", "cooling holder"), it is necessary to ensure that the specimen is in good contact with the stage. Inadequate contact may cause a large temperature difference (tens of degrees) between specimen and holder, and may give rise to more severe damage, radiolytic and mechanical, due to charging effects (Knapek et al. 1984).

Earlier work was covered by Glaeser and Taylor (1978). Of this one should note the discrepancies between the findings of various groups. For example, whereas Venables and Basset (1967) found no significant improvement in the stability of crystalline polyethylene (PE) as specimen temperature was reduced from 300 K to 20 K, Grubb and Groves (1971) did observe an improvement by a factor of three in the time of decay of electron diffraction spots under the beam upon cooling to 18 K (most of the improvement was achieved by cooling to 77 K). Glaeser and Hobbs (1975) even reported an increase in the radiation sensitivity of negatively stained catalase at liquid helium temperature, although later Taylor and Glaeser (1976) reported that the radiation sensitivity of ice-embedded catalase decreased by a factor of ten compared to room temperature, and Hayward and Glaeser (1980) showed that glucose-embedded catalase was more resistant to the beam at about 180 K.

Special mention should be made of the work of Siegel (1970, 1972), who elegantly demonstrated that electron beam radiation damage is composed of two steps, the second of which, diffusion of the free radicals, can be slowed down by cooling. Siegel showed that the survival of *Bacillus subtilis* after irradiation at (nominal) 4 K was only slightly better than at 300 K, and that organic crystals of tetracene irradiated at 4 K and showing very little damage, were badly damaged upon warming to 300 K without additional irradiation. This is the so-called latent dose effect. He also found that most of the low temperature protective effect in paraffin occurred already at 77 K, whereas in tetracene the more significant "cryoprotection" took place between 77 K and 4 K.

The development of a superconducting objective lens several years ago (Dietrich et al. 1979) made it possible to examine biological specimens very close to liquid helium temperature of 4 K. (In other designs of liquid helium cooled stages a temperature gradient of 4 to 15 K between specimen and cryogen is inevitable.) Early radiation damage experiments with this superconducting lens showed reduction of the rate of radiation damage, as measured by fading of diffraction spots of organic crystals, by one to two orders of magnitude (Dietrich et al. 1979; Dubochet and Knapek 1979; Müller et al. 1981; Chiu et al. 1981; Knapek and Dubochet 1980). The excitement about these promising results later turned to disappointment when it was found (Lepault et al. 1983c) that the earlier results were erroneous. Careful work on three different microscopes equipped with superconducting lenses showed that cryoprotection effects are much smaller than previously reported. It seems now that the most that can be expected is an improvement by a factor of three to five, and that this can usually be achieved at liquid nitrogen temperatures when precautions are taken to limit beam heating and charging effects (Knapek et al. 1984; Wade 1984).

Recently, a group of 12 investigators in several laboratories in Europe and in the USA conducted a careful study in an attempt to resolve the questions surrounding the cryoprotection factor afforded by cooling specimens to liquid nitrogen and liquid helium temperatures (Chiu et al. 1986). The study involved two standardized crystalline samples, paraffin and bacterial purple membrane, examined following the same standard protocol in all laboratories. The results of this study agree well with previous observations (Knapek et al. 1984; Wade 1984) that the electron diffraction patterns of crystalline protein and hydrocarbon specimens last approximately three to five times longer when cooled to about 150 K from room temperature. Whether further cooling to the vicinity of 4 K gives further cryoprotection, could not be determined in that study with any certainty. However, a later report by Zemlin et al. (1986) claimed that the sensitivity of a glucose-embedded purple membrane is six times less at 143 K and twenty times less at 4.5 K than at room temperature.

Another aspect of radiation damage is mass loss. This aspect is especially important for X-ray microanalysis, where the electron exposures used are several orders of magnitude higher than those used in the experiments described above. One would expect that despite the damage on the molecular scale, the chemical species formed could be localized by cooling the specimens, and thus more accurate chemical analyses could be performed. Indeed, Ramamurti et al. (1975) and Dubochet (1975) have reported negligible mass loss from irradiated amino acids and bacteria respectively, after cooling the specimen to liquid helium temperature. This was later verified by Egerton (1980, 1982) for mass loss of light elements from polycarbonate, PMMA, Formvar and collodion; cooling the specimen from 300 to 100 K reduced the rate of removal of these light elements by factors up to 100.

The above discussion was limited to dry electron microscope specimens. The presence of water, usually in one of its solid forms, adds a new dimension to the problems of radiation damage to organic materials at low temperature; problems that are discussed in the next two sections.

4 Radiation Damage to Ice

Damage to ice by ionizing radiation has been mostly studied outside the electron microscope. These were studies of fairly thick specimens of hexagonal ice, irradiated by electrons, gamma-rays or ion beams. Although in many of these studies the energies involved were comparable to or even exceeded those of the beam in the electron microscope, the doses were smaller than normally encountered in electron microscopy. Nevertheless, these experiments have provided important information on the physical processes that take place when an electron beam interacts with ice in the microscope. The advantage of experiments outside the microscope is the possibility of chemical analysis of radiolysis products, especially by electron spin resonance (ESR) spectroscopy. Studies of ice radiolysis in the microscope have been few and involved no chemical analysis of radiolysis products. However, these studies have provided useful microstructural information on all three forms of ice [hexagonal (I_h), cubic (I_c), and vitreous (I_v)] that can exist under the conditions in the microscope, and about radiation-induced phase transformation between various forms of I_v, and between vitreous and crystalline ice.

4.1 Radiolysis of Crystalline Ice

The major, primary event in radiolysis of ice is ionization of the water molecules, i.e. the ejection of an electron:

$$H_2O \rightarrow H_2O\cdot^+ + e^- \ . \qquad (1)$$

The dot to the right of H_2O denotes an unpaired electron. The probability ("cross-section") for displacement of light atoms such as hydrogen or oxygen is very small (Cosslett 1978), and thus such "knock-on" processes are negligible at low and moderate electron beam acceleration voltages (< 1 MeV).

The products of ionization interact with the water molecules and with products of the so-called primary event of radiolysis to form a large number of active chemical species (Taub and Eiben 1968; Symons 1982a). The most important of these primary events following water ionization is the reverse reaction which forms an excited water molecule (H_2O^*), that may either return to its ground state, dissipating energy as heat, or decompose and form free radicals:

$$H_2O\cdot^+ + e^- \rightarrow H_2O^* \rightarrow H\cdot + \cdot OH \ . \qquad (2)$$

Other probable reactions are proton transfer:

$$H_2O\cdot^+ + H_2O \rightarrow H_3O^+ + \cdot OH \qquad (3)$$

and interaction of an ejected electron with the hydronium ion:

$$H_3O^+ + e^- \rightarrow H\cdot + H_2O \ . \qquad (4)$$

At very low temperatures (10 K), a reversible process has been identified by Symons (1982a):

$$\cdot OH + H_2O \leftrightarrow \cdot O^- + H_3O^+ \ . \qquad (5)$$

The G value for ice radiolysis at cryogenic temperatures (< 100 K) is fairly low, about 0.9 for ·H or ·OH formation (Siegel et al. 1961). This indicates that the reactions that lead to reformation of H_2O are quite important, and explains the relative stability of hexagonal ice in the electron microscope.

Secondary radiolysis processes involve annealing of the H· atoms to form hydrogen; this takes place above 20 K, annealing of the ·OH radicals above 77 K:

$$\cdot OH + \cdot OH \rightarrow H_2O_2 \ , \qquad (6)$$

and formation of $HO_2\cdot$ radicals:

$$H_2O_2 + \cdot OH \rightarrow H_2O + HO_2 \ . \qquad (7)$$

These last two processes are especially important at fairly high temperatures (above 110 K), which are not uncommon in cryoelectron microscopy. As high concentrations of ·OH build up at high electron doses, the following reactions may also become important (Symons 1982a):

$$\cdot OH + e^- \rightarrow OH^- \tag{8}$$

and

$$H_2O_2 + e^- \rightarrow \cdot OH + OH^- \ . \tag{9}$$

These "double-event" reactions and the recombination reaction

$$H\cdot + \cdot OH \rightarrow H_2O \tag{10}$$

combined with charge neutralization by the mobile H_3O^+ and OH^- ions may explain the relative stability of hexagonal ice to high electron doses.

When pure hexagonal ice is irradiated by an electron beam, it loses mass quite uniformly from the outer surface of the specimen. The sample retains its crystallinity until all ice is etched away. This mass loss is a result of radiolysis, not beam heating (Talmon et al. 1979; Talmon 1982). I_h has a thermal conductivity of 6 $Wm^{-1}K^{-1}$ at 100 K, and beam-heating calculations have shown (Talmon and Thomas 1977a, b) that temperature rise under normal microscope operating conditions is too low to induce sublimation.

The fact that mass is lost only from the outer surfaces is peculiar. It may be explained by the high probability of reactions such as (2), (4), (7) and (10), which restore the original structure of the water molecule. Only radiolysis products which are formed at or diffuse to the interface have a high probability of escaping into the microscope column. Another attempt to explain mass loss from the sample surface is through the "pre-knock-on" concept suggested by Symons (1982b). This involves a local shock wave, created by the impinging electron/sample interaction, that travels through the specimen to the surface where it is able to eject an atom from the surface layer.

The rate of etching of ice by a 100 kV electron beam at about 100 K was found to be approximately 1 μg C^{-1} (Talmon et al. 1979; Dubochet et al. 1982a; Heide 1982). Heide (1984) has also shown that the rate of etching is almost constant between 10 and 90 K and increases by a factor of 2.8 between 90 and 150 K (Fig. 2). Above 150 K sublimation becomes appreciable. He also found that at 200 kV the rate of etching was reduced by a factor of 1.6 (from 0.8 to 0.5 μg C^{-1}), and that the ice removal rate by the beam is independent of specimen thickness. These

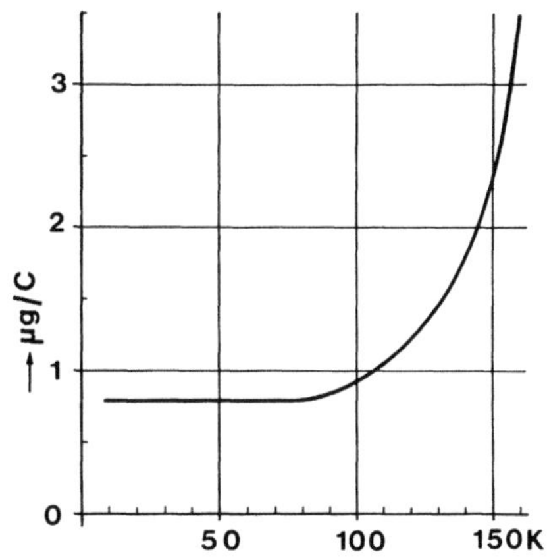

Fig. 2. Mass loss of pure ice by electron irradiation, depending on the temperature (Heide 1984)

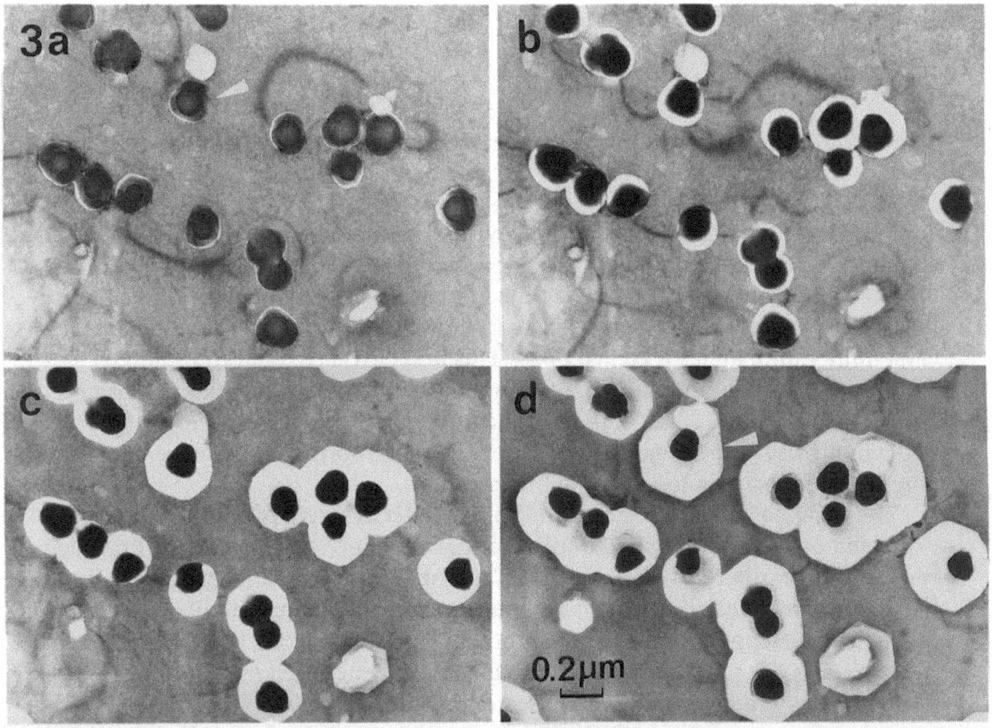
3a
b
c
d
0.2µm

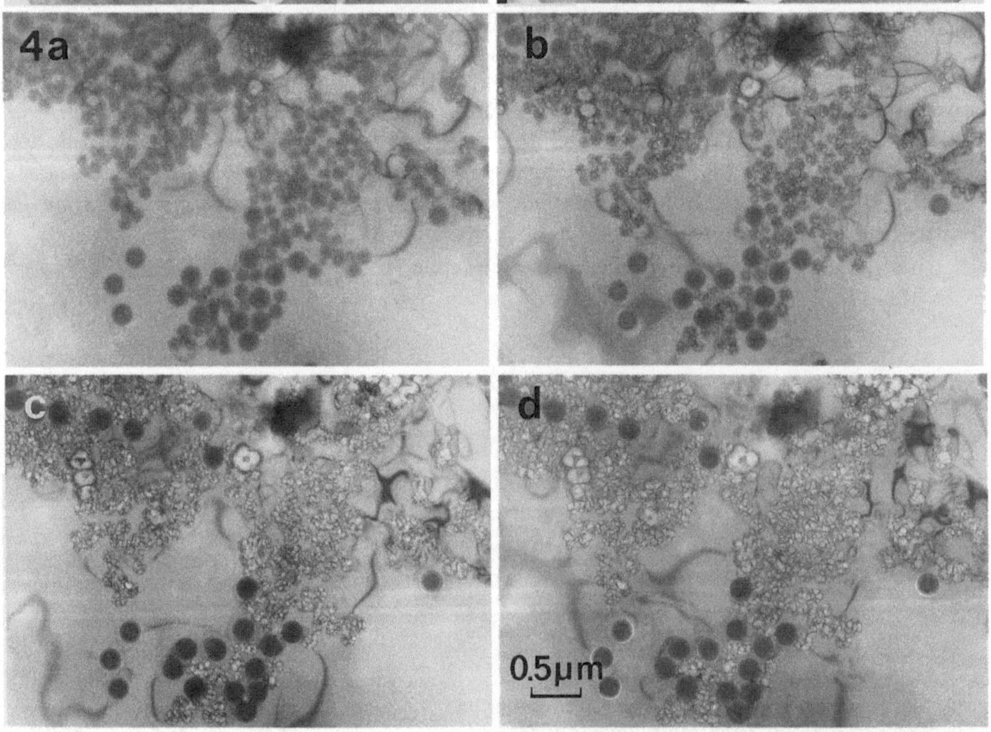
4a
b
c
d
0.5µm

observations seem to indicate that ionization, not knock-on processes, causes ice mass loss.

Another phenomenon observed upon electron beam irradiation of hexagonal ice is the formation of cavities and "bubbles", especially following high electron exposures (Unwin and Muguruma 1971; Glaeser and Taylor 1978). Also, small visible holes in the specimen, a result of trapped air bubbles or mechanical damage, grow when exposed to the beam (Fig. 3; Narkis et al. 1985; Talmon et al. 1985). It seems that the smaller cavities form in the ice from small, initially invisible defects in the ice lattice, or around small organic contamination particles. These holes, when exposed long enough to the electron beam, attain a shape characteristic of the orientation of the crystal with respect to the beam: hexagons for the [0001] zone axis (Fig. 3), or elongated rectangles for the [1$\bar{2}$10] zone axis (Talmon, unpubl. results). The growth of these holes along preferred crystallographic planes supports the notion of the importance of free ice interface in the loss-of-mass process; it may also indicate that, at least in some situations, atom or molecule displacement (and redeposition) does play a role in ice etching by the electron beam.

Below liquid nitrogen temperature, crystalline ice becomes vitreous ("amorphous") when irradiated by the electron beam (Lepault et al. 1983a; Heide 1984; Heide and Zeitler 1985). In the range of 30 to 70 K, I_h or I_c are transformed under the beam to the low density form of vitreous ice ($I_{v,l}$); this in turn may undergo another transformation, depending on the temperature within the range (the lower the temperature, the higher the conversion), to the high density form of vitreous ice ($I_{v,h}$). In the 8 to 20 K range all forms of ice tend to transform upon irradiation to $I_{v,h}$. This process may take place through an intermediate state, i. e. $I_c \rightarrow I_{v,l} \rightarrow I_{v,h}$, the stable form of ice at that temperature at low pressures. Heide and Zeitler (1985) also found that between 80 and 90 K no irradiation-induced phase transformation occurs in any of the three forms of ice, $I_{v,l}$, I_c, I_h, which may coexist in this temperature range.

4.2 Radiation Damage to Vitreous Ice

Our knowledge of the processes that take place in electron-irradiated vitreous ice is much more limited than our knowledge about crystalline ice. Most of what we do know comes from the careful study of Heide (1984) and Heide and Zeitler (1985). All three groups of phenomena which have been recorded in crystalline

◄

Fig. 3a–d. TEM of a frozen specimen (95 K) of a two-stage latex made by adding styrene to PBA seeded latex (PS/PBA = 1:1) after increasing electron exposures: **a** 10 kC m^{-2}; **b** 40 kC m^{-2}; **c** 80 kC m^{-2}; **d** 110 kC m^{-2}. Note the change in the shape of cavities around the particles as they grow; *arrowheads* in **a** and **d** point at the same growing cavity. Note also the shape and direction of the final cavities; compare to the cavity formed around a random hole in the ice (*upper left* part of the micrographs) (Talmon et al. 1985)

Fig. 4a–d. Transmission electron micrographs of a frozen specimen (95 K) of mixed latexes, PMMA (*smaller spheres*) and PS (*larger spheres*) at increasing exposures to the electron beam: **a** 2 kC m^{-2}; **b** 7 kC m^{-2}; **c** 35 kC m^{-2}; **d** 52 kC m^{-2}. Note the different damage patterns to the two polymers (Talmon et al. 1985) ice, namely mass loss, "bubble" formation and phase transformation have also been seen in vitreous ice.

Mass loss from vitreous ice in the temperature range of 75 to 90 K is about 1.5 times higher than that of hexagonal ice, whereas cubic ice seems to have a somewhat higher rate of etching than I_v, probably because of its small grain size. Above 100 K, etching experiments are not practicable because of rapid phase transformation under the electron beam to I_c. No data are available yet for mass loss of I_v below 70 K.

Bubble formation in vitreous ice condensed from the vapour phase was reported by Dubochet et al. (1982b) at 110 K. Heide and Zeitler (1985) reported very pronounced void formation in $I_{v,h}$ layers thicker than 20 nm between 8 and 20 K. In the transition zone of 30 to 70 K void formation is rarer; between 80 and 90 K in pure $I_{v,l}$, void formation was not observed at all. Above 100 K voids were formed only in condensed layers of I_v which always contain foreign gas molecules adsorbed from the microscope column.

Below 70 K, irradiation of $I_{v,l}$ transforms it to the high density form of vitreous ice, $I_{v,h}$. Below 30 K this transformation is complete. Between 30 and 70 K the conversion is temperature-dependent. In the 80 to 90 K range no irradiation-induced phase transformations have been recorded. Above 100 K vitreous ice ($I_{v,l}$) is transformed into I_c when exposed to the electron beam. The exposure needed for this transformation was found to be temperature-dependent (Heide 1984).

5 Radiation Damage to Frozen-Hydrated and Vitrified-Hydrated Specimens

There are two important conclusions to be drawn from the discussion of radiolysis of ice presented in Section 4. The first is that ice is a fairly stable substance under electron irradiation. Electron exposures well above those used in low-dose techniques are required to induce appreciable damage (mass loss or phase transformation) to ice. In other words, ice is a much more stable compound under the beam than most organic substances that an electron microscopist would like to examine embedded in ice, i. e. in their frozen-hydrated state. But electron beam-irradiated ice is also a very active source of free radicals, and these, if produced in high enough concentrations, may attack organic materials embedded in the ice, and initiate free-radical chain reactions that involve both the ice and the material in contact with it. This may lead to much enhanced radiation damage in the ice and in the organic material. Such reactions may take place at electron exposures from just above the low-dose range ($>100\,e^-\,nm^{-2}$) to normal viewing conditions ($>10^5\,e^-\,nm^{-2}$). The following is a discussion of the phenomena, their study, possible mechanisms and the parameters affecting radiation damage in low temperature water-containing specimens.

5.1 The Modification of Radiolysis by the Presence of Water or Ice

That water may affect the sensitivity of organic specimens to the electron beam was noted by Heide (1965), who suggested that water molecules absorbed on an organic specimen surface, especially at low temperatures, are ionized by the electron beam, producing reactive chemical species that oxidize the organic substrate. Such etching by the product of radiolysis of ice condensed on cooled biological specimens was later observed by Hartman and Hartman (1971) and Somlyo et al. (1976). In most electron microscopes available today condensation of an ice layer on the specimen surface is inevitable. Water in small concentrations (which are sufficient for radiolysis enhancement) may also be found in the sample as physically bound water, water of hydration (in crystals) or due to incomplete drying during specimen preparation.

First observations of unique and massive radiation damage to a biological frozen-hydrated specimen were published by Heide and Grund (1974), who reported bubble and hole formation in their specimens. Bubble and hole formation in ice in contact areas with organic matter were reported to occur at relatively high electron exposures by Glaeser and Taylor (1978). Talmon et al. (1979) noticed severe interactions between the electron beam and polymer films in the presence of ice as a continious layer or as isolated ice crystallites. These interactions led to the formation of a high-contrast network snapping into small balls of (probably) polymerized organic material. Talmon (1982, 1984) noted violent and rapid bubbling of vitrified sucrose and glycerol aqueous solutions under the electron beam. These bubbles were constantly changing under the beam until water was completely consumed by the radiolytic reactions. Similar bubbling in pure vitreous ice was observed by Dubochet et al. (1982b). This bubbling effect is similar to the even more severe effects observed in wet organic specimens in an environmental chamber (Bourgeois et al. 1980).

The importance of the organic material/ice interface as the locus where the radiolytic events are initiated and propagate at the highest rate, emerged from some of the above mentioned works, and will be discussed in more detail below. Heide and Zeitler (1985) have added to these observations the notion of the importance of a free surface or a small gap between the ice and the organic substrate. Such a gap, they observed, allows free radicals to move and act freely, while in its absence, as in the case of a completely ice-embedded organic particle, radiation damage is slow.

It seems that, in general, the presence of ice has no measurable effect on the lifetime of highly sensitive organic or biological crystals in the electron beam, as may be determined by the fading of spots in diffraction patterns (Hui 1980; Lepault et al. 1983b). This is a result of the extreme radiation sensitivity of these crystals, whose microstructural order is destroyed before the deleterious effects of the presence of ice can be manifested.

5.2 Controlled Experiments: Radiation Damage at Ice/Organic Material Interface

These experiments were devised to study carefully the complex phenomena involved in radiation damage of frozen-hydrated specimens. Electron exposure, the nature of the organic material and ice, and the location and relative amount of interfacial area were well defined. Through such experiments it has been possible to assess the effects and importance of the various factors involved.

One approach is to use an organic polymer film on which small ice crystallites are deposited by condensation outside the microscope. Talmon et al. (1979) used Formvar [poly(vinylformal)] and polyimide films to demonstrate excessive damage, network and hole formation at areas of contact between ice and polymer. Such experiments demonstrated the high sensitivity of Formvar, collodion and other films commonly used in electron microscopy specimen preparation to radiation damage in the presence of ice, and suggested an alternative: polyimides or carbon films. [Another useful alternative, SiO_x films, had been used by Taylor and Glaeser (1976)].

A similar experiment by Heide (1982) involved a layer of vitreous ice deposited in the TEM on a thin carbon film. No damage to the substrate was observed during irradiation by the electron beam; the ice layer was simply etched away. Only when almost all the ice was gone did some etching of the carbon occur, suggesting the importance of a free surface for massive radiation damage by the radiolysis products of ice. A different approach is to use polymer latexes, i. e. dispersions of polymer particles in water, stabilized by surfactants or by surface charges. Latexes, the products of emulsion polymerization, may be acquired in a wide range of narrow distribution sizes of different polymer species. Thus, they provide chemically and structurally well-defined systems in which the chemical nature and surface area-to-volume ratio can be easily varied. Polymers are very suitable systems for radiation damage studies because their behaviour under ionizing radiation varies from high sensitivity of scission-type polymers that undergo mostly scission reactions, to high stability of crosslinked polymers or crosslinking-type polymers that undergo mostly crosslinking in the electron microscope.

Polymer latexes as model systems for radiation damage to frozen-hydrated systems have been used by Talmon (1984) and by Talmon et al. (1986). These studies have shown that radiolytic processes in ice-embedded latexes are different for scission-type and crosslinking-type polymers, and that there is a marked difference depending on whether these processes take place in crystalline or vitreous ice.

Under electron beam irradiation, dry polymer latex particles lose mass quite uniformly all around their outer surfaces. Scission-type polymer particles, such as poly(methylmethacrylate) (PMMA), may lose up to 50% of their mass under the beam. The rest of the material becomes a crosslinked polymer. Crosslinking-type polymers, polystyrene (PS) for example, lose less mass (typically less than 10%). However, when embedded in crystalline ice, different phenomena are observed (Fig. 4). Even relatively small electron exposures ($\sim 5\ ke^-\ nm^{-2}$) are sufficient to start hole formation in scission-type polymer particles (PMMA, for example, the smaller particles in Fig. 4). Higher electron exposures lead to extensive mass loss until the particles become sponge-like in appearance. Quite often such badly damaged particles change shape and expand during the process. Neighbouring particles may be observed to fuse in I_h under the electron beam. The behaviour of crosslinking-type latex (PS, for example, the larger particles in Fig. 4) is quite different. No damage is discernible in the particles up to fairly high electron exposures ($> 40\ ke^-\ nm^{-2}$), but cavities form in the ice around them. Only after higher exposures can mass loss from the particles also be detected. In small particles (< 100 nm) the situation is different (Fig. 5); mass loss in quite fast even in the case of a crosslinking polymer, such as PS.

PMMA and PS represent the two extremes of behaviour of organic material in ice under electron beam radiation. Other polymers exhibit variations on the pattern of the above described behaviour, e.g. formation of a crust and fusion of neighbouring particles as in the case of poly(vinylchloride) (PVC) (Talmon 1984).

Radiation damage to polymer latex particles in vitreous ice is rather different (Talmon et al. 1986). The most striking differences have been observed in the most radiation-sensitive test system, PMMA. The electron exposure needed to observe any discernible damage on an electron micrograph negative taken at 12000 x was found to be about $4\times10^4\,e^-\,nm^{-2}$ in I_v, compared to about $5\times10^3\,e^-\,nm^{-2}$ in crystalline ice, cubic or hexagonal. The pattern of the damage has also been found to be different. In I_v, most of the damage took place in the ice around the test particles, and only after much higher electron exposures was etching seen in the polymer, whereas in I_c or I_h damage was throughout the particles without any visible cavities around the latex spheres. In other polymers such as PVC the pattern of damage was similar to that of PMMA (most damage in the ice), but initial damage has been observed at about the same electron exposures as in I_v or I_h, about $4\,ke^-\,nm^{-2}$.

The surprising results of these experiments in vitreous ice suggest that, at least for some systems of organic material suspended in water, examining vitreous rather than crystalline ice specimens may have the important advantage of reducing electron beam radiation damage, in addition to affording better preservation of the original microstructure of the system and avoiding electron-optical effects associated with a crystalline matrix. Indeed, the much reduced radiation damage in vitreous ice has been demonstrated in the study of frozen-hydrated suspensions and micellar solutions of various natural and synthetic surfactants, such as phospholipids, sulphonates and quaternary ammonium surfactants (Talmon 1986; J.R. Bellare, H.T. Davis, L.E. Scriven, Y. Talmon, in preparation; Miller et al. 1987). This effect seems to be less pronounced in protein systems (Talmon, unpubl. results), but more systematic work needs to be done in this area.

5.3 Postulated Mechanism of Radiation Damage to Organic Inclusions in Ice

The experimental data reported in the literature and the results of the above described controlled experiments suggest that radiation damage to organic materials embedded in ice involves high concentrations of free radicals formed in the ice and in the organic material at the organic material/ice interface. These free radicals from both chemical species give rise to a chain of free-radical reactions that cause mass loss and some mass rearrangement in both ice and inclusions. Because ice is a good source of small, very mobile free radicals, these free-radical chain reactions are much faster in its presence than in the case of dry organic molecules under electron beam irradiation. This free-radical attack affects not only etching of the inclusions, but also of the ice matrix.

A clear demonstration of this may be seen in Fig. 3 (Talmon et al. 1985), that shows polymer latex particles, made of 50% poly(butylacrylate) (core), and 50% polystyrene forming the particle shells. Since PS is concentrated in the shells, radiation behaviour of these particles is indistinguishable from that of pure PS

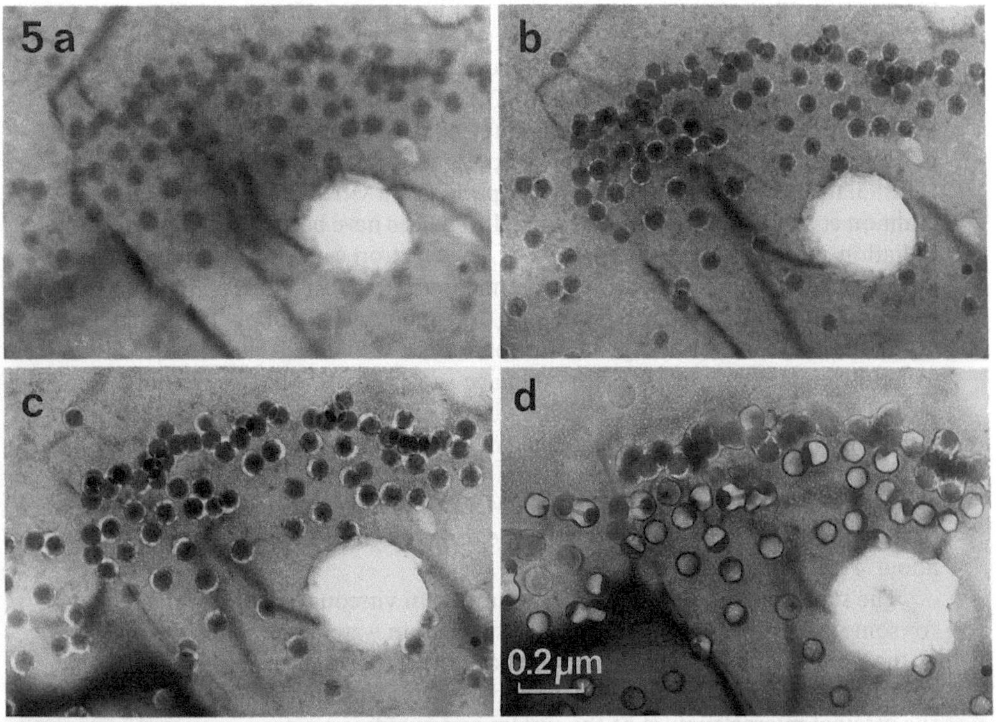
5 a
b
c
d
0.2 µm

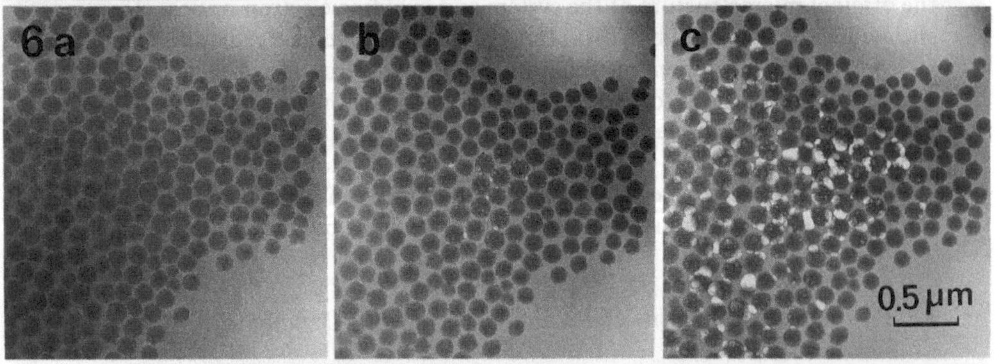
6 a
b
c
0.5 µm

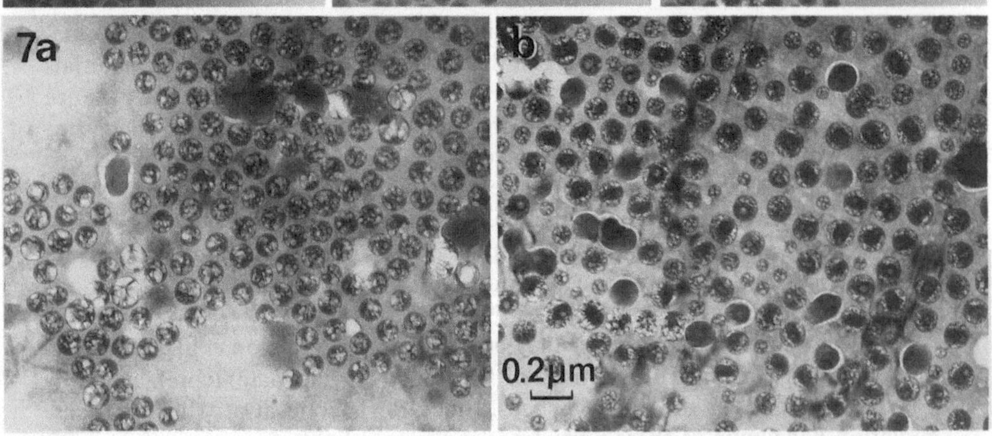
7a
b
0.2µm

spheres (cf. Fig. 4). At low electron exposures, cavities form in the hexagonal ice matrix. These cavities (Fig. 3a, b) are more or less round without any preferred orientation of growth. At higher electron exposures (Fig. 3c, d), as the gap between the ice and the polymer widens, the cavities turn into hexagons, all with corresponding sides parallel. This form of cavity is imposed by the crystalline structure of the ice ([0001] zone axis in this case), and is similar to the pattern of cavities growing around a random defect in the ice. As long as the gap between the ice and the polymer surface is small, free radicals from the ice can diffuse to and attack the polymer and vice versa. But as the gap widens, diffusion rates go down and the effects of mutual attack are diminished. The etching process continues as in pure ice and (almost) dry inclusions.

The importance of a gap between ice and an organic inclusion during radiation damage was emphasized by Heide and Zeitler (1985). H.G. Heide (private commun. 1985) also suggested that in the case of PS in I_h, radiation damage is initially slow, until enough PS is etched to form a free ice surface which becomes a source of free radicals. This does not explain the very severe and fast radiation damage of PMMA in I_h, where no gap can be observed. But gaps may form by mass loss of the polymer, and refill immediately by repolymerization.

More support for the free-radical mechanism suggested here is drawn from the enhancement of radiation damage when small polymer particles are irradiated in ice (Fig. 5). The straightforward explanation for this phenomenon is the high surface area-to-volume ratio for small spheres (this ratio is inversely proportional to the diameter of the particles). This means more free radicals per unit mass of polymer, and thus more rapid radiolysis. Repolymerization as manifested by fusion of neighbouring polymer particles, crust formation at the end of radiation damage to small PS (Fig. 6) or PVC particles, and a stable polymer left over after the termination of radiolysis of scission-type polymers in ice, also support the free-radical hypothesis of radiolysis of organic inclusions in ice.

A fully satisfactory explanation for the marked difference of radiolysis in vitreous and crystalline ice is not available yet. One explanation connects it to faster diffusion of free radicals away from the particle/ice interface through vitreous ice, in essence a high viscosity liquid. In crystalline ice, the rigid matrix of the true solid "cages in" the free radical formed at that interface and makes radiolysis more efficient (Talmon et al. 1986). This is supported by the finding that in a large cluster of polymer particles in I_v, radiolysis starts in the middle of the cluster (Fig. 6), where the concentration of free radicals is high despite their

◄

Fig. 5a–d. A frozen sample of PS latex (nominal size 0.091 μm) at 95 K after increasing exposure to 100 kV electrons: **a** exposure of 400 C m^{-2}; **b** 6 kC m^{-2}; **c** 12 kC m^{-2}; **d** 24 kC m^{-2} (Talmon 1984)

Fig. 6a–c. A large aggregate of PMMA spheres in vitreous ice after increasing electron exposures: **a** 1.0 kC m^{-2}; **b** 5.3 kC m^{-2}; **c** 11 kC m^{-2} (Talmon et al. 1986)

Fig. 7a, b. Frozen-hydrated specimens of two-stage latex of poly(butylacrylate) and polystyrene (PBA/PS = 3:1): **a** latex prepared by adding styrene to a PBA seed-latex, after exposure to 88 kC m^{-2}; **b** latex prepared by adding butylacrylate to PS seed-latex, after exposure to 60 kC m^{-2}. Radiation damage reveals the formation of an interpenetrating network in the former case, and a core-and-shell structure in the latter high diffusivity through vitreous ice. Another explanation by Heide and Zeitler (1985) asserts that the finite viscosity of I_v makes it relatively probable for a forming gap to close around an etched particle, and thus reduce radiolysis.

As there is little doubt now that free radicals and their diffusion rates play an important role in radiolysis of organic particles in ice, and since diffusion rates are strongly temperature-dependent, one would expect to be able to reduce these radiolytic effects by cooling the specimen close to liquid helium temperature. This should reduce radiolysis in ice to levels encountered in dry specimens at the same temperature. A careful study of the effect of very low temperatures on well-controlled model systems is obviously necessary.

5.4 Applications of Radiation Damage to Organic Inclusions in Ice

The best known application of radiation damage is electron beam lithography, an important technique in the fabrication of microelectronic devices. This is done at room temperature. Recently, Narkis et al. (1985) and Talmon et al. (1985) have demonstrated the use of radiation damage at low temperature and in the presence of ice to elucidate the microstructure of certain polymer systems. The technique makes a judicious use of the significant difference between radiolysis of scission-type and crosslinking-type organic materials in hexagonal ice. The technique has been demonstrated for polymer systems only, but it should be applicable to other organic or biological systems as well.

The problem solved by the technique is whether a polymer produced by two-stage emulsion polymerization, i.e. by polymerizing one monomer in emulsion and then adding a second monomer, has a core (of the first polymerized monomer) and a shell (of the second polymerized monomer), or whether the two polymerized monomers form an interpenetrating network of small, connected, sample-spanning polymer domains.

Frozen-hydrated specimens of a two-stage latex, made by first polymerizing polyacrylate (a mixture of 70% ethylacrylate and 30% butylacrylate) and then adding styrene so that the styrene-to-acrylate ratio was 30:70, showed radiolytic behaviour quite similar to that of pure butylacrylate (Fig. 7a). This behaviour excludes the possibility of PS shell formation. However, when the ratio of styrene-to-acrylate is increased to 50:50, the situation is different, as can be seen in Fig. 3. Here, the particles undergo radiolysis in a way similar to the way pure PS does, indicating that the outer layer is at least very PS-rich, if not pure PS.

When the order of polymerization is reversed, namely, acrylate monomer is added to a PS seed-latex, the microstructure formed is different (Fig. 7b). In this case the outer regions of the particles are damaged like a scission-type polymer in I_h, while a seemingly intact core becomes visible. This shows that here a core (PS) and shell (polyacrylate) geometry is formed. In addition to the two-stage polymer particles, one can easily detect small, pure polyacrylate particles characterized by their sponge-like appearance, and large PS particles characterized by the cavities in the ice around them. Without radiolysis it would have been impossible to distinguish between the three kinds of particles present in this system.

Another application of radiation damage in ice is to distinguish between organic particles embedded in ice and particles located above or below the ice. Those embedded in ice will be damaged under the beam much faster and in a different way than those not embedded in the ice matrix (Talmon 1984). Such a test should be conducted, of course, only after useful microstructural data has been recorded from the area in question at low electron exposures.

6 Conclusions

At low temperatures, typically below 150 K, electron beam radiation processes are slowed by a factor of three to five in comparison to room temperature. Although the primary ionization processes are not affected by lowering the temperature, the amount of structural damage is reduced by lowering the diffusivities of free radicals and other products of the primary events by cooling the specimen. Most of the gain in this so-called cryoprotection effect is already achieved at 150 K, and certainly at 77 K (liquid nitrogen). At least at the present it seems that only marginal benefits can be gained by using a liquid helium-cooled specimen holder, or a superconducting lens (also liquid helium-cooled), which are very expensive and technically difficult to operate.

An important aspect of cryomicroscopy is the presence of water in the specimen, either deliberately, as an integral and usually a major component of frozen-hydrated specimens, or as an adsorbed layer from the microscope vacuum, or as a residual substance due to incomplete drying during specimen preparation. The presence of water modifies radiation damage to electron-irradiated organic specimens. Because ice is an excellent source of free radicals when exposed to ionizing radiation such as electrons, radiation damage can be much more severe in its presence than in its absence. But since it takes a rather high electron dose to produce enough free radicals to start a chain of free-radical reactions at the ice/organic material interface (on the order of $1000\,e^- \,nm^{-2}$ [$10\,e^-\,Å^{-2}$]), very radiation-sensitive organic crystals are damaged in most low-dose experiments before the effect of surrounding ice can be noticed. However, at moderately high or at high electron exposures the presence of ice may lead to massive structural changes (bubbling, polymerization, network formation) and mass loss.

Different organic materials show different sensitivity to electron beam irradiation in the presence of ice. Although crystallinity is lost after rather small exposures (10^2 to $10^5\,e^-\,nm^{-2}$), some organic materials tend to crosslink and form a more stable compound, thus preserving supramolecular microstructure. But in most biological specimens, the predominant radiolytic process is scission, giving rise to massive structural changes and mass loss in the presence of ice.

The structure of ice surrounding the organic material also affects the pattern and the extent of damage. For certain highly radiation-sensitive polymers and smaller organic and biological molecules, such as synthetic surfactants and natural phospholipids, the radiation sensitivity in vitreous ice is about ten times less than in crystalline ice. This is probably the result of the rigidity of crystalline ice that "cages" the forming free radicals at the organic inclusion/ice interface, bringing their concentrations to a level where chain reactions are very fast. This recently observed phenomenon adds one more good reason for examining vitrified-hydrated (rather than crystallized, frozen-hydrated) specimens by cryoelectron microscopy techniques.

Acknowledgement. The author's research on cryoelectron microscopy has been supported in part by a grant from the United States – Israel Binational Science Foundation (BSF), Jerusalem, Israel.

References

(References marked by an asterisk are review papers or books respectively)

Bourgeois CA, Zanchi G, Khin Y, Lacaze JC, Zalta J, Bouteille M (1980) Hydrated cell nucleus fractions as studied by energy filter in a 1.2 MeV electron microscope. In: Brederoo P, de Priester W (eds) Electron microscopy 1980, vol 2. Proc 7th Eur Congr Electron Micros, Leiden, pp 118–119

Charlesby A (1972) Radiation effects in polymers. In: Jenkins AD (ed) Polymer science, vol 2. Elsevier/North-Holland Biomedical Press, Amsterdam New York, pp 1544–1559

Charlesby A (1981) Crosslinking and degradation of polymers. Radiat Phys Chem 18:59–66

Chiu W, Knapek E, Jeng TW, Dietrich I (1981) Electron radiation damage of a thin protein crystal at 4 K. Ultramicroscopy 6:291–296

Chiu W, Downing KH, Dubochet J, Glaeser RM, Heide HG, Knapek E, Kopf DA, Lamvik MK, Lepault J, Robertson JD, Zeitler E, Zemlin F (1986) Cryoprotection in electron microscopy. J Microsc (Oxford) 141:385–391

* Cosslett VE (1978) Radiation damage in high resolution electron microscopy of biological materials: a review. J Microsc (Oxford) 113:113–129

Dietrich I, Formanek H, Fox F, Knapek E, Weyl R (1979) Reduction of radiation damage in an electron microscope with a superconducting lens system. Nature (London) 277:380–381

Dubochet J (1975) Carbon loss during irradiation of T4 bacteriophages and *E. coli* bacteria in electron microscopes. J Ultrastruct Res 52:276–288

Dubochet J, Knapek E (1979) Use of very low temperature to reduce electron beam damage in biological specimens. Chem Scr 14:267–269

Dubochet J, Lepault J, Freeman R, Berriman JA, Homo JC (1982 a) Electron microscopy of frozen water and aqueous solutions. J Microsc (Oxford) 128:219–237

Dubochet J, Chang JJ, Freeman R, Lepault J, McDowall AW (1982 b) Frozen aqueous suspensions. Ultramicroscopy 10:55–62

Egerton RF (1980) Chemical measurements of radiation damage in organic samples at and below room temperature. Ultramicroscopy 5:521–523

Egerton RF (1982) Organic mass loss at 100 K and 300 K. J Microsc (Oxford) 126:95–100

Glaeser RM (1971) Limitations to significant information in biological electron microscopy as a result of radiation damage. J Ultrastruct Res 36:466–482

*Glaeser RM (1979) Radiation damage with biological specimens and organic materials. In: Hren JJ, Goldstein JI, Joy DC (eds) Introduction to analytical electron microscopy. Plenum, New York London, pp 423–436

Glaeser RM, Hobbs LW (1975) Radiation damage in stained catalase at low temperature. J Microsc (Oxford) 103:209–214

*Glaeser RM, Taylor KA (1978) Radiation damage relative to transmission electron microscopy of biological specimens at low temperature: a review. J Microsc (Oxford) 112:127–138

Grubb DT, Groves GW (1971) Rate of damage of polymer crystals in the electron microscope: dependence on temperature and beam voltage. Philos Mag 24:815–828

Hartman RE, Hartman RS (1971) Residual gas reaction in the electron microscope: IV. a factor in radiation damage. In: Arceneaux CJ (ed) Proc 29th Annu EMSA Meet Boston Mass. Claitors, Baton Rouge, pp 74–75

Hayward SB, Glaeser RM (1980) High resolution cold stage for the JEOL 100B and 100C electron microscopes. Ultramicroscopy 5:3–8
Heide HG (1965) Contamination and irradiation effects and their dependence on the composition of residual gas in the electron microscope. Lab Invest 14:396/1134–401/1139
Heide HG (1982) On the irradiation of organic samples in the vicinity of ice. Ultramicroscopy 7:299–300
Heide HG (1984) Observations on ice layers. Ultramicroscopy 14:271–278
Heide HG, Grund S (1974) Eine Tiefkühlkette zum Überführen von wasserhaltigen biologischen Objekten ins Elektronenmikroskop. J Ultrastruct Res 48:259–268
Heide HG, Zeitler E (1985) Physical behaviour of solid water at low temperatures and the embedding of electron microscopical specimens. Ultramicroscopy 16:151–160
Hobbs LW (1979a) Radiation damage in electron microscopy of inorganic solids. Ultramicroscopy 3:381–386
*Hobbs LW (1979b) Radiation effects in analysis of inorganic specimens by TEM. In: Hren JJ, Goldstein JI, Joy DC (eds) Introduction to analytical electron microscopy. Plenum, New York London, pp 437–480
Hui SW (1980) Radiation damage of phosphatidylcholine bilayers: effects of temperature and hydration. Ultramicroscopy 5:505–512
Knapek E, Dubochet J (1980) Beam damage to organic material is considerably reduced in cryo-electron microscopy. J Mol Biol 141:147–161
Knapek E, Formanek H, Lefranc G, Dietrich I (1984) The interpretation of radiation damage measurements with electron diffraction of organic materials at very low temperatures. Ultramicroscopy 14:253–264
Lepault J, Freeman R, Dubochet J (1983a) Electron beam induced "vitrified ice". J Microsc (Oxford) 132:RP3–RP4
Lepault J, Booy FP, Dubochet J (1983b) Electron microscopy of frozen biological suspension. J Microsc (Oxford) 129:89–102
Lepault J, Dubochet J, Dietrich I, Knapek E, Zeitler E (1983c) Amendment to: Electron beam damage to organic specimens at liquid helium temperature. J Mol Biol 163:511
Miller D, Bellare JR, Evans DF, Talmon Y, Ninham BA (1987) On the meaning and structure of amphiphilic phases: inferences from video enhanced microscopy and cryo-transmission electron microscopy. J Phys Chem 91:674–685
Müller KH, Zemlin F, Zeitler E (1981) Cryoprotection of electron-irradiated organic crystals. In: Bailey GW (ed) Proc 39th Annu EMSA Meet, Atlanta. Claitor, Baton Rouge, pp 26–27
Narkis M, Talmon Y, Silverstein M (1985) Properties and structure of elastomeric two-stage emulsion interpenetrating networks. Polymer 26:1359–1364
Ramamurti K, Crewe AV, Isaacson MS (1975) Low temperature mass loss of thin films of L-phenylalanine and L-tryptophan upon electron irradiation. Ultramicroscopy 1:156–158
Reimer L (1984) Methods of detection of radiation damage in electron microscopy. Ultramicroscopy 14:291–304
Siegel G (1970) The influence of low temperature on the radiation damage of organic compounds and biological objects by electron irradiation. In: Favard P (ed) Proc 7th Int Congr Electron Microsc, vol 2, Grenoble, pp 221–222
Siegel G (1972) Der Einfluß tiefer Temperaturen auf die Strahlenschädigung von organischen Kristallen durch 100 keV-Elektronen. Z Naturforsch 27a:325–332
Siegel S, Flournoy JM, Baum LH (1961) Irradiation yields of radicals in gamma-irradiated ice at 4.2° and 77° K. J Chem Phys 34:1782–1788
Somlyo AP, Shuman H, Somlyo AV (1976) Quantitation, minimal detectable levels and application of biological electron probe analysis. In: Silcox J (ed) Analytical electron microscopy: report of a specialist workshop, August 1976, Cornell Univ, pp 114–117
Symons MCR (1982a) Radiation processes in frozen aqueous systems. Ultramicroscopy 10:97–104
Symons MCR (1982b) The pre-knock-on concept. Ultramicroscopy 10:41–44
Talmon Y (1982) Thermal and radiation damage to frozen hydrated specimens. J Microsc (Oxford) 125:227–237
Talmon Y (1984) Radiation damage to organic inclusions in ice. Ultramicroscopy 14:305–316
Talmon Y (1986) Imaging surfactant dispersions by electron microscopy of vitrified specimens. Colloids Surfaces 19:237–248

Talmon Y, Thomas EL (1977a) Temperature rise and sublimation of water from thin frozen hydrated specimens in cold stage microscopy. Scanning Electron Microsc 1977/I:265–272

Talmon Y, Thomas EL (1977b) Beam heating of a moderately thick cold stage specimen in the SEM/STEM. J Microsc (Oxford) 111:151–164

Talmon Y, Davis HT, Scriven LE, Thomas EL (1979) Mass loss and etching of frozen hydrated specimens. J Microsc (Oxford) 117:321–332

Talmon Y, Narkis M, Silverstein M (1985) Electron beam radiation damage as an analytical tool for polymer science. J Electron Microsc Tech 2:589–596

Talmon Y, Adrian M, Dubochet J (1986) Electron beam radiation damage to organic inclusions in vitreous, cubic, and hexagonal ice. J Microsc (Oxford) 141:375–384

Taub IA, Eiben K (1968) Transient solvated electron, hydroxyl, and hydroperoxy radicals in pulse-irradiated crystalline ice. J Chem Phys 49:2499–2513

Taylor KA, Glaeser RM (1976) Electron microscopy of frozen hydrated biological specimens. J Ultrastruct Res 55:448–456

Unwin PNT, Muguruma J (1971) Transmission electron microscopy of ice. J Appl Phys 42:3640–3641

Venables JA, Basset DC (1967) Electron microscopy of polyethylene below 20 °K. Nature (London) 214:1107–1108

Wade RH (1984) The temperature dependence of radiation damage in organic and biological materials. Ultramicroscopy 14:265–270

Zemlin F, Reuber E, Beckmann E, Dorset D (1986) High-resolution electron microscopy of beam-sensitive specimens: results with paraffin. In: Bailey GW (ed) Proc 44th Annu EMSA Meet, Albuquerque. San Francisco Press, pp 10–13

Part II General Methodology

Chapter 4 Cryofixation Without Pretreatment at Ambient Pressure

HELLMUTH SITTE, LUDWIG EDELMANN, and KLAUS NEUMANN[1]

1 Introduction

Almost a century ago biological or medical objects were used in their frozen state in order to accelerate pathological diagnosis (immediate section) and to better maintain their chemical constitution (Altmann 1890). This alternative to chemical fixation gained in importance during the first half of this century due to the availability of liquefied air used as coolant ("cryogen") as well as to the fact that effective cooling systems for cryostats had been developed (see e.g. Gersh 1932; Simpson 1941; Eränkö 1954; Kulenkampff 1955; Neumann 1958). Attempts to freeze wet objects for electron microscopy revealed that at normal atmospheric pressure and under the most favourable conditions only an approx. 30- µm border zone can be perfectly frozen (see e.g. Sitte 1979; Plattner and Bachmann 1982; Robards and Sleytr 1985). At greater depths the mixed plasmatic phases segregate. The size of these segregation compartments formed by the growing ice crystals within the specimen increases so rapidly that deeper layers cannot be used for electron microscopy. The depth of the well-preserved border zone can be increased without chemical pretreatment to, at the most, 300 µm by applying high pressures of about 2100 bar (Müller and Moor 1984). These limits can be considerably extended by the use of anti-freezing agents ("cryoprotectants"): glycerol has proved to be an effective anti-freeze for the freeze-fracture/freeze-etch method (Moor and Mühlethaler 1963) and saccharose in cryoultramicrotomy (see e.g. Bernhard and Leduc 1967; Tokuyasu 1973; Griffiths et al. 1984).

Chemical prefixation is usually required in order to render the cell membranes more permeable for an effective penetration by these cryoprotectants and to reduce detrimental side effects on cell preservation.

After fixation and chemical cryoprotection, objects of approx. 1 mm diameter can be easily frozen (e.g. in liquid nitrogen, hereafter referred to as LN_2) without producing disturbing ice-crystal artefacts. Thus, sections may be obtained for histochemical studies without dehydration in polar media and without resin embedding in the cryoultramicrotome. However, this pretreatment distinctly alters the state of these specimens from the in vivo condition. Any cryoprotection of this type is not applicable if optimal preservation of the ultrastructure and chemical

1 Fachrichtung 3.5, Medizinische Biologie, Universität des Saarlandes, D-6650 Homburg-Saar, FRG

Cryotechniques in Biological Electron Microscopy
Edited by R. A. Steinbrecht and K. Zierold

composition of wet biological specimens is required. The topic of this chapter is the presentation of methods enabling cryofixation without prior chemical treatment. Particular emphasis is placed on the practical execution of cryofixation. Information on the physicochemistry of water and hydrous mixed phases as well as on freezing processes is supplied by Bachmann and Mayer (Chap. 1, this Vol.). Moor (Chap. 8, this Vol.) competently deals with high pressure cryofixation and this technique is only briefly compared and discussed in Section 4.5.

Theoretically, the basic problem of cryofixation without cryoprotection is to extract the heat from the object as rapidly as possible, because the result is determined by the speed of cooling ("cooling rate" measured in $K\,s^{-1}$ or $°C\,s^{-1}$; see e.g. van Venrooij et al. 1975; Kopstad and Elgsaeter 1982; Bald 1983; Bald 1984; Bald 1985; Mayer 1985; Bald 1986; see also Chap. 1, this Vol.). It may be assumed that cooling rates above some $10^6\,K\,s^{-1}$ cause real "vitrification" of the usual biological specimens during which even water molecules are not able to assume lattice-like arrangements. Under the same conditions, cooling rates of approx. $10^4\,K\,s^{-1}$ seem to be necessary to limit the segregation and the size of the ice crystals so that they do not interfere with the electron microscope image. Due to the poor thermal conductivity of ice these high cooling rates can only be realized in the previously mentioned border zones of 30 μm depth at maximum: however, at normal pressure this depth is only then attained if the temperature at the interface of specimen and cryogen is kept continuously at a minimum during the decisive initial phase of cryofixation.

In practice, other factors also influence the quality of preservation. For example, for freezing compact tissue not only the method of cryofixation used is decisive for the preservation of structures, but also the method of specimen excision. Layers of liquid on the specimen surface drastically reduce the depth of the well-preserved zone to be examined: the cryogen has to absorb the heat through this layer of liquid. This leads to a decrease in the steepness of the decisive temperature gradient. Water evaporation has to be avoided as well, since it considerably changes the appearance of free specimen surfaces. Limitations due to the very restricted depth of the well-preserved border layer after cryofixation can be minimized by special methods of preparation: thus, most parts of larger organs may be laid bare by careful cutting and the preparation may be obtained by subsequent impact cryofixation (Bernard and Krigman 1974; Van Harreveld and Fifkova 1975). Cryopreparation therefore may be applied to almost any category of fresh specimens.

2 Cryofixation with Liquid Cryogen

Rapid freezing by contact with a cold liquid is, without doubt, the simplest method. This is due to the fact that for objects of any shape intimate surface contact between specimen and cryogen is easy to establish. The choice of cryogen is of utmost importance: several authors have investigated the efficiency of various coolants by comparing cooling rates of model objects under standard conditions (see Table 1). Liquefied gases with a low boiling point (LHe, LN_2) are only suitable for use as "primary cryogens", i.e. to cool "secondary cryogens" such as

Table 1. Cooling efficiency (CE) and freezing point (Fp) of cryogens in comparison with liquid propane[a]

Cryogen	Fp (K)	Fp (°C)	CE
Ethane	102	−171	1.3
Propane	83	−190	1.0
Propane-Isopentane (3:1)			0.8
Freon 13	88	−185	0.8
Freon 22	118	−155	0.7
Freon 12	121	−152	0.5
Isopentane	113	−160	0.5
LN_2 cooled to Fp	63	−210	0.2
Boiling LN_2	77	−196	0.1

[a] CE is roughly calculated by comparison of the cooling rates obtained with bare thermocouples by Costello and Corless (1978) and Ryan et al. (1987). CE values >1 indicate higher cooling rates, CE values <1 lower cooling rates than propane (CE = 1). It should be mentioned that cooling depends to a considerable extent on the kind and shape of the specimen. As far as the primary cryogen LN_2 is concerned, CE >1.3 will be reached by suppression of the Leidenfrost phenomenon (Bald 1984). The 3:1 mixture of propane und isopentane is often used to inhibit freezing at LN_2 temperature. See discussion in Sections 2 and 4.2.

propane, ethane or halogenated hydrocarbons. Boiling LHe or LN_2 are not appropriate for direct use as a cryogen because they immediately envelop any warmer object with a thermally insulating layer of gas (Leidenfrost phenomenon). The only exception here is pressurized LN_2 (Bald 1984), although its practical application presents some problems. LN_2 undercooled to 63 K (−210 °C; Umrath 1974) is only suitable for very small specimens as its heat capacity for cooling larger specimens is not adequate without additional measures being taken. Isopentane often used in the past provides inferior results. The same applies to Freon 12 and Freon 22. Propane and ethane provide optimal results. Ethane, apart from the best cooling rates, also has the advantage of a comparatively high vapour pressure at low temperatures. Ethane residues are therefore easy to eliminate from vitrified specimens, which must not be heated to above 133 K (−140 °C; see Sect. 4.3) to avoid devitrification. For all other work propane has sufficient cooling power as well as the advantage of being easy and inexpensive to acquire. The use of propane in quantities of less than 50 ml presents no danger, if the appropriate safety regulations are carefully observed (see Sitte et al., Chap. 16, this Vol.).

2.1 Immersion Cryofixation (Plunging)

The immersion cryofixation method may be successfully applied to very small or thin specimens and for the cryofixation of suspensions. Equipment as illustrated in Fig. 1 a is often used for this procedure. In practice, this configuration suffers from the disadvantage that the temperature of the LN_2 (77 K; −196 °C) is below the freezing point of propane, ethane or halogenated hydrocarbons. Either the LN_2 level, initially raised for liquefaction of the secondary cryogen, has to be

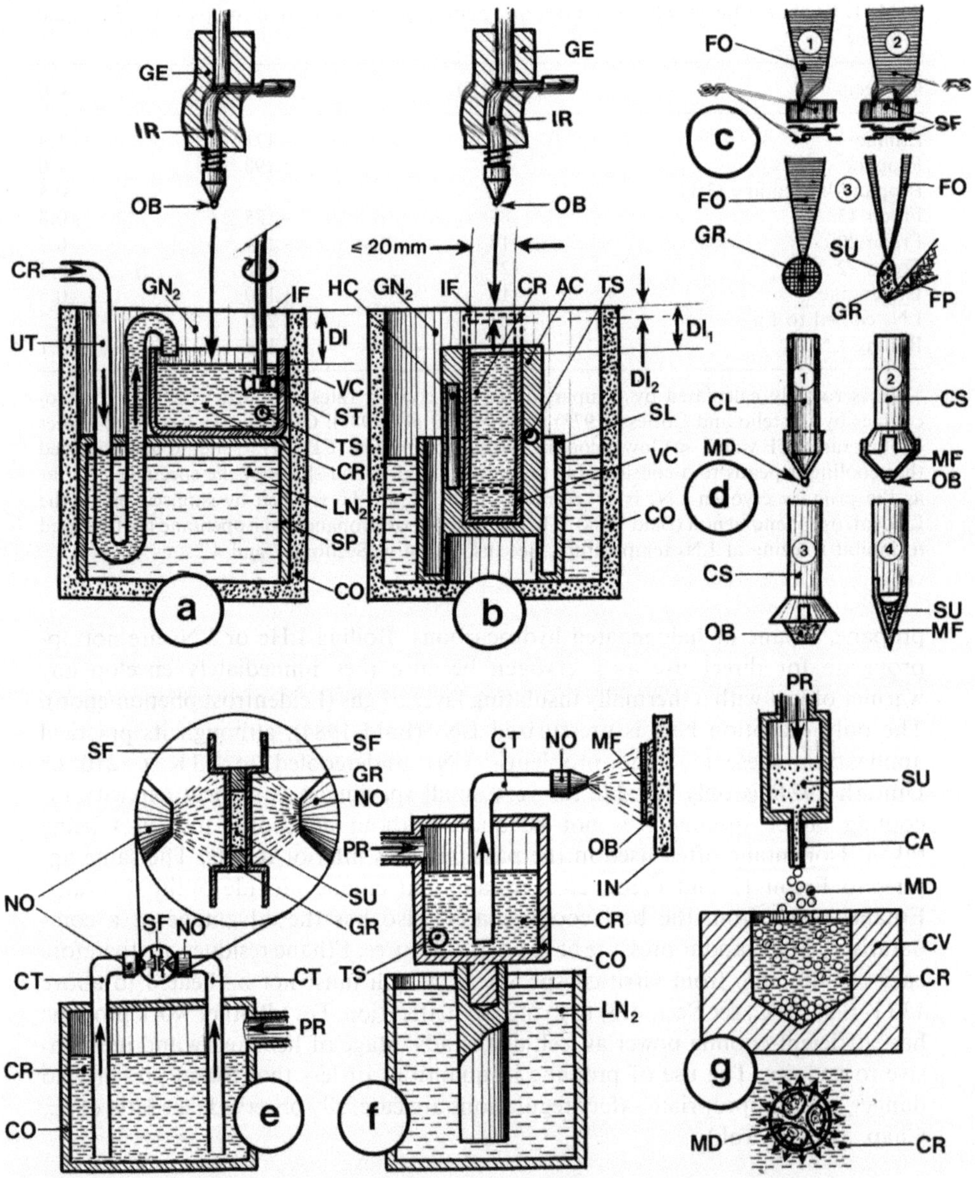

Fig. 1a–g. Procedures and instrumentation for cryofixation in liquid secondary cryogens (e.g. liquefied propane, ethane or halogenated hydrocarbons; see explanations in Sect. 2).
a Simple laboratory setup: the thermally insulating container *CO* is partially filled with LN_2. Gaseous secondary cryogen *CR* is liquefied in the U-tube *UT* and collected in the vessel *VC* positioned on the stage plate *SP*. The liquid *CR* is moved by the stirrer *ST*, its temperature is measured by the sensor *TS*. The distance *DI* between the surface of the cryogen *CR* and the interface *IF* between the GN_2 inside the container *CO* and the room atmosphere has to be kept >5 mm with respect to condensation of H_2O and O_2 on the cryogen. The object *OB* is mounted to the injector rod *IR* guided by the element *GE*. The *arrow* indicates injection movement.

lowered or another medium has to be added to the cryogen in order to decrease the freezing point (e.g. addition of isopentane to propane). However, in the latter case cooling efficiency suffers. Another disadvantage is that the distance from the cryogen level to the border between the atmosphere and the gaseous nitrogen (hereafter referred to as GN_2) usually exceeds 5 mm. This causes precooling of the specimen before it comes into contact with the cryogen (Ryan and Purse 1984, 1985; Ryan et al., in press). If the vessel is lifted in order to reduce this distance, the cryogen warms up. It is therefore necessary to keep up convection in the cryogen with the help of a stirrer and to control the cryogen temperature with a sensor. Working with the system illustrated in Fig. 1 b (Sitte 1984; Sitte et al. 1986) is more straightforward: the cryogen container is made of a metal with a high thermal conductivity. If the diameter of the cryogen container is only approx. 20 mm, temperature deviations in the cryogen column remain within ±2 K and no stirrer is required. This also applies if the distance between the level of the cryogen and the air/GN_2 interface is reduced. By using a thermostat, the hollow

---

b Immersion system according to Sitte et al. (1986; see also Sitte 1984): a sleeve *SL* prevents continuous contact between LN_2 and the secondary cryogen *CR* inside the vessel *VC* surrounded by the aluminium cylinder *AC*. The temperature of the cylinder *AC* is thermostatically controlled by the heating cartridge *HC* and the thermal sensor *TS*. With respect to the small diameter (20 mm) of the vessel *VC* a convection of the cryogen *CR* by a stirrer is not necessary. The distance DI_1 between the level of the cyrogen *CR* and the interface *IF* between GN_2 and room atmosphere can be reduced by lifting the vessel *VC* (*dotted lines*) to $DI_2 \leq 5$ mm to protect the specimen from precooling. This is done only for the injection procedure. For initial cooling and liquefaction of the secondary cryogen *CR* the LN_2 level may be raised above the top of the sleeve *SL* to cool down the cylinder *AC* directly.
c, d Different specimen carriers for plunging: $\mathbf{c_1}$ Mounting of sandwich foils *SF* with a forceps *FO* for subsequent freeze-cleavage. The forceps *FO* is fixed to the injector rod *IR* (see **a**). $\mathbf{c_2}$ Mounting by a special forceps *FS* forming a fork: only for freezing of slippery liquids, e.g. lipids, to avoid dislocation of the sandwich foils *SF* by the cryogen. $\mathbf{c_3}$ Mounting of a grid *GR* in the forceps *FO* for "bare-grid peparation" according to Adrian et al. (1984): the excess of the particle suspension *SU* is drained away by filter paper *FP*.
$\mathbf{d_1}$ Low weight carrier *CL* for subsequent cryoultramicrotomy of microdroplets *MD* or small objects in diameters ≤ 0.5 mm: the overall wall thickness is kept ≤ 0.1 mm to reduce heat capacity and thermal conductivity. $\mathbf{d_2}$ Standard carrier CS of a cryoultramicrotome adapted with a V-shaped thin metal foil *MF* of low heat capacity and thermal conductivity to take up the small object *OB* (see Ryan and Purse 1985). $\mathbf{d_3}$ Standard carrier CS of a cryoultramicrotome: only suited for cryoprotected objects *OB* with respect to heat capacity and thermal conductivity. $\mathbf{d_4}$ Use of a sharply edged V-shaped metal foil *MF* mounted on the injector rod for high-speed immersion of a suspension *SU* according to Handley et al. (1981).
e Double jet according to Müller et al. (1980): the secondary cryogen *CR* in the container *CO* is pressed through two tubes *CT* and nozzles *NO* against the sandwich specimen, e.g. the suspension *SU* between two sandwich foils *SF*. The thickness of the suspension layer is reproduced by a spacer, e.g. the grid *GR*.
f Single jet according to Burstein and Maurice (1978; see also Pscheid et al. 1981): the jet is directed vertically to the object *OB*, which is covered by a thin metal foil *MF* and mounted on a thermally insulating support *IN*. The secondary cryogen *CR* in the container *CO* is cooled by LN_2. The temperature is controlled by sensor *TS*.
g Spray freezing according to Bachmann and Schmitt (1971): the particle suspension *SU* is pressed through a thin capillary *CA* and forms microdroplets *MD* with diameters below 100 µm. The droplets are frozen and collected in liquified propane *CR (upper part)*. The heat is extracted in all directions (lower part: arrows). With respect to the small size the cooling is very efficient. *PR* pressure

metal cylinder can be kept precisely at a temperature just above the freezing point of the cryogen used. Additional agents for decreasing the freezing point are therefore not necessary. Heavy boiling of the primary cryogen LN_2 is avoided by a sleeve, which prevents direct contact between LN_2 and the surface of the metal container during standby operation. For initial cooling of the unit the LN_2 level can be raised so that the container is directly cooled by LN_2[1].

The system depicted in Fig. 1 b is suitable for immersion cryofixation of sandwich specimens for freeze-etching (Fig. 1 c_{1-2}; see also Costello et al. 1984) of thin suspension films ("bare-grid" method or "ice embedding" as in Fig. 1 c_3; see also Lepault et al. 1983; Adrian et al. 1984; Lepault 1985; Mandelkow et al. 1986; Trinick et al. 1986), for freezing microdroplets of suspensions for subsequent cryoultramicrotomy (Fig. 1 d_1; see also Dubochet et al. 1983) as well as for vitrifying microspecimens (Fig. 1 d_2; see also Ryan et al., in press). The minimum heat capacity of the various specimen carriers and maximum thermal resistance to the injector system are of decisive importance. The specimen carriers normally employed for cryoultramicrotomy of cryoprotected specimens (Fig. 1 d_3) are unsuitable for plunging: their heat capacity and thermal conductivity are too high for fresh specimens. In this case, after the specimen has come to a standstill, recrystallization may take place due to the inflow of heat from the carrier into the object and to insufficient cryogen convection for carrying off this heat.

Heat capacity and thermal resistance of the specimen holder are not the only decisive factors. The quality of immersion cryofixation depends to a large extent on the speed with which the object enters into and is moved within the cryogen. The path length has to be chosen in such a way that by the time the specimen reaches a halt, it is completely frozen and the inside of the specimen has attained a temperature which excludes recrystallization processes due to redistribution of heat after stopping (Robards and Crosby 1983; Steinbrecht 1985). At one time manual dipping of specimens affixed to a thermally insulating rod or held by tweezers was common practice, but the results obtained with this method are not reproducible: because of insufficient cryogen convection a coat of heated cryogen may form around the specimen, thus preventing rapid cooling. With increased specimen velocity as achieved by using mechanical injector systems (Costello and Corless 1978; Handley et al. 1981; Costello et al. 1982; Robards and Crosby 1983; Costello et al. 1984; Sitte 1984; Sitte et al. 1986) usually the yield of well-frozen specimens is increased. For reasons which are not yet completely understood, this is true only for velocities up to approx. 3 m s^{-1}. Further increase of the specimen velocity often leads to the seemingly paradox observation, that cooling rates are not increased in the most important temperature range down to 223 K (−50 °C) as presented in Fig. 2 (see Robards and Crosby 1983). It seems from microthermocouple measurements that the immersion speed has no significant influence on the cooling rates during the motion of the specimen through the cryogen. At higher specimen velocities a steeper decrease of temperatures occurs after the specimen has stopped. Probably the inertia of the measurement (diameter and inert mass of the thermocouple bead) and the limited speed of the heat extraction cause this delay. Possibly also turbulences and/or air pockets contribute to these

[1] An apparatus working according to Fig. 1 b is available from Reichert-Jung (KF 80).

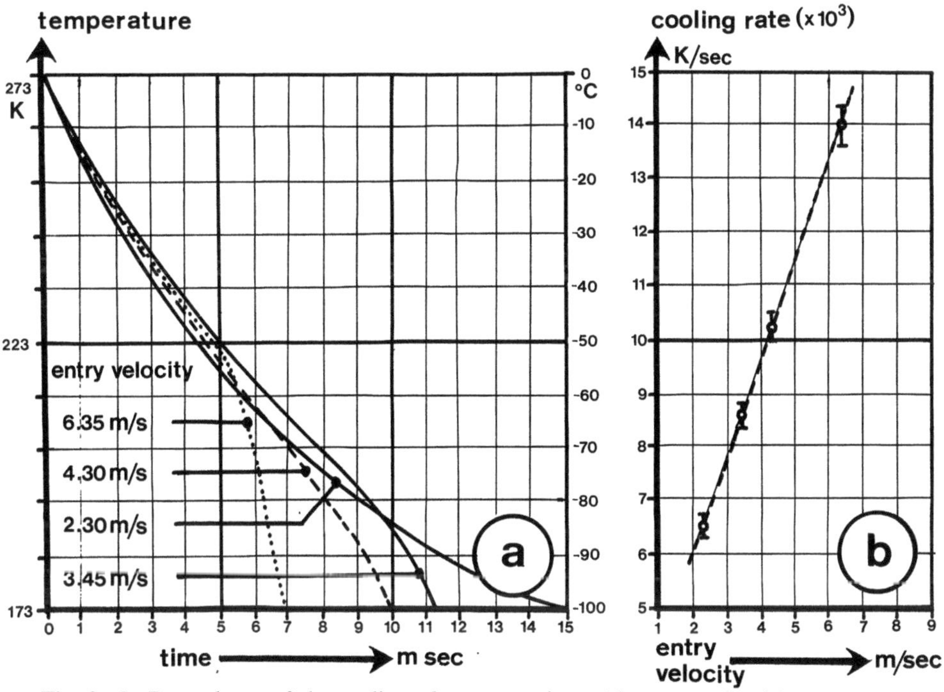

Fig. 2a, b. Dependence of the cooling of a test specimen (thermocouple with approx. 60 µm bead diameter, sandwiched by two 50-µm-thick copper foils) on the entry velocity into liquid propane according to Robards and Crosby (1983). **a** Cooling profiles between 273 and 173 K (0 and $-100\,°C$) at different entry velocities: it is of interest that the cooling profiles are similar down to 223 K ($-50\,°C$) and that the differences occur after the standstill of the specimen. **b** Cooling rates show a linear dependence from the entry velocity. By definition the cooling rates are calculated from the time needed for the temperature decrease from 273 K (0 °C) to 173 K ($-100\,°C$). Compare discussion in Section 2.1 and Fig. 3.

phenomena. One may conclude that for most purposes with specimens shaped as illustrated in Fig. 1c, the best cooling rates are already obtained with injection velocities of approx. $3\ m\ s^{-1}$. Immersion with these limited speeds are more convenient as dangerous cryogen splashing does not occur. An exception to this rule are specimens which for aerodynamic or hydrodynamic reasons have been given the shape of sharp blades or pointed needles (Fig. 1 d_4): Handley and co-workers (1981) were able to freeze blood perfectly using an injection velocity of $15\ m\ s^{-1}$ (Fig. 3). The blood was protected against the impact in the cryogen by a V-shaped, 4-µm-thick titanium sheet with an edged entering profile. This method, however, cannot be easily applied to most tissue samples as they cannot be pressed into an edged profile without causing mechanical damage.

2.2 Cryofixation by Cryogen Jet

Theoretically it does not make any difference whether the object is moved at a high velocity within resting cryogen or whether the object is stationary and is cooled with a moving jet of cryogen. In practice, however, this reversal of the procedure has some important consequences. To begin with, it is much easier to ac-

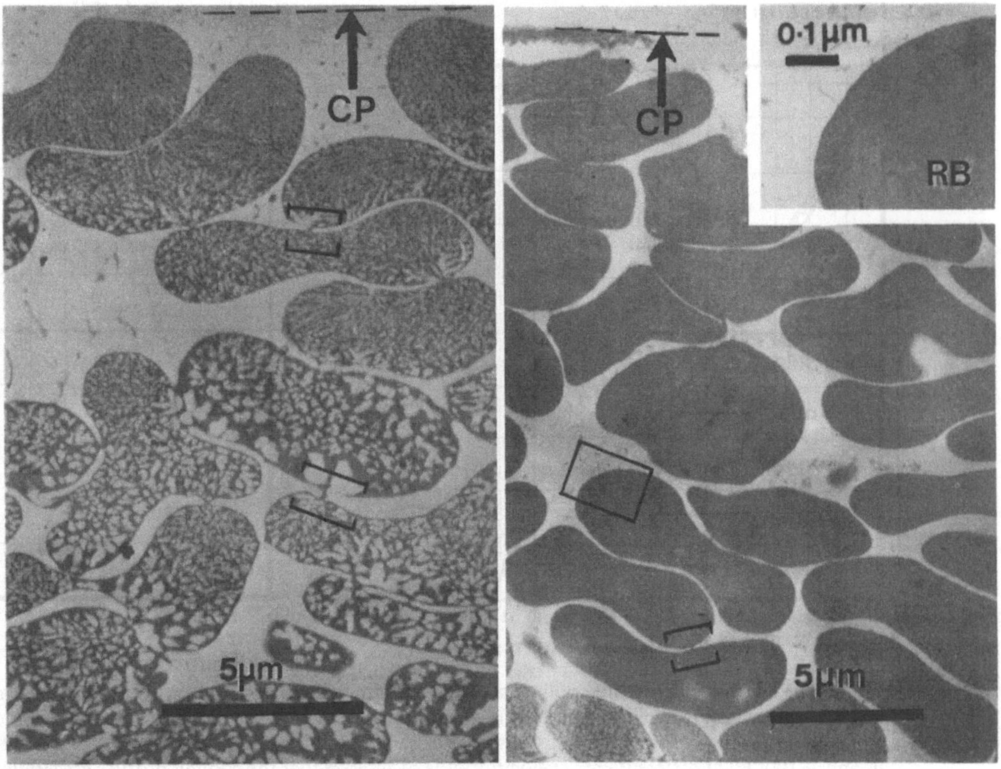

Fig. 3. Striking difference of the state of preservation of red cell suspensions after immersion cryofixation with slow (*left*: 1.25 m s^{-1}) and high entry velocity (*right*: 10 m s^{-1}) into the secondary cryogen. (Electron micrographs courtesy of D. A. Handley). The blood suspension is covered by a sharply edged V-shaped 4-µm titanium foil and plunged into subcooled Freon 22 at 109 K (−164 °C). Note the different shape and size of the segregation compartments ("ice-crystal artefacts") even in neighbouring cells (*brackets*). Contact plane *CP* of the blood suspension with the titanium foil indicated by *dashed line.* Freeze-substitution in OsO_4-acetone at 193 K (−80 °C) for 18 h, epoxide embedding, ultrathin sections stained with uranyl acetate (Handley et al. 1981)

celerate a jet of liquid cryogen to an extremely high velocity with the help of a nozzle and a compressor than it is to accelerate a specimen (see e.g. Burstein and Maurice 1978; Müller et al. 1980; Pscheid et al. 1981; van Venetie et al. 1981; Knoll et al. 1982; Kopstad and Elgsaeter 1982; Plattner and Knoll 1984; Haggis 1986). On the other hand, it is more difficult to perform cryogen-jet cryofixation with the small quantities of cryogen (5−20 ml) which are usually sufficient for a series of immersion procedures. It is therefore important during cryogen-jet cryofixation that the relevant safety regulations are strictly observed (see Sitte et al., Chap. 16, this Vol.). In addition, the exposed specimen surfaces must be protected from the impulse of the jet with sufficiently strong metal foils.

Nowadays different jet systems are used for thin sandwich specimens with a thickness of ≤ 0.1 mm (Fig. 1 e) and for thicker specimens (Fig. 1 f). The double jet, introduced by Müller and co-workers (1980) is a symmetrical arrangement (Fig. 1 e) where the specimen (preferably a cell suspension) is placed in the plane of symmetry between two thin sandwich foils[2]. The minimum layer thickness is reproduced, e.g. by a specimen grid used as a spacer. Two nozzles direct jets of liquid propane onto the specimen if pressure is applied to the container vessel. If all elements are correctly adjusted, this system provides very high cooling rates. This is probably due to the fact that heat is extracted synchronously from both sides. Commercially available jet systems of this design do not, however, always attain such high cooling rates. It is possible to freeze thicker specimens with a single jet (Burstein and Maurice 1978; Pscheid et al. 1981; Knoll et al. 1982; Plattner and Knoll 1984) which can be easily produced in a workshop (Fig. 1 f; Knoll 1984). Thus, monolayers of tissue cultures can be cryofixed in situ without removal from the support. In order to obtain the optimum effect it is important that the jet gun be exactly of the described dimensions. Freeze-fracturing of sandwich specimens is especially easy by simply taking off the sandwich supports.

2.3 Spray Freezing

A particularly elegant method for cryofixation of aqueous cell or particle suspensions was described by Bachmann and Schmitt (1971, for details see Bachmann and Schmitt-Fumian 1973). They spray suspensions through thin capillaries or orifices (see Fig. 1 g). On ejection the suspensions form fine droplets with diameters of 10–50 μm. These immediately freeze on contact with the propane used as cryogen. The small volume of the droplets is optimally suited for heat extraction. The cooling rate at the centre of each sphere is even increased for the given dimensions (Van Venrooij et al. 1975). Unfortunately, this technique is only suitable for spraying suspensions of low viscosity. These are optimally frozen in this way, even if containing cells with an extremely high water content (Plattner et al. 1972). Highly viscous liquids cannot be pressed through the narrow capillaries or orifices of the device with optimum pressure.

3 Impact Cryofixation (Slamming)

3.1 Basic Concept of Van Harreveld

Cooled metal surfaces were previously applied (Simpson 1941) and later successfully used (Eränkö 1954) for histochemical preparations. They were first used for electron microscopy by Van Harreveld and Crowell (1964), who cooled a highly polished silver surface down to 63 K (−210 °C) with N_2 partly solidified by vacuum ("N_2 slush") and froze natural organ surfaces on this mirror surface ("metal-mirror cryofixation" also called "slamming"). Frost precipitates in the relatively small cooling chamber were avoided by flushing with helium gas

---

[2] Equipment for double-sided jet-freezing according to Fig. 1 e is commercially available from Balzers Union, Hexland and RMC.

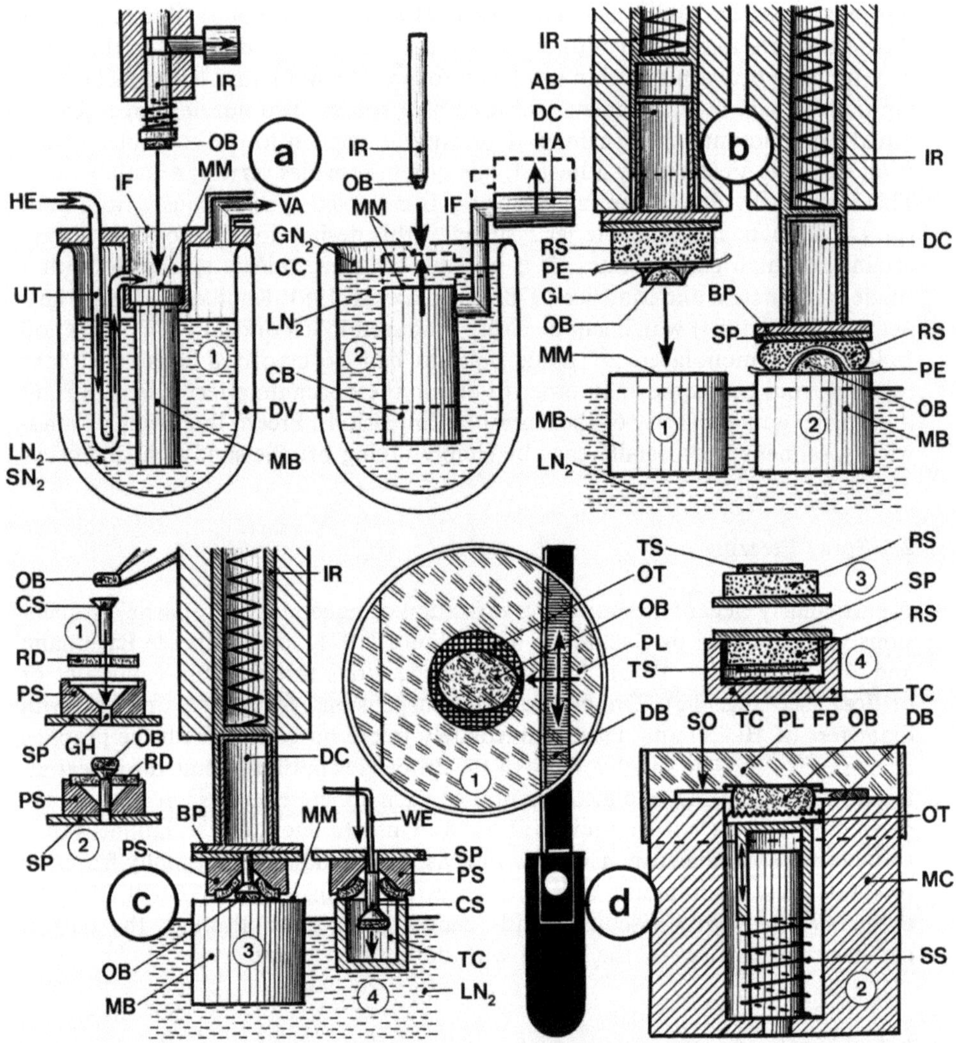

Fig. 4a–d. Impact cryofixation (**a–c**) and preparation of tissue slices for this procedure (cf. discussion in Sects. 3 and 4.3). $\mathbf{a_1}$ Simplified schematic drawing of the Van Harreveld system (Van Harreveld and Crowell 1964): a metal block *MB* (silver) with the highly polished "mirror" surface *MM* is cooled by N_2 slush, a mixture of LN_2 and solid nitrogen (SN_2) obtained in the Dewar vessel *DV by vacuum VA*. The cooling chamber *CC* is flushed with gaseous helium *He* precooled by N_2 slush in the U-tube *UT*. The object *OB* is mounted on an injector rod *IR* and moves to the mirror *MM* as indicated by the *arrow*. $\mathbf{a_2}$ Simplified impact procedure according to Dempsey and Bullivant (1976): after cooling the copper block *CB* with the mirror surface *MM* to LN_2 temperature, it is lifted with the thermally insulated handle *HA* as indicated by the *dotted lines*. The mirror surface *MM* in this case remains in GN_2 below the interface *IF* between GN_2 and room atmosphere without frosting. The object *OB* mounted on a thermally insulating rod *IR* is gently pressed against the mirror *MM* and afterwards moved below the LN_2 level together with the copper block *CB*. The results are not reproducible (see discussion in Sect. 3.3).

(Fig. $4a_1$). The particular attraction of impact cryofixation is that the large heat capacity and good thermal conductivity of some cooled solids appears to be superior a priori to any liquid cryogen. In addition, a solid can be cooled to lower temperatures with LHe than any liquid secondary cryogen. It is also apparent that a highly polished, smooth metal surface produces, by replication, a complementary surface on the specimen eminently suitable for subsequent section preparation in the cryoultramicrotome (Sitte 1979): its geometry is comparable to the precut surface of a resin embedding. The well-preserved border zone can be used optimally if sectioned parallel or slightly oblique to the specimen surface.

In numerous publications van Harreveld and co-workers (see van Harreveld et al. 1974) clearly documented the superiority of this new method of cryofixation over chemical fixation and dehydration at normal temperatures used up to that time. They always used freeze-substitution in OsO_4/acetone at 193 K (−80 °C) as follow-up procedure (Van Harreveld and Crowell 1964; see also Steinbrecht and Müller, Chap. 7, this Vol.). Despite the fact that these results were interesting, the new method has been used comparatively little. There are several reasons for this: the relatively complicated apparatus was not commercially available, specimen preparation for metal-mirror cryofixation was too difficult without additional tools and the overall results were too variable (see Heuser et al. 1976; Boyne 1979; cf. also Sects. 3.3 and 3.4).

◀———

b Air-cushioned, low-weight injection system developed by Sitte, Hässig and Neumann (will be published in detail elsewhere): the object *OB* is mounted by white office adhesive *GL* (e.g. Pritt glue-stick manufactured by Henkel) to a soft foam-rubber support *RS* (intermediate foil of thin polyethylene *PE* for easy separation after freezing) and moved with the injector rod *IR* towards the mirror plane *MM* (*arrow! left*). After contact the soft foam *RS* deforms, surrounds the object *OB* and hardens superficially by freezing (*right*). The shape of the object is only slightly changed because in the initial phase of freezing pressure is built up slowly during emptying of the air bag *AB* of the damping cylinder *DC*.

c Impact freezing of tissue samples on a specimen carrier *CS* of a cryoultramicrotome for subsequent cryosectioning. (After Phillips and Boyne 1985). **c_1, c_2** The carrier *CS* is mounted together with a rubber foam disc *RD* on the plastic support *PS* with a steel plate *SP*. The steel plate *SP* serves for simple magnetic mounting on the base plate *BP* (magnet) of the injector rod *IR*. The cylindrical part of the carrier *CS* slides into the guiding hole *GH*. This movement is possible by a cone-shaped opening of the support *PS*. **c_3** After initial contact between the object *OB* and the metal mirror *MM* the rubber foam *RD* is deformed. The gentle but continuous contact between mirror *MM* and object *OB* results in perfect freezing without disturbing compression artefacts in the object. **c_4** After complete freezing the object *OB* on the carrier *CS* is ejected by a wire *WE* into a transfer container *TC* precooled (or filled) with LN_2 used for cryotransfer to the sectioning chamber of the cryoultramicrotome.

d Manufacturing and transfer of tissue slices (e.g. Fig. 6) for subsequent impact cryofixation. **d_1, d_2** Tissue slicer modified after Stadie and Riggs (1944) in top view (**d_1**) and cross-section side view (**d_2**). The slicer consists of a heavy weight metal cylinder *MC* with springloaded object table *OT* (soft coil spring *SS*). The object *OB* on the table *OT* is gently pushed downwards by the transparent plexiglass lid *PL* which flattens the object surface. The slice is produced by slow movements of a disposable microtome blade *DB* (e.g. Reichert Histostat, 76 mm × 8 mm) introduced into a slit opening *SO* between the lid *PL* and the metal cylinder *MC*. **d_3, d_4** After removal of the lid PL in a "humid chamber" (e.g. glove box with 100% humidity by a wet cloth), the tissue slice *TS* is positioned on the foam rubber support *RS* for slamming (**d_3**; see also scheme **b**). After turning the support *RS* is positioned on a transfer chamber *TC* moistened by wet filter paper *FP* preventing evaporation of H_2O from the sectioned surface. The slice remains in this small humid chamber up to the start of the slamming motion

3.2 Further Development of the Van Harreveld System – Use of LHe as Coolant

The Van Harreveld system was further developed by two of his colleagues (see Heuser et al. 1976; Heuser 1977; Boyne 1979; Heuser et al. 1979; Phillips and Boyne 1984). Both Heuser and Boyne realized that bounce phenomena drastically reduce the quality of metal-mirror cryofixation. Heuser tried to suppress bouncing with strong retaining magnets. Boyne solved the problem by installing hydraulic damping into the injector system and was the first to investigate the effect of this arrangement. To this end he used a simple electrical circuit to indicate continuous contact between the injector (specimen) and the surface of the metal mirror necessary for optimum metal-mirror cryofixation. Similar measurements in our laboratory have shown that tests to check the absence of bouncing are necessary even if the system used appears to be extremely stable: bouncing movements practically interrupt the thermal contact between the specimen and the metal block and drastically reduce the quality of cryofixation, even if the bouncing amplitude is only a few micrometers. Both authors used a block of extremely pure copper for the metal mirror instead of the silver block employed by van Harreveld[3].

In addition to bounce suppression, Heuser used LHe cooling instead of the usual LN_2 cooling. Retrospectively, the latter delayed rather than encouraged the use of metal-mirror cryofixation as the impression was created that the excellent demonstration of exocytosis of synaptic vesicles (Heuser 1977) was only or at least mainly due to the use of LHe as cryogen. This impression was supported by calculations and indirect comparative measurements to assess the freezing process on a capacitive basis (Heuser et al. 1979). Publications on other LHe-cooled slammers (e.g. Escaig et al. 1977; Escaig 1982; Usukura et al. 1983; Escaig 1984) and the information that the results could be further improved by using LHe II as cryogen (Akahori et al. 1980, 1983) had the same effect. Anyone interested in using impact cryofixation was therefore confronted with an expensive and complicated LHe technology as "a LN_2-cooled block is approx. 50 to 100% slower than a LHe-cooled one" (Heuser et al. 1979). Apart from the fact that 100% slowing down of the cooling process is not imaginable, considerable doubts arose as to the real advantages of LHe compared to LN_2. Based on theoretical analyses Kopstad and Elgsaeter (1982) pointed out that the cooling rates could at the most be improved by 10 to 20% if LHe was used instead of LN_2. Moreover, Bald (1983) drew attention to the fact that the mirror surfaces of extremely pure copper blocks as used practically exclusively for LHe-cooled slammers, due to their temperature-dependent material properties, could not be expected to provide essentially better cooling rates at LHe temperatures than at LN_2 temperatures. If an additional effect is to be achieved with the considerable technical effort involved in LHe cooling, a silver mirror should be used as silver provides optimum conditions at 15 K (−258 °C). To our knowledge an LHe-cooled silver mirror was never correctly tested in comparison with an LN_2-cooled copper mirror

[3] Equipment for impact cryofixation can be obtained from Eiko, Hexland, Polaron, Reichert-Jung and RMC.

using a demonstrably bounce-free injection system according to Boyne (1979). In the present situation a final judgement on the potential advantages of LHe cooling requires that this experiment be first performed.

The chances that the use of LHe provides drastic advantages are small, because even the lowest temperatures at the mirror or contact surface do not alter the poor thermal conductivity of water-containing biological specimens. In addition, the temperature gradient between the front of the cold wave penetrating the specimen and the mirror surface, decisive for further cooling, decreases steeply. At 1 μm distance between the front of the cold wave and the contact surface between mirror and specimen, this gradient reaches at maximum 288 K μm^{-1} if LHe is used. This value is given by the optimum mirror temperature of 15 K (−258 °C) and the temperature of the specimen of approx. 303 K (+30 °C) before slamming. Under the same conditions the gradient on an LN_2-cooled metal mirror of 77 K (−196 °C) reaches at maximum 226 K μm^{-1}. At 10 μm distance from the mirror surface these maximum gradients drop to 28.8 K μm^{-1} if LHe or to 22.6 K μm^{-1} if LN_2 is used. It is evident that both these gradients are very poor and that even these poor values will never be reached in reality. A large difference between LHe and LN_2 on this basis cannot be expected. According to our present knowledge, there is no indication that more than negligible differences are obtained in the deeper parts of the well-preserved zones if LHe is used instead of LN_2. Skeleton muscle, often referred to as a difficult specimen to work with, is clearly no better frozen with LHe (see Fig. 5 in Usukura et al. 1983) than with LN_2 (cf. Fig. 5a). Generally, the same also applies to the depth and quality of the well-preserved border zones of other specimens (see Figs. 1 and 2 in Sitte et al. 1986; as well as Figs. 5 and 6, this Chap.). It should always be remembered when making comparisons of this sort that the mirror material and the mirror temperature are not the only parameters affecting the quality of preservation. It is also determined, often to a far greater extent, by the specimen used, in particular its water content, its metabolic state, the way it is excised and by the way the specimen is applied to the metal mirror (see Sect. 3.4).

3.3 Simplified Impact Procedures

A number of attempts have been made to replace the relatively complicated setup shown in Fig. $4a_1$ by simpler systems (see Christensen 1971; Dempsey and Bullivant 1976; Sitte 1979). The method illustrated in Fig. $4a_2$ appeared to be particularly suitable. A solid copper block is cooled in a Dewar filled with LN_2 and then, with the help of an insulated manipulator lifted just far enough above the LN_2 level that it still remains below the air/GN_2 interface, thus avoiding frost precipitation. The specimen mounted onto a rod is placed on the mirror surface and is lowered into the LN_2 together with the copper block. However, during this procedure the rod is unavoidably tilted relative to the block. As a result a wedge-like space between the already frozen specimen surface and the mirror surface is formed from which LN_2 evaporates due to the specimen's residual heat. This may lead to the recrystallization of the originally perfectly frozen border zone of the specimen. The edge remaining in contact with the mirror surface may also be

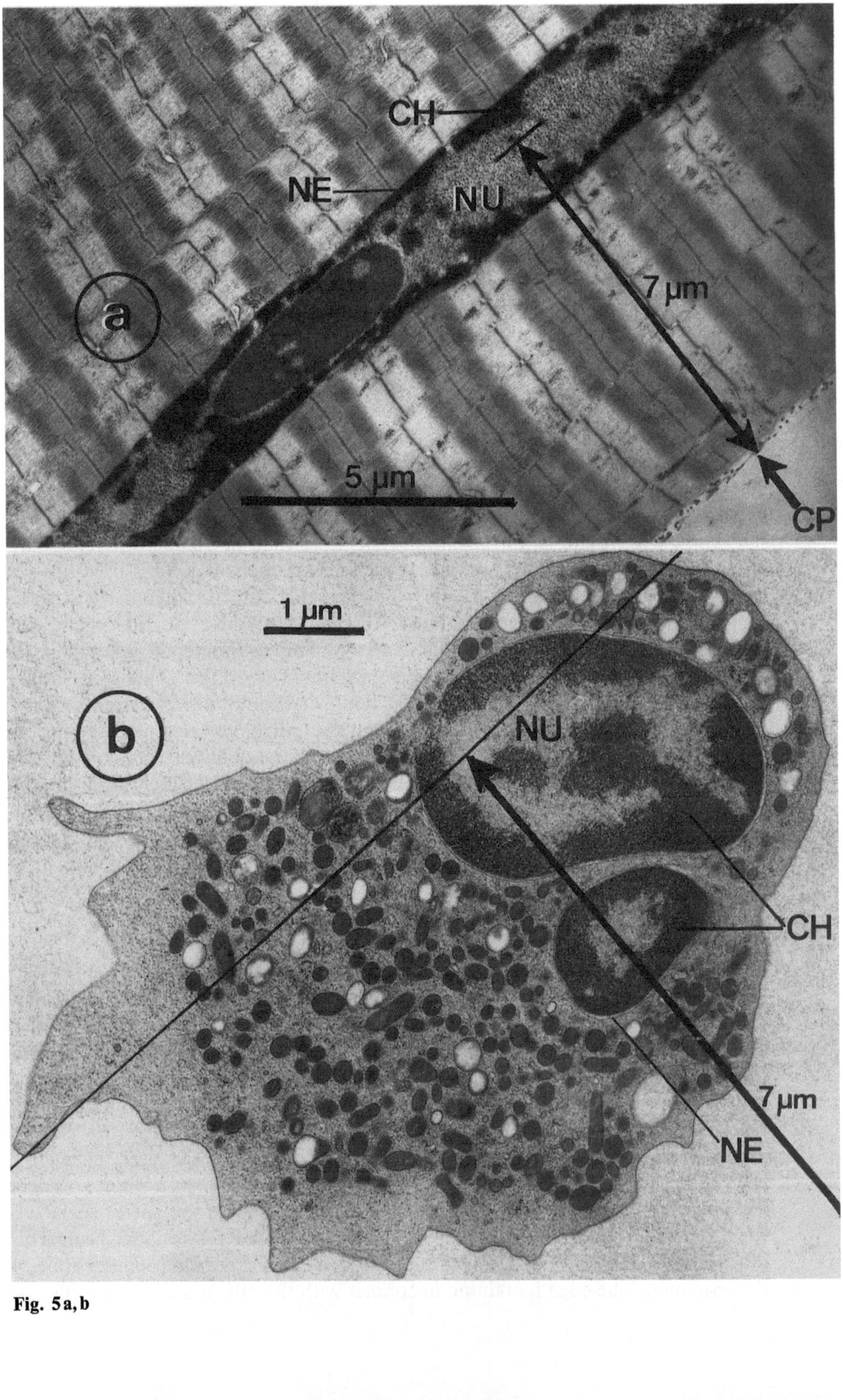

Fig. 5a,b

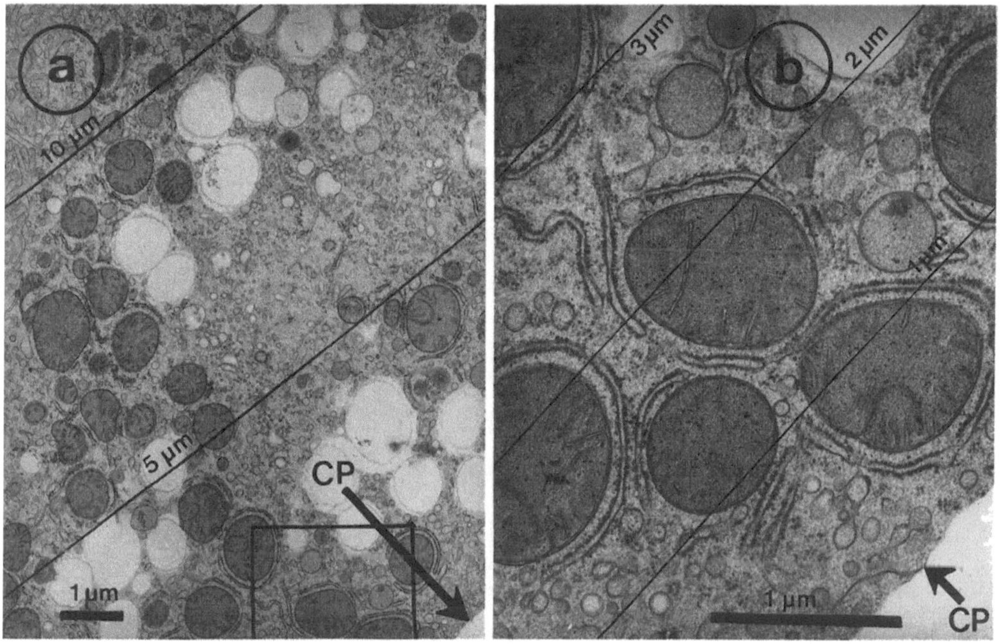

Fig. 6 a, b. Quality of preservation of the area in the immediate vicinity of the sectioned surface *CP* of a liver tissue slice produced with a modified Stadie-Riggs slicer (see Fig. 4d). Impact cryofixation on an LN_2-cooled copper mirror. Freeze-substitution, embedding and staining as in Fig. 5. The ultrathin sections are cut exactly perpendicular to the contact plane *CP*. **a** Survey picture: distances to the contact plane *CP* indicated by *parallel lines*. **b** Enlarged area as indicated in **a**

compressed. On the whole, the results are not reproducible. Similarly, the numerous types of freezing pliers (see e.g. Hagler and Buja 1984) provide comparable results only for particularly suitable specimens and require considerable skill from the user.

3.4 Own Investigations and Further Developments

Our own investigations showed that not only the speed at which the specimen slams against the metal mirror is of importance: the thrust developed by the

Fig. 5 a, b. Quality of preservation by impact cryofixation on an LN_2-cooled copper mirror. Note the compact and dense appearance of the sharply delineated chromatin *CH* attached to the nuclear envelope *NE*. Sections are cut exactly perpendicular to the contact plane of the specimen with the metal mirror. The distance between the contact plane *CP* and the nucleus *NU* is indicated. Freeze-substitution in OsO_4-acetone at 193 K (−80 °C) for 48 h, epoxide embedding, ultrathin sections stained with uranyl acetate and lead citrate. **a** Frog sartorius muscle: the contact plane *CP* is visible and parallel to the myofibrils and the longitudinal extension of the cell nucleus *NU*. **b** Neutrophil granulocyte frozen according to Fig. 8 onto the metal mirror. Note the gradient of chromatin preservation which depends on the distance to the contact plane (outside the area shown in the micrograph). (Electron micrograph courtesy of E. Morgenstern)

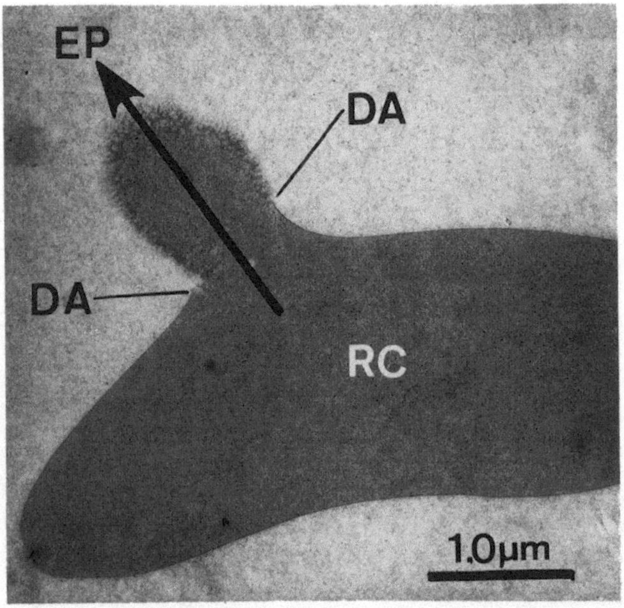

Fig. 7. Artefact by slamming with high velocities $\geq 5\,m\,s^{-1}$ and/or high inert masses. The damage *DA* of the membrane of the red blood cell *RC* and the exit of red cell plasm *EP* (*arrow*) probably result from a pressure wave created by the hard impact of the specimen to the mirror surface. Impact velocity approx. $5\,m\,s^{-1}$, slamming without air-bag damping (see AB in Fig. 4b and c) according to the method shown in Fig. 8. Compare this artefact with the excellent results obtainable with lower impact velocities and the air-cushioned system (Figs. 5b and 9). Freeze-substitution, embedding and staining as in Fig. 5. (Electron micrograph courtesy of E. Morgenstern)

system after initial contact also considerably affects cryofixation quality. The best results are obtained with relatively slow specimen contact (slamming speed $<5\,m\,s^{-1}$), if the specimen is mounted onto a support of extremely low mass (mass of the fully loaded support with specimen carrier and specimen below 10 g; Fig. 4b, c). The support slides inside a receiver sleeve in the metal mirror injector similar to the air cushioning of a telescopic arrangement. Therefore, as the specimen contacts the metal mirror surface, it stops its motion without build up of a detrimental compression wave, the pressure gradient of which often causes cell membranes to rupture (Fig. 7). In addition, this method limits the destruction and compression of the not yet frozen and, therefore, not stabilized deeper specimen layers as well as the horizontal displacement of parts of the specimen, if the object surface and mirror surface are not exactly plane and parallel. Another point of importance proved to be the increase of the deliberately selected weak initial contact pressure to a secondary pressure with values >0.5 kp after emptying the air cushion. This is done in order to prevent bending or twisting of the specimen during the process of complete freezing, and is necessary because the specimen surface may detach from the mirror surface, so that the extraction of residual heat from the specimen is interrupted and recrystallization may take place. Finally, in the system described, bounce phenomena are eliminated by a

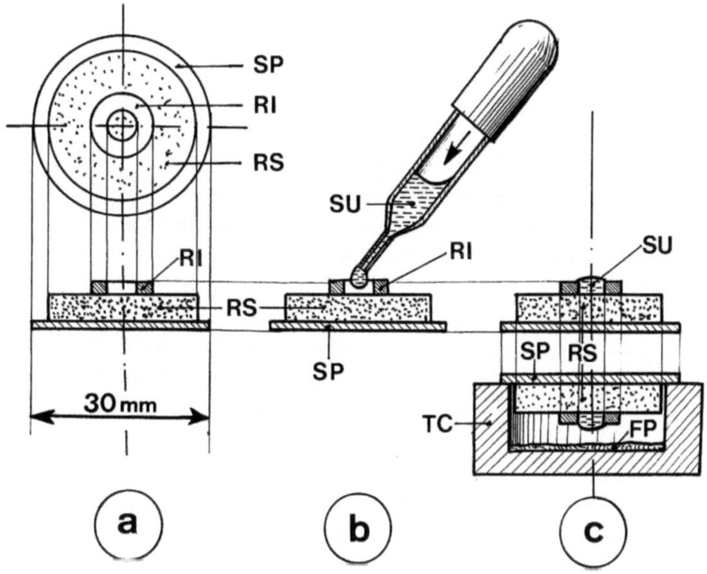

Fig. 8a–c. Impact cryofixation of suspensions *SU* on the metal mirror. Compare results shown in Figs. 5b and 9. The method is mainly suited for suspensions containing protein which forms a matrix during freeze-substitution and prevents the loss of incorporated cells or particles. **a** Specimen carrier *RS/SP* as shown in Fig. 4b, d in top view (*above*) and sectioned side view (*middle*). A ring *RI* of anti-adhesive plastic is placed on the surface of the foam rubber support RS. **b** Filling of the suspension *SU* into the ring *RI*. **c** The loaded carrier is transferred upside down into a transfer chamber *TC* moistened by wet filter paper *FP* to the slammer as shown and described in Fig. 4d (3–4)

simple reverse motion stop lock. This lock and the device for secondary specimen contact pressure are not included in the simplified diagrams in Fig. 4b, c[4].

Slamming at speeds above 5 m s^{-1} leads to deformations which may cause the total destruction of the specimen, if large moving masses or momentums and soft specimens are used. Experience has shown that these effects can be reduced by an appropriately designed specimen carrier (e.g. cushion of soft foam rubber according to Escaig 1984; see also Fig. 4b, c). A tissue slicer as illustrated in Fig. 4d often proves to be of help when preparing slices: this apparatus has been developed in our laboratory on the basis of the principle described by Stadie and Riggs in 1944, and yields tissue slices between 0.5 and 2.0 mm thick. Excellent cutting surface quality is obtained (Fig. 6) if water evaporation is prevented, as illustrated in Fig. 4d. The whole preparation procedure of tissue slicing, transfer of slices to the slammer in a humid chamber and impact cryofixation takes not much more than 10 s and provides good, reproducible results.

Not only compact tissues and tissue slices, but also suspensions may be vitrified on a metal mirror (see Fig. 8): the suspension to be examined is placed into a ring made of anti-adhesive plastic positioned on a foam rubber support

[4] This system is available from Reichert-Jung (KF80-MM80) and will be described in detail elsewhere.

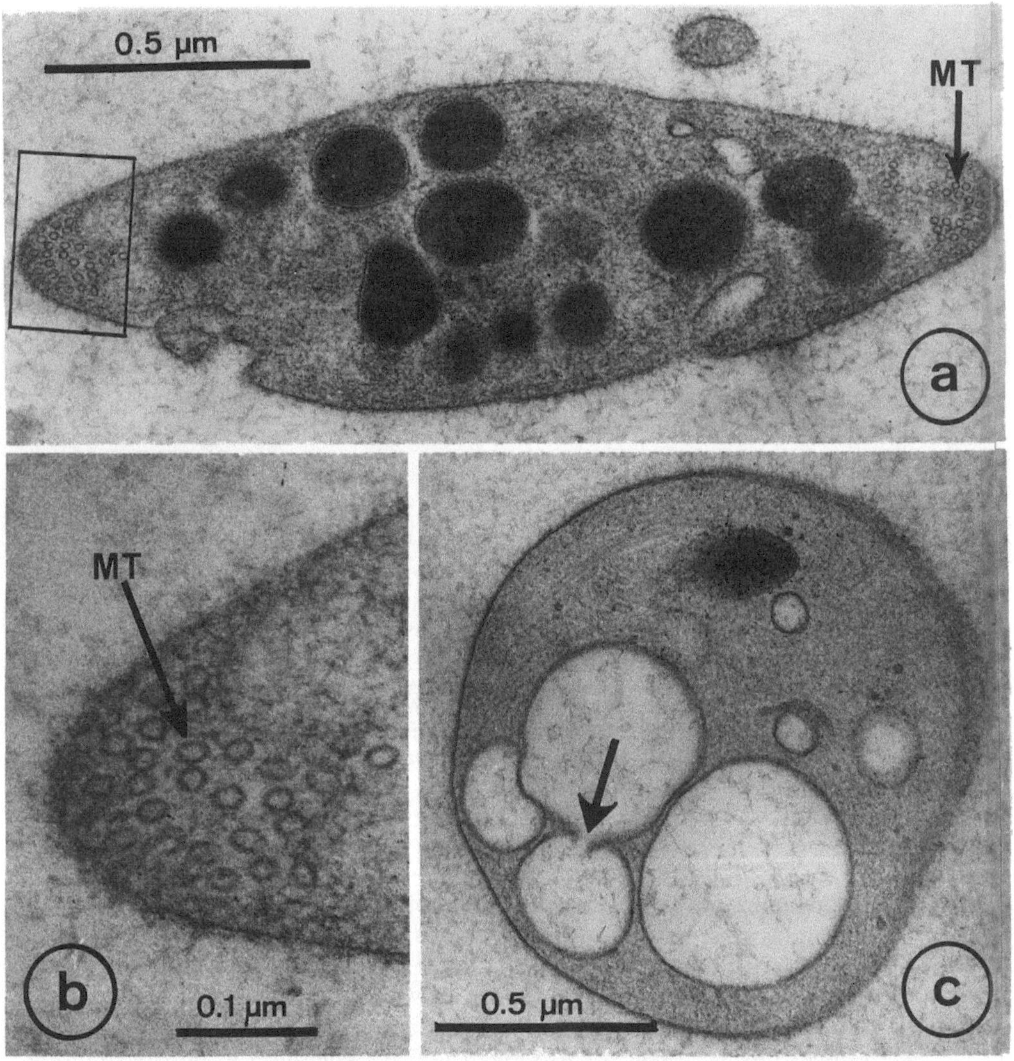

Fig. 9a–c. Blood platelets prepared by impact cryofixation on the LN_2-cooled copper mirror following the method described in Fig. 8. Freeze-substitution, embedding and staining as in Fig. 5. All micrographs are by courtesy of E. Morgenstern (see Morgenstern et al. 1987). **a** Survey picture: section of a discoid resting platelet. **b** Enlarged area from **a** shows clearly well-preserved microtubules *MT.* **c** Membrane fusion processes (*arrow*) in thrombin-stimulated platelets during exocytosis can exclusively be demonstrated by cryofixation. They may be useful to show the quality of preservation (cooling rate; see discussion in Sect. 4.4)

(Fig. 8a). This method proved particularly suitable for examining delicate thrombocyte fractions (Fig. 9; Morgenstern et al. 1987). Metal-mirror cryofixation has also become particularly important for the cryoultramicrotomy of tissue blocks for subsequent electron-probe microanalysis or examination of frozen-hydrated sections in the TEM. In 1985, Phillips and Boyne were the first to present a holder for subsequent cryoultramicrotomy where the tissue blocks could be frozen directly on a low weight specimen carrier. An alternative arrangement of this kind is exemplified in Fig. 4c.

4 Discussion

4.1 General Aspects and Advantages of Cryofixation

The specific advantages of cryofixation over a normal chemical fixation and standard processing at temperatures between 273 K and 373 K (0 and 100 °C) are now undisputed (see Shotton 1980, Plattner and Bachmann 1982, Menco 1986). These advantages are mainly due to the extremely rapid physical stabilization of the specimen. According to calculations (Bald 1983), the attainable cooling rates in a 10-μm border zone of a compact specimen may exceed 3×10^6 K s^{-1}. Conversely, the time required for vitrification in these border zones is certainly less than 0.1 ms. This "fixation time" is shorter by a factor of 10000 than any chemical fixation by immersion or perfusion if a minimum of 1 s is assumed for the effect of aldehydes or heavy metal compounds. In fact, the diffusion of the fixative through an intact cell membrane and the chemical reactions required for permanent chemical stabilization probably result in times in excess of the assumed value of 1 s for the fixation of a single cell. Rapid processes can therefore not be captured by chemical crosslinking of proteins, lipids or nucleic acids. Van Harreveld and co-workers (1974) and Heuser (1977; see also Heuser et al. 1976, 1979) provided impressive evidence for this. Plattner and Bachmann (1982) quite rightly pointed out that time resolution is as important for the registration and interpretation of rapid processes in the submicroscopic region as spatial resolution. This is particularly apparent in endocytotic and exocytotic processes (see Heuser 1977; Morgenstern et al. 1987; Fig. 9c; see also Knoll et al., Chap. 14, this Vol.).

In contrast to the extremely rapid cryofixation of the samples on contact with the secondary cryogen or with the metal mirror, a sample preparation itself requires times of at least several seconds. Moreover, this usually includes processes (e.g. specimen excision and slicing, specimen transfer into the cryogen or onto the metal mirror) where artificial alterations in the specimen cannot be excluded. The same applies to the follow-up procedures (Kellenberger et al. 1986; see also Chap. 2, this Vol.). Specimens may also undergo considerable changes compared to their in vivo status during freeze-substitution, freeze-drying and subsequent embedding at normal or reduced temperature (Edelmann 1986; Steinbrecht and Müller, Chap. 7, this Vol.). Changes may also occur during cryosectioning (Zierold 1986; Chap. 6, this Vol.) and during the examination of frozen-hydrated samples in the electron beam (Taylor and Glaeser 1976; Taylor 1978; Dubochet et al. 1983;

Lepault et al. 1983; McDowall et al. 1983; Adrian et al. 1984; Dubochet and McDowall 1984; see also Dubochet et al., Chap. 5, this Vol.).

Particular care should be taken to avoid mechanical damage during specimen excision and the subsequent procedures up to freezing. Sectioning with a tissue slicer, as described above, in a great variety of applications provides good results. The same may apply to slicing with a blade oscillating in the direction of the knife edge ("Vibratome" principle), but the authors have no first-hand experience. In some cases (e.g. physiological experiments with skeleton or heart muscles), it is of course impossible to slice these objects before cryofixation so that cryofixation has to be carried out through the intact connective tissue layer. In these cases the quality of cryofixation is strongly dependent on the thickness of this layer (see Heuser et al. 1976). The same applies if the specimen is coated with a layer of liquid (e.g. lymph, blood). Conversely, serious artefacts result if the surface layer of cells dries out due to exposure to ambient air. As it is important that the border layers of 10 to 20 µm are always well preserved, any loss of moisture must be prevented by using sandwich preparations (Fig. 1 e, f) or "humid chambers" (Figs. $4c_4$ and 8c).

4.2 Comparison of the Efficiency of Different Liquid Cryogens

Arguments relating to the merits of a particular cryofixation method are often associated with the question of the measurable cooling rates. It is an established fact that higher cooling rates guarantee better results. But it is often forgotten that cooling curves only permit statements to be made on a given experimental setup for exactly defined standard conditions. Using test specimens, Costello and Corless (1978; see also Costello et al. 1982, 1984) in this way compared various secondary cryogens. The resulting order of cryogens has been confirmed by experience and has now to be supplemented by ethane (Silvester et al. 1982; Bald 1984; Ryan et al., in press; see also Table 1). In order to avoid misunderstandings, it must be pointed out that Table 1 only includes those secondary cryogens, which are routinely used and appropriate for specimens of interest. Of these liquids propane is most commonly used today (for safety rules see Sitte et al., Chap. 16, this Vol.): the results obtained are good and reproducible. Liquid ethane provides somewhat better results than propane. However, the higher costs of this method are only justified if problems are encountered during the removal of propane covering the specimens in the course of subsequent cryoultramicrotomy at temperatures below 133 K (−140 °C). LN_2 would certainly provide better results than propane and ethane, because of its comparatively high density and low viscosity (Bald 1984), if the Leidenfrost phenomenon is avoided by cooling down to 63 K (−210 °C; see Umrath 1974) and/or by increasing the pressure. This is, however, not easily and reproducibly possible for larger specimens: forced convection of the cryogen and a very low heat capacity of the specimen are necessary in order to achieve this. The same applies to an even greater extent to the use of LHe or LHe II for directly cooling samples without the intermediary of a metal mirror. The latter cryogen put forward by Fernández-Morán in 1960 has never attained any practical degree of importance for the reasons mentioned.

4.3 Choice of the Most Suitable Method of Cryofixation

One reason for the relatively infrequent use of cryofixation in routine TEM work is undoubtedly the confusing variety of methods to choose from. An attempt will be made below, from the authors' standpoint and with the simplification inevitable in such a comparison, to point out the most appropriate method for different applications. In this section only methods presented in this chapter applying ambient pressure are dealt with. Moreover, only methods are discussed which are described in detail either in this chapter or in other publications and which have been sufficiently well tested in routine use. In most cases these methods may be applied with commercially available apparatus, so that instruments do not have to be built. A final comparison of these methods with high pressure cryofixation will be given in Section 4.5.

Larger organs or organisms (volumes in excess of 0.5 cm^3) are usually neither suitable for immersion in secondary cryogen nor for double-jet cryofixation because of their considerable heat capacity. Cryofixation of superficial regions is possible with either a metal-mirror system or single-jet cooling. Specimen preparation is necessary in both cases. For the metal mirror, tissue slices should be prepared after the method presented in Section 3.4 (Fig. 4d). If the natural organ surface is of interest, it may be directly applied to the metal mirror. Metal-mirror cryofixation is particularly recommended for preparations with subsquent freeze-substitution or freeze-drying followed by resin embedding, because this enables the comparatively large surfaces to be easily processed. Jet cryofixation as seen in Fig. 1f requires that the relevant surface area be covered with a very thin foil of a material with high thermal conductivity. The nozzle of the jet, after prior cooling, is directed vertically onto the covered part of the specimen. Approximately 20 ml propane is required for initial freezing of the border zones. Freezing may then be completed with LN_2. The jet method is particularly suitable for cryofixation of cell monolayers or organ surfaces in situ prior to freeze-fracturing/freeze-etching or freeze-substitution.

In contrast to the methods available for freezing compact specimens, a number of different methods may be used to freeze suspensions. The most suitable method depends on the follow-up procedure applied. If a thin film of the frozen suspension is to be investigated in the TEM, immersion, as presented in Section 2.1 (Fig. 1 c_3), is the only suitable method ("ice embedding" or "bare-grid preparation"). Propane or ethane are usually used for freezing. To this end an immersion system as well as a cryotransfer system and a compatible electron microscope with a cold stage are necessary. Cryotransfer and examination in the TEM must be carried out at temperatures below 133 K (−140 °C) to avoid devitrification. Subsequent freeze-fracturing/freeze-etching may preferentially be carried out using either the double-jet method (Sect. 2.2, Fig. 1e), the immersion-cryofixation method (Sect. 2.1, fig. 1 c_{1-2}) or the spray-freezing method (Sect. 2.3, Fig. 1g). Our own experience has shown that the immersion method is the simplest and safest, since commercially available double-jet systems do not always provide the results optimally obtainable with the double-jet method. The spray-freeze method may only be applied to suspensions of low viscosity and is more difficult to combine with freeze-substitution. Blood or similar suspensions with suspension media containing proteins can, in our experience, also be frozen without difficulty on a metal mirror and substituted as usual (see Sect. 3.4, Figs. 5b, 8 and 9). After impact cryofixation there are few restrictions as to the follow-up procedures.

Special requirements have to be met for the cryofixation of specimens to be sectioned in the cryoultramicrotome. In the case of suspensions or specimens of minute dimensions (diameter below 0.5 mm) freezing is best achieved by immersion in propane or ethane on a low weight specimen carrier as illustrated in Figure 1 d_1 (see Sect. 2, Table 1). Larger specimen blocks are most easily frozen on a normal specimen carrier on the metal mirror (see Sect. 3.4, Fig. 4c). Sufficiently thin specimens may be frozen as sandwich specimens by immersion or with the cryogen jet, and sectioned once the sandwich foil has been removed (see Fig. 2 in Zierold 1986). It is also possible to glue arbitrarily frozen specimens to the specimen carrier of the cryoultramicrotome with n-heptane (T_m 182 K; −91 °C) and to cut them afterwards (Steinbrecht and Zierold 1984; see also Zierold, Chap. 6, this Vol.). Border layers of thin and rigid specimens [e. g. insect antennae (Steinbrecht 1984) or leaves of plants or fine hairs in diameters $\ll 0.5$ mm] may be optimally frozen by direct immersion in secondary cryogen.

4.4 Criteria for the Quality of Cryofixation

Considerable misunderstandings were and still are encountered in discussions on quality criteria of cryofixation, in particular if different methods are to be compared. Strictly speaking a statement concerning the vitrification of an electron microscope specimen is only possible when it is investigated in the frozen-hydrated state using electron diffraction within the frame of a quantitative evaluation. After freeze-substitution or freeze-drying as well as after freeze-fracturing/freeze-etching, it is not possible to distinguish between "amorphous" and "crystalline" (see Bachmann and Mayer, Chap. 1; Dubochet et al., Chap. 5, this Vol.). Reference is often made to the existence or the lack of visible segregation compartments (often called "ice crystals" or "ice-crystal ghosts") or the preservation of different cell organelles and cells. However, these criteria are not equally valid for different procedures of preparation and investigation. Different metabolic states of the cells may also cause different reactions to the freezing process in otherwise similar specimens.

Our own experiment confirms the suggestion made by Phillips and Boyne (1984; see Fig. 8, p. 18), that chromatin adjacent to the nuclear envelope is one of the most sensitive indicators of freezing damage. If the chromatin is compact and dense, sharply delineated and, therefore, clearly contrasted to the karyoplasm, the state of preservation is optimal within this range of depth. Otherwise the chromatin appears honeycombed and broken up and not distinctly separated from the karyoplasm. As cell nuclei have diameters of approx. 5 μm often both of these manifestations may be observed in one nucleus. In the case of good metal-mirror cryofixation preserved chromatin structures of this type may be found normally in depths of up to approx. 10 μm, at the most up to approx. 15 μm (see Fig. 5).

If the sensibility of different methods to the representation of the above mentioned freezing artefacts is compared, ultrathin sections after cryosubstitution appear to provide the most certain way of detection. Our own experience has shown that during freeze-drying, probably due to thermal collapse (see Kellenberger et al. 1986; see also Gross, Chap. 10, and Kellenberger, Chap. 2, this Vol.), further artefacts appear in addition to those already present and these cannot always be clearly distinguished from freezing artefacts. This applies to both freeze-dried cryosections and, to a lesser extent, to ultrathin plastic sections after resin embedding (see Edelmann 1986). Any form of assessment using the nuclear chromatin of a freeze-fracture/freeze-etching preparation may prove to be relatively difficult, but we have no first-hand experience. Information of interest on the quality of the state of preservation is also provided by observing exocytotic and endocytotic processes (see also Knoll et al., Chap. 14, this Vol.). Membrane fusions of the kind illustrated in Fig. 9c have up till now only been seen in well cryofixed material.

4.5 Cryofixation at Ambient Pressure vs High Pressure Work

In addition to the cryofixation methods at atmospheric pressure as described in this chapter, cryofixation may also be carried out at high pressure, approx. 2100 bar (Müller and Moor 1984; see also Moor, Chap. 8, this Vol.). High pressure cryofixation enables considerably thicker specimens to be frozen (maximum thickness 600 μm with double-sided cooling) without the formation of artificial segregation compartments ("ice crystals"). This method undoubtedly opens up completely new perspectives for the cryofixation of larger specimens. When comparing this procedure to the methods at ambient pressure discussed in this chapter it should be remembered that freezing with this high pressure system is not necessarily implemented at the high cooling rates (approx. $10^6\,Ks^{-1}$), which are required for vitrification at ambient pressure. In the centre of thick specimens comparatively low cooling rates of approx. $500\,Ks^{-1}$ prevail. Thus, high pressure cryofixation is a "slow" process with poor time resolution. It will be of interest when examining rapid processes (e. g. membrane fusions), whether and how this clear difference is manifested.

If we compare the potential advantages provided by high pressure cryofixation with those of ambient pressure methods we find, on the one hand, preparations which cannot be produced with the high pressure system. This applies for example to the vitrification of thin suspension films ("bare-grid method", Sect. 2.1) for direct observation of the frozen-hydrated structures in the TEM. On the other hand, the potential of the high pressure system has been impressively demonstrated, e. g. with the cryofixation of cartilage. It should also be possible to obtain good results with all specimens which can be placed directly into the freezing chamber: the natural dimensions of these specimens should be less than 1 mm diameter × 0.6 mm thickness. Plant specimens (e. g. leaves or roots of appropriate dimensions) should also be well suited for high pressure cryofixation as in these cases no serious artefacts are to be expected during cutting. Soft animal tissue, however, is often severely damaged during cutting down to the size of the specimen chamber before cryofixation: here, the production of larger and thicker tissue slices followed by metal-mirror cryofixation appears to be the better method. The future will show whether the high pressure method will prove as practicable for routine work as expected: the specific advantages and disadvantages of this system have to be identified.

Summing up, the various ambient pressure methods presented in this chapter are still useful and appropriate for the cryofixation of a large variety of biological and medical specimens in addition to the now available high pressure technique.

Acknowledgements. The technical developments presented in this chapter would not have been possible without the engaged assistance of H. Hässig, head of the workshop of the Biophysikalisches Institut der Universität des Saarlandes, Homburg-Saar. The Deutsche Forschungsgemeinschaft (Sachbeihilfe Si 86/6 "Kryomethoden") and the C. Reichert-Jung, Optische Werke AG in A-1170 Vienna, have actively supported the work. The valuable advice provided by H. Kleber and A. Lang from the Design Department and G. Kappl from the Electronics Department of Reichert-Jung substantially contributed towards the success of the project. We thank our colleagues E. Morgenstern and G. Werner for helpful hints and discussions. Mrs. Ch. Rosinus expertly carried out numerous cryopreparations. The manuscript was translated into English by Mrs. E. Menzel and typed by Mrs. B. Reiland. We sincerely thank them all.

References

(References marked by an asterisk are review papers or books respectively)

Adrian M, Dubochet J, Lepault J, McDowall AW (1984) Cryo-electron microscopy of viruses. Nature 308:32–36

Akahori H, Yamada E, Usukura J, Takahashi H (1980) Development of a rapid freezing device. I. J Clin Electr Micr 13:5–6

Akahori H, Takahashi H, Usukura J, Yamada E (1983) Rapid freezing apparatus for biological materials at the temperature of helium II. J Electr Micr 31:335

Altmann R (1890) Die Elementarorganismen und ihre Beziehungen zu den Zellen. Veit und Co., Leipzig

Bachmann L, Schmitt WW (1971) Improved cryofixation applicable to freeze etching. Proc Nat Acad Sci USA 68:2149–2152

Bachmann L, Schmitt-Fumian WW (1973) Spray-freezing and freeze-etching. In: Benedetti EL, Favard P (eds) Freeze-etching, techniques and applications. Soc Franc Micr Electr, Paris pp 73–79

Bald WB (1983) Optimizing the cooling block for the quick freeze method. J Microsc (Oxford) 131:11–23

Bald WB (1984) The relative efficiency of cryogenic fluids used in the rapid quench cooling of biological samples. J Microsc (Oxford) 134:261–270

Bald WB (1985) The relative merits of various cooling methods. J Microsc (Oxford) 140:17–40

Bald WB (1986) On ice crystal size and cooling rate. J Microsc (Oxford) 143:89–102

Bernard SA, Krigman MR (1974) Ultrastructural analysis of the deeper layers of freeze-substituted guinea pig cortex. Brain Res 76:325–329

Bernhard W, Leduc EH (1967) Ultrathin frozen sections. I. Methods and ultrastructural preservation. J Cell Biol 34:757–771

Boyne AF (1979) A gentle, bounce-free assembly for quick-freezing tissues for electron microscopy: Application to isolated torpedine ray electrocyte stacks. J Neuro Sci Methods 1:353–364

Burstein NL, Maurice DM (1978) Cryofixation of tissue surfaces by a propane jet for electron microscopy. Microsc 9:191–198

Christensen AK (1971) Frozen thin sections of fresh tissue for electron microscopy, with a description of pancreas and liver. J Cell Biol 51:772–804

*Costello MJ, Corless JM (1978) The direct measurement of temperature changes within freeze-fracture specimens during quenchin in liquid coolants. J Microsc (Oxford) 112:17–37

Costello MJ, Fetter R, Hoechli M (1982) Simple procedures for evaluating the cryofixation of biological samples. J Microsc (Oxford) 125:125–136

*Costello MJ, Fetter R, Corless JM (1984) Optimum conditions for the plunge freezing of sandwiched samples. In: Revel J-P, Barnard T, Haggis GH (eds) The science of biological specimen preparation. SEM Inc, AMF O'Hare, IL 60666, pp 105–115

Dempsey GP, Bullivant S (1976) A copper block method for freezing non-cryoprotected tissue to produce ice-crystal-free regions for electron microscopy. I. Evaluation using freeze-substitution. J Microsc (Oxford) 106:251–260

Dubochet J, McDowall AW (1984) Frozen hydrated sections. In: Revel J-P, Barnard T, Haggis GH (eds) The science of biological specimen preparation. SEM Inc, AMF O'Hare, IL 60666, pp 147–152

Dubochet J, McDowall AW, Menge B, Schmid EN, Lickfeld KG (1983) Electron microscopy of frozen-hydrated bacteria. J Bacteriol 155: 381–390

*Edelmann L (1986) Freeze-dried embedded specimens for biological microanalysis. Scanning Electron Microsc 1986 IV:1337–1356

Fränkö O (1954) Quenching of tissues for freeze-drying. Acta Anat 22:331–336

Escaig J (1982) New instruments which facilitate rapid freezing at 83 K and 6 K. J Microsc (Oxford) 126:221–229

Escaig J (1984) Control of different parameters for optimal freezing conditions. In: Revel J-P, Bernard T, Haggis GH (eds) The science of biological specimen preparation. SEM Inc, AMF O'Hare, IL 60666, pp 117–122

Escaig J, Géraud G, Nicolas G (1977) Congélation rapide de tissus biologiques. Mesure des températures et des vitesses de congélation par thermocouple en couche mince. CR Acad Sc Paris, Serie D 284:2289–2292

Fernández-Morán H (1960) Low temperature preparation techniques for electron microscopy of biological specimens based on rapid freezing with liquid helium II. Ann NY Acad Sci 85:689–713

Gersh I (1932) The ALTMANN technique for fixation by drying while freezing. Anat Rec:309–337

*Griffiths G, McDowall AW, Back R, Dubochet J (1984) On the preparation of cryosections for immuno cytochemistry. J Ultrastruct Res 89:65–78

Haggis GH (1986) Study of the conditions necessary for propane-jet freezing of fresh biological tissues without detectable ice formation. J Microsc (Oxford) 143:275–282

*Hagler HK, Buja LM (1984) New techniques for the preparation of thin freeze dried cryosections for X-ray microanalysis. In: Revel J-P, Barnard T, Haggis GH (eds) The science of biological specimen preparation. SEM Inc, AMF O'Hare, IL 60666, pp 161–166

Handley DA, Alexander JT, Chien S (1981) The design and use of a simple device for rapid quench-freezing of biological samples. J Microsc (Oxford) 121:273–282

Heuser JE (1977) Quick-freezing to catch the membrane changes that occur during exocytosis. In: Bailey GW (ed) Proc 35 Annu EMSA Meet, Boston Mass. Claitor's Publ Div, Baton Rouge, pp 676–679

Heuser JE, Reese TS, Landis DMD (1976) Preservation of synaptic structures by rapid freezing. Cold Spring Harbour Symp Quant Biol 40:17–24

Heuser JE, Reese TS, Dennis MJ, Jan Y, Jan L, Evans L (1979) Synaptic vesicle exocytosis captured by quick freezing and correlated with quantal transmitter release. J Cell Biol 81:275–300

*Kellenberger E, Carlemalm E, Villiger W (1986) Physics of the preparation and observation of specimens that involve cryoprocedures. In: Müller M, Becker RP, Boyde A, Wolosewick JJ (eds) The science of biological specimen preparation 1985. SEM Inc, AMF O'Hare, IL 60666, pp 1–20

Knoll G (1984) Dynamische Interaktionen der mitochondrialen Hüllmembranen. Eine ultrastrukturelle Analyse auf der Basis schneller Kryofixation. Dissertation Universität Konstanz

Knoll G, Oebel G, Plattner H (1982) A simple sandwich-cryogen-jet procedure with high cooling rates for cryofixation of biological materials in the native state. Protoplasma 111:161–176

Kopstad G, Elgsaeter A (1982) Theoretical analysis of specimen cooling rate during impact freezing and liquid-jet freezing of freeze-etch specimens. Biophys J 40:163–170

Kulenkampff H (1955) Zur Technik der Gefriertrocknung histologischer Präparate. I. Die Frage der Strukturerhaltung. Z wiss Mikrosk 62:427–438

Lepault J (1985) Cryo-electron microscopy of helical particles TMV and T4 polyheads. J Microsc (Oxford) 140:73–80

Lepault J, Booy FP, Dubochet J (1983) Electron microscopy of frozen biological suspensions. J Microsc (Oxford) 129:89–102

Mandelkow E-M, Rapp R, Mandelkow E (1986) Microtubule structure studied by quick freezing: Cryo-electron microscopy and freeze fracture. J Microsc (Oxford) 141:361–373

Mayer E (1985) Vitrification of pure liquid water. J Microsc (Oxford) 140:3–15

McDowall AW, Chang JJ, Freeman R, Lepault J, Walter CA, Dubochet J (1983) Electron microscopy of frozen hydrated sections of vitreous ice and vitrified biological samples. J Microsc (Oxford) 131:1–9

*Menco BPhM (1986) A survey of ultra-rapid cryofixation methods with particular emphasis on applications to freeze-fracturing, freeze-etching, and freeze-substitution. J Electron Microsc Technique 4:177–240

Moor H, Mühlethaler K (1963) Fine structure in frozen-etched yeast cells. J Cell Biol 17:609–628

Morgenstern E, Neumann K, Patscheke H (1987) The exocytosis of human blood platelets. A fast freezing and freeze-substitution analysis. Eur J Cell Biol 43:273–282

Müller M, Meister N, Moor H (1980) Freezing in a propan jet and its application in freeze-fracturing. Mikroskopie (Wien) 36:129–140

Müller M, Moor H (1984) Cryofixation of thick specimens by high pressure freezing. In: Revel J-P, Barnard T, Haggis GH (eds) The science of biological specimen preparation. SEM Inc, AMF O'Hare, IL 60666, pp 131–138

*Neumann KH (1958) Anwendung der Gefriertrocknung für histochemische Untersuchungen. In: Graumann W, Neumann KH, Handbuch der Histochemie. Band I/1 (see pp 17–24, Das Einfrieren des Gewebes) G. Fischer, Stuttgart

Phillips TE, Boyne AF (1984) Liquid nitrogen-based quick freezing: Experiences with bounce-free delivery of cholinergic nerve terminals to a metal surface. J Electron Microsc Technique 1:9–29

*Plattner H, Bachmann L (1982) Cryofixation: A tool in biological ultrastructural research. Int Rev Cytol 79:237–304

Plattner H, Fischer WM, Schmitt W.W, Bachmann L (1972) Freeze etching of cells without cryoprotectants. J Cell Biol 53:116–126

Plattner H, Knoll G (1984) Cryofixation of biological material for electron microscopy by the methods of spray-, sandwich-, cryogen-jet and sandwich-cryogen-jet-freezing: A comparison of techniques. In: Revel J-P, Barnard T, Haggis GH (eds) The science of biological specimen preparation. SEM Inc, AMF O'Hare, IL 60666, pp 139–146

Pscheid P, Schudt C, Plattner H (1981) Cryofixation of monolayer cell cultures for freeze-fracturing without chemical pretreatments. J Microsc (Oxford) 121:149–167

Robards AW, Crosby P (1983) Optimisation of plunge freezing: Linear relationship between cooling rate and entry velocity into liquid propane. Cryo-Letters 4:23–32

*Robards AW, Sleytr UB (1985) Low temperature methods in biological electron microscopy. In: Glauert AM (ed) Practical methods in electron microscopy, Vol 10. Elsevier, Amsterdam New York Oxfordord

Ryan KP, Purse DH (1984) Rapid freezing: Specimen supports and cold gas layers. J Microsc (Oxford) 136:RP5–6

Ryan KP, Purse DH (1985) Plunge-cooling of tissue blocks: determinants of cooling rates. J Microsc (Oxford) 140:47–54

Ryan KP, Purse DH, Robinson SG, Wood JW (1987) The relative efficiency of cryogens used for plunge-cooling biological specimens. J Microsc (Oxford) (in press)

*Shotton D (1980) Quick-freezing – the new frontier in freeze-fracture. Nature 283:12–14

Silvester NR, Marchese-Ragona S, Johnston DN (1982) The relative efficiency of various fluids in the rapid freezing of protozoa. J Microsc (Oxford) 128:175–186
Simpson WL (1941) An experimental analysis of the Altmann technic of freezing-drying. Anat Rec 80:173–189
*Sitte H (1979) Cryofixation of biological material without pretreatment – A review. Mikroskopie (Wien) 35:14–20
Sitte H (1984) Instruments for cryofixation, cryo-ultramicrotomy and cryosubstitution for biomedical TEM. Zeiss Information MEM magazine for electron microscopists 3:25–31
Sitte H, Neumann K, Edelmann L (1986) Cryofixation and cryosubstitution for routine work in transmission electron microscopy. In: Müller M, Becker RP, Boyde A, Wolosewick JJ (eds) The science of biological specimen preparation 1985. SEM Inc, AMF O'Hare, IL 60666 pp 103–118
Stadie WC, Riggs BC (1944) Microtome for the preparation of tissue slices for metabolic studies of surviving tissues in vitro. J Biol Chem 154:687–690
Steinbrecht RA (1984) Cryofixation and follow-up techniques in biological electron microscopy – outline and demonstration. Zeiss Information MEM magazine for electron microscopists 3:9–17
Steinbrecht RA (1985) Recrystallization and ice-crystal growth in a biological specimen as shown by a simple freeze substitution method. J Microsc (Oxford) 140:41–46
Steinbrecht RA, Zierold K (1984) A cryoembedding method for cutting ultrathin cryosections from small frozen specimens. J Microsc (Oxford) 136:69–75
Taylor KA (1978) Structure determination of frozen, hydrated, crystalline biological specimens. J Microsc (Oxford) 112:115–125
Taylor KA, Glaeser RM (1976) Electron microscopy of frozen hydrated biological specimens. J Ultrastruct Res 55:448–456
Tokuyasu KT (1973) A technique for ultracryotomy of cell suspensions and tissues. J Cell Biol 57:551–565
Trinick J, Cooper J, Seymour J, Egelman EH (1986) Cryo-electron microscopy and three-dimensional reconstruction of actin filaments. J Microsc (Oxford) 141:349–360
Umrath W (1974) Cooling bath for rapid freezing in electron microscopy. J Microsc (Oxford) 101:103–105
Usukura J, Akahori A, Takahashi H, Yamada E (1983) An improved device for rapid freezing using liquid helium. J Electron Microsc (Tokyo) 32:180–185
Van Harreveld A, Crowell J (1964) Electron microscopy after rapid freezing on a metal surface and subsitution fixation. Anat Rec 149:381–385
Van Harreveld A, Fifkova E (1975) Rapid freezing of deep cerebral structures for electron microscopy. Anat Rec 182:377–386
Van Harreveld A, Trubatch J, Steiner J (1974) Rapid freezing and electron microscopy for the arrest of physiological processes. J Microsc (Oxford) 100:189–198
Van Venetie R, Hage WJ, Bluemink JG, Verkleij AJ (1981) Propane jet-freezing: A valid ultra-rapid freezing method for the preservation of temperature dependent lipid phases. J Microsc (Oxford) 123:287–292
Van Venrooij GEPM, Aertsen AMHJ, Hax WMA, Ververgaert PHJT, Verhoeven JJ, van der Vorst HA (1975) Freeze-etching: Freezing velocity and crystal size at different locations in samples. Cryobiol 12:46–61
*Zierold K (1986) Preparation of cryosections for biological microanalysis. In: Müller M, Becker RP, Boyde A, Wolosewick JJ (eds) The science of biological specimen preparation 1985. SEM Inc, AMF O'Hare, IL 60666 pp 119–127

Chapter 5 Cryoelectron Microscopy of Vitrified Specimens

JACQUES DUBOCHET, MARC ADRIAN, JIIN-JU CHANG[2],
JEAN LEPAULT, and ALASDAIR W. MCDOWALL[1]

1 Introduction

Water is the most abundant component of biological material, but it is systematically excluded from conventional electron microscopy. This is because water evaporates rapidly under the vacuum conditions of an electron microscope. Cryoelectron microscopy has long been seen as a possible avenue to overcome this limitation, but until recently the direct observation of frozen-hydrated specimens was relatively unsuccessful because of a number of serious difficulties. These were, in particular, due to the absence of a good cryospecimen holder, the inherently low contrast of hydrated specimens and the structural damage due to ice crystals formed during freezing. As a consequence, the cryomethods which have flourished in electron microscopy during the last 20 years were not aimed at preserving the hydration of the specimen in the electron microscope. Freezing was only used as an aid to preparation. The objects ultimately observed in the electron microscope were dry and at room temperature. Such cryomethods have recently been reviewed in detail (Robards and Sleytr 1985).

The development of cryoelectron microscopy of vitrified specimens has radically changed the situation during the last few years. Water is now a suitable substance for electron microscopy and high resolution investigations of aqueous material are now done routinely in a rapidly increasing number of laboratories. This development was made possible by the work of a number of pioneers: Fernández-Morán has fought since the 1950s for the advancement of cryoelectron microscopy. In 1960 he had already succeeded in observing ice in an electron microscope (Fernández-Morán 1960) and since then has continued to work towards the same goal (Fernández-Morán 1985). It is frequently forgotten that P.N.T. Unwin who, later, was among the first to apply cryoelectron microscopy to high resolution study of biological specimens (Unwin and Ennis 1984), also started his career by observing ice (Unwin and Muguruma 1971). The Cambridge group of T.A. Hall and B.L. Gupta also began in the early 1970s with the study of frozen-hydrated biological specimens by X-ray microanalysis in a scanning microscope (cited by Echlin 1971; Moreton et al. 1974). At the same time, tech-

---

1 European Molecular Biology Laboratory, Meyerhofstr. 1, Postfach 102209, D-6900 Heidelberg, FRG
2 Present address: Institute of Biophysics, Academia Sinica, Peking, China

Cryotechniques in Biological Electron Microscopy
Edited by R.A. Steinbrecht and K. Zierold

niques were developed for cutting, mounting and observing frozen-hydrated sections (Hutchinson et al. 1974, 1978).

A breakthrough took place in 1975, when Taylor and Glaeser at Berkeley demonstrated by electron diffraction that the structure of catalase crystals is superbly preserved in frozen-hydrated specimens (Taylor and Glaeser 1975) and that surprisingly good direct images of such specimens can also be obtained (Taylor and Glaeser 1976; Glaeser and Taylor 1978). A second breakthrough was when vitrification of pure liquid water was found possible (Brüggeller and Mayer 1980; Dubochet and McDowall 1981) and applicable to electron microscopy (Dubochet et al. 1982b). This discovery coincided with the time when the instrumentation for cryoelectron microscopy became commercially available and easy to operate.

The discovery of vitrification opened the door to a number of further developments which greatly facilitated cryoelectron microscopy. Simple methods were found for preparing a thin vitrified layer of aqueous specimens (Dubochet et al. 1982a, b, c; Lepault et al. 1983a; Adrian et al. 1984; Jaffa and Glaeser 1984; Milligan et al. 1984). The cutting of vitrified sections, thin enough for high resolution observation was also found possible (Dubochet et al. 1983; McDowall et al. 1983). Among the various kinds of cryospecimen holders, those which are the simplest and the easiest to manipulate, turned out to give the best results (Chanzy et al. 1977; Dubochet et al. 1982b; Unwin and Ennis 1984). It was also observed that an optimal use of phase contrast (Unwin and Henderson 1975) could compensate to a large extent for the inherent low contrast of unstained vitrified specimens (Adrain et al. 1984; Lepault and Pitt 1984; Vogel et al. 1986). Furthermore, low temperature helps to reduce the effect of electron beam damage [Lepault et al. 1983a; Jeng and Chiu 1984; Chiu et al. 1986 (Int. Exp. Study Group)].

Cryoelectron microscopy of vitrified specimens is described in detail in the articles cited above and in a recent review (Dubochet et al., in press). Other reviews by Stewart and Vigers (1986) and Chiu (1986) focus on results obtained with this method. Since then, important contributions have been made on clathrin cages and coated vesicles (Vigers et al. 1986a, b) enveloped viruses (Vogel et al. 1986, Fuller 1987), on the packaging of DNA in bacteriophage heads (Lepault et al. 1987), on the assembly of microtubules (Mandelkow et al. 1986), on actin (Trinick et al. 1986) and myosin filaments (Milligan and Flecker 1986) and on the structure of chromosomes and chromatin (Dubochet et al. 1986; McDowall et al. 1986; P. Schultz and J. Dubochet, in prep.). All these subjects will not be discussed again here. In this chapter we present, however, a critical discussion of the method by focussing on the various phenomena which govern its potentials and limitations. In the first part we try to provide an answer to the questions: How is vitrification possible, and what is its effect on biological structures? The fact that liquid solutions have to be formed into a thin film and that vitrified bulk material has to be cut into thin sections is discussed in the second part. The chapter ends with a discussion on beam damage and the conditions for optimal image formation.

2 Vitrification

2.1 Vitrification vs Freezing

Vitrification of a liquid consists of making it solid without allowing it to crystallize. Achieving vitrification was the aim of the early cryobiologists (Luyet and Gehenio 1940). They saw the possibility of stopping movement without changing the structure, of realizing the old dream of "freezing time". The idea was nice. Its realization turned out to be difficult. Over more than 40 years, it was found that, except for the most concentrated aqueous solutions, crystallization of water always takes place before cryofixation is completed. Furthermore, theoretical arguments were given in support of the claim that vitrification of water is fundamentally impossible (Franks 1982; Rasmussen 1982). Consequently, from the 1950s to the early 1980s, the idea of vitrification became less and less fashionable. Most of the efforts of cryobiologists were redirected towards development of methods in which ice crystals are allowed to form without producing lethal cellular damage (Mazur 1970, 1984) and electron microscopists looked towards freezing conditions under which the damage caused by ice crystals is not recognized in the specimen.

The discovery of vitrification came as a surprise, but it had immediate consequences. A new breeze is now blowing over cryobiology, where vitrification is once more recognized as the primary goal (Fahy et al. 1984; Rall and Fahy 1985; Takahashi et al. 1986). Vitrification has also allowed the rapid development of cryoelectron microscopy, thus providing the reason for this chapter. The physical phenomena underlying vitrification are discussed in a preceding chapter (Bachmann and Mayer, Chap. 1, this Vol.). We will focus here on two questions: What is the largest volume of material that can be vitrified, and what structural changes take place when biological samples are vitrified?

2.2 Cryoprotectant

Small liquid droplets or thin layers of pure water can be vitrified by rapid immersion in an efficient liquid cryogen or by projection on a cold surface (reviewed in Mayer 1985; Bachmann and Mayer, Chap. 1, this Vol.). However, the size of a pure water sample that can be vitrified by these methods is maximally 3 μm. This may explain why vitrification was not discovered before and put a severe limitation on the usefulness of the method for cryopreservation. Larger volumes can also be vitrified, but this requires the help of a cryoprotectant. Fortunately, biological material or organic solutes themselves have some cryoprotective ability. In many cases, however, cryoprotectant must be added to the sample.

In the past, the effect of a cryoprotectant was generally characterized by how much it lowered the equilibrium freezing temperature. Many studies have been made on this effect (reviewed by MacKenzie 1977). For our purposes, however, the only parameter of importance is how much the cryoprotectant retards the onset of crystallization during rapid cooling. These two effects are not expected to be simply related and preliminary experiments indeed show that they are not. The consequence is that all previous studies on cryoprotectants are not directly applicable for defining optimal vitrification conditions. Results obtained with methods like freeze-drying or freeze-substitution, which require the specimen to be warmed much above the devitrification temperature before observation, are not relevant either. The only sources of reliable information are therefore X-ray diffraction at low temperature and cryoelectron microscopy as described in this chapter. The results of preliminary studies with these methods suggest different conclusions about freezing and vitrification than hitherto accepted (Dubochet and McDowall 1984a).

2.3 Size of Vitrified Samples

In principle, there is no limit to the size of a sample that can be vitrified provided enough cryoprotectant is added. Rall and Fahy (1985) have recently been able to vitrify mouse embryos by using a mixture of cryoprotectant at a total concentration of about 50% and to preserve their viability. As shown by this group, the problem of cryoprotection is then to find a cryoprotectant which does not kill the organism by its toxicity at the concentration required for vitrification. The cryosectioning method of Tokuyasu (1973, 1980), which is successfully used for immunolabelling studies in electron microscopy, also relies in part on vitrification for preserving biological structures (Griffiths et al. 1984). In this case, the cellular fluid is replaced by a saturated sugar solution. Vitrification of samples in the mm^3 size is then possible by direct immersion in liquid nitrogen, but the penetration of sugar requires that the sample be permeabilized by chemical fixation which also preserves labile structures.

High concentrations of cryoprotectant are, however, harmful for unfixed biological structures and all efforts must be made to maintain conditions which are closer to the native state. In our work with vitrified sections, we have found, in general, that uncryoprotected tissues, like liver or kidney, can be vitrified up to a depth of 5 – 10 μm. This is less than that generally accepted as the depth of good preservation as judged by freeze-substitution, but it is sufficient for cryosectioning. This value of 5 to 10 μm is, however, only possible because the cell interior contains a high concentration of soluble material which plays the role of cryoprotectant. Vitrification of interstitial fluid is more difficult. In general, it has a freezing behaviour which is not very different from pure water. Consequently, biological samples containing large extracellular spaces, as is the case, for example, in washed glycerinated muscle or bacterial pellets, cannot be vitrified over more than about 1 μm. This is neither enough for representing the bulk of the sample, nor for allowing good cryosectioning. Cryoprotectant must therefore be added to the interstitial fluid.

The choice of cryoprotectant and its concentration is a weighted compromise. Low concentration keeps the specimen close to its native state, but makes the vitrified volume very small. The yield of good sections thus becomes prohibitively small. High concentration of cryoprotectant makes bulk vitrification possible, but disturbs its native state. It also reduces the contrast of biological structures (contrast matching). The molecular weight of the cryoprotectant must also be taken into consideration. Some molecules added to the interstitial fluid may have a strong osmotic effect on cells or organelles. Some may also enter the cell. Long polymers present neither of these disadvantages. Last, but not least, the cryoprotectant sometimes changes the cutting properties of the vitrified sample. Frequently, it makes it more prone to cutting artefacts. For example, vitrified DMSO solution tends to become exceptionally crevassed upon cryosectioning. The hope still exists that a cryoprotectant will be found which would confer better cutting properties to the vitrified sample. For most of our work we have used sucrose at 15 to 20% concentration (about 0.5 M). It provides adequate vitrifying conditions and it has a low toxicity. The sucrose polymer dextran has also been found adequate for many applications.

2.4 Preservation of Vitrified Samples

Water does not crystallize when an aqueous specimen is vitrified. This means that the first-order exothermic phase transition between the liquid and the solid does not take place and that the basic nature of a liquid is preserved. According to the classical nucleation theory (see Franks 1982), crystallization takes place in two steps. First, a group of molecules forms an ice nucleus. This is a stochastic process based on local entropy fluctuation. It corresponds in classical chemistry to the jump over a free energy barrier. Once a nucleus is formed, crystallization can proceed very rapidly. This second step corresponds to a steep fall in a deep energy well. According to this theory, a vitrified sample is obtained if the molecules are put to a thermal rest before they have time to form an ice nucleus. We know by experiment that about 1 μm^3 of pure water can be vitrified. This corresponds to about 3×10^{10} water molecules. The time required for vitrifying this volume is not known, but a rough estimation can be made: cooling speed measurements made with thin thermocouples, extrapolated to a μm-size water layer, suggest that vitrification time is in the 10^{-4} s range. This duration may seem short. It is very long on the molecular time scale. The length of this time at the molecular level and the enormous complexity of the nucleation process is illustrated by molecular dynamics calculations. In a typical run lasting for about 1 h in a medium-sized computer, the dynamics of 250 molecules are calculated by steps of 10^{-15} s during 10^{-10} s. According to the numbers cited above for the vitrification conditions, the molecular dynamics calculation should be upgraded by a factor of 10^{14} in order to have a chance to simulate a nucleation event. This would last for a duration comparable to the age of the universe. This example is not very serious. It emphasizes, however, that at the molecular level, nucleation is a rare event and many other things can eventually take place in a biological sample before crystallization has a chance to start. In particular, the question must be raised if, during cooling, the distribution of macromolecules and the conformation of biological structures has time to change. Some elements to answer these questions can be found in the works on the behaviour of enzymes and other biological systems at low temperature (Douzou 1977). Others come from the various physical chemical studies of fast reaction kinetics in biological systems (Hiromi 1979). Direct imaging or X-ray diffraction during the cooling process is, however, not yet possible at the desirable resolution in space and time. The only source of information about the transformation suffered by biological structures during vitrification is therefore to compare the native structure at room temperature and the vitrified specimen. Up to now, this has only been done by cryoelectron microscopy, but recently developed methods will also allow vitrification of large enough volumes for X-ray studies (Mayer 1985).

The overwhelming majority of observations of vitrified specimens point to a faithful preservation of the biological structures. This has been illustrated qualitatively in many studies (reviewed by Stewart and Vigers 1986) and this is what makes cryoelectron microscopy of vitrified specimens attractive. The first of these observations was the historical contribution of Taylor and Glaeser (1975) showing that the structure of catalase crystals seems to be preserved in frozen-hydrated specimens as well as when it is observed in a humid chamber at room temperature (Parsons 1974). At that time it was, however, not fully realized that this good result can only be obtained when the water in the interior of the catalase crystal is truly vitrified (Lepault et al. 1983a). Catalase crystals have also been used for a precise calibration of the size of vitrified samples. The period of the crystals measured by X-ray diffraction in liquid water was compared to that measured by electron diffraction in vitrified preparations. In both cases, the measurements can be made with great precision by comparison with internal standards. The unpublished result we have obtained in collaboration with P.N.T Unwin shows that, in both cases, the lattice period is the same to within less than 1%. This close agreement is surprising. Indeed, direct measurement of the density of vitreous water gives the value of 0.93 g cm^{-3} (Dubochet et al. 1982a), which, as expected from the study of supercooled water (Angell 1982) is close to that of the ice. From this value it is deduced that water expands linearly by about 2% during vitrification without affecting in a comparable amount the period of catalase. Other, though less precise, measurements on various viruses or components thereof, led to the same conclusion (Adrian et al. 1984; Lepault and Leonard 1985). Another quantitative assessment of the preservation of biological samples during vitrification has been provided by a cryosection study of washed glycerinated insect flight muscle (McDowall et al. 1984) which has also been studied extensively by X-ray in liquid medium. In both cases, they seem identical. The technical difficulties due to sectioning have, however, limited this study to dimensions larger than 10 nm. The only example we know of structural effects of cryofixation on a biological specimen, although the water had been vitrified, was demonstrated in the study of lipid vesicles by electron diffraction (Lepault et al. 1985). Vesicles made of phosphatidylcholine were studied in these experiments. This lipid is known to have an α to β transition when it is cooled below 24 °C. When the liquid sample is vitrified from a starting temperature below phase transition, its electron diffractogram is in agreement with X-ray diffractograms obtained in the liquid. Discrepancies appear when the starting temperature is above the transition point. In this case, the rapid cooling causes a partial rearrangement of the aliphatic chains, but it does not give time for the full α to β transition to take place. The effect is noticeable on the electron diffractogram which depends on the precise arrangement of the molecules. It is, however, not directly detectable on the image where the information concerns larger dimensions.

Altogether it seems that vitrification is an excellent method for preserving biological structures. It must, however, not be forgotten that the vitrification time is very large on the scale of atomic or molecular movement. Until more is known about the transformation that biological specimens can suffer during a cold shock of this duration, it is important that cryoelectron microscopists remain aware and attentive to this possible artefact.

3 Preparation of Thin Specimens

One of the basic requirements for transmission electron microscopy is that the specimen be very thin. This can be achieved by the technique of cryosectioning after the bulk sample has been vitrified. Alternatively, liquid biological suspensions can first be prepared in the form of a thin film before they are vitrified. The first method is of general use, but is very demanding. The second is simple and rapid, but limited to small particles in liquid suspensions of not too high viscosity. Both methods introduce their own artefacts.

3.1 Cryosections

Cryosectioning of vitrified specimens is, in principle, an almost ideal preparation method: the bulky sample is vitrified, cut and observed. All the usual artefacts due to chemical fixation, dehydration, staining and embedding are ruled out. Trouble, unfortunately, comes with cutting the sample below 135 K and indeed the cutting artefacts are more pronounced than with conventional plastic sections. Nevertheless, many of the micrographs obtained are good. An example is shown in Fig. 1. It represents the nucleus of the dinoflagellate *Porocentrum micans*. Because the specimen is not stained or chemically fixed, only the most massive features, for example the chromosomes, are visible in the low magnification print. Some enlarged views of the same sections, however, show that they really contain a great deal of information. Unfortunately, some specimens never produce good sections, even well-vitrified specimens do not always give good sections.

Among the limitations of the cryosection method with vitrified specimens, the difficulties imposed by the vitrification requirement have been discussed in the previous paragraph. The damage due to the cutting process has been described previously in detail (Chang et al. 1983). Typical is the compression along the cutting direction, and the network of crevasses. The low contrast of unstained hydrated specimens also makes the observation more difficult. The result of all this is that several experienced cryoelectron microscopists regard vitrified sections as inadequate for ultrastructural studies. Obviously, we do not share this point of view and enough articles disprove it (Chang et al. 1983; Dubochet et al. 1983; McDowall et al. 1983, 1984, 1986; Dubochet and McDowall 1984a, b; Griffiths et al. 1984). However, more work still needs to be done before the method becomes reliable and can be routinely used in any laboratory. For the moment, some remarks may be useful.

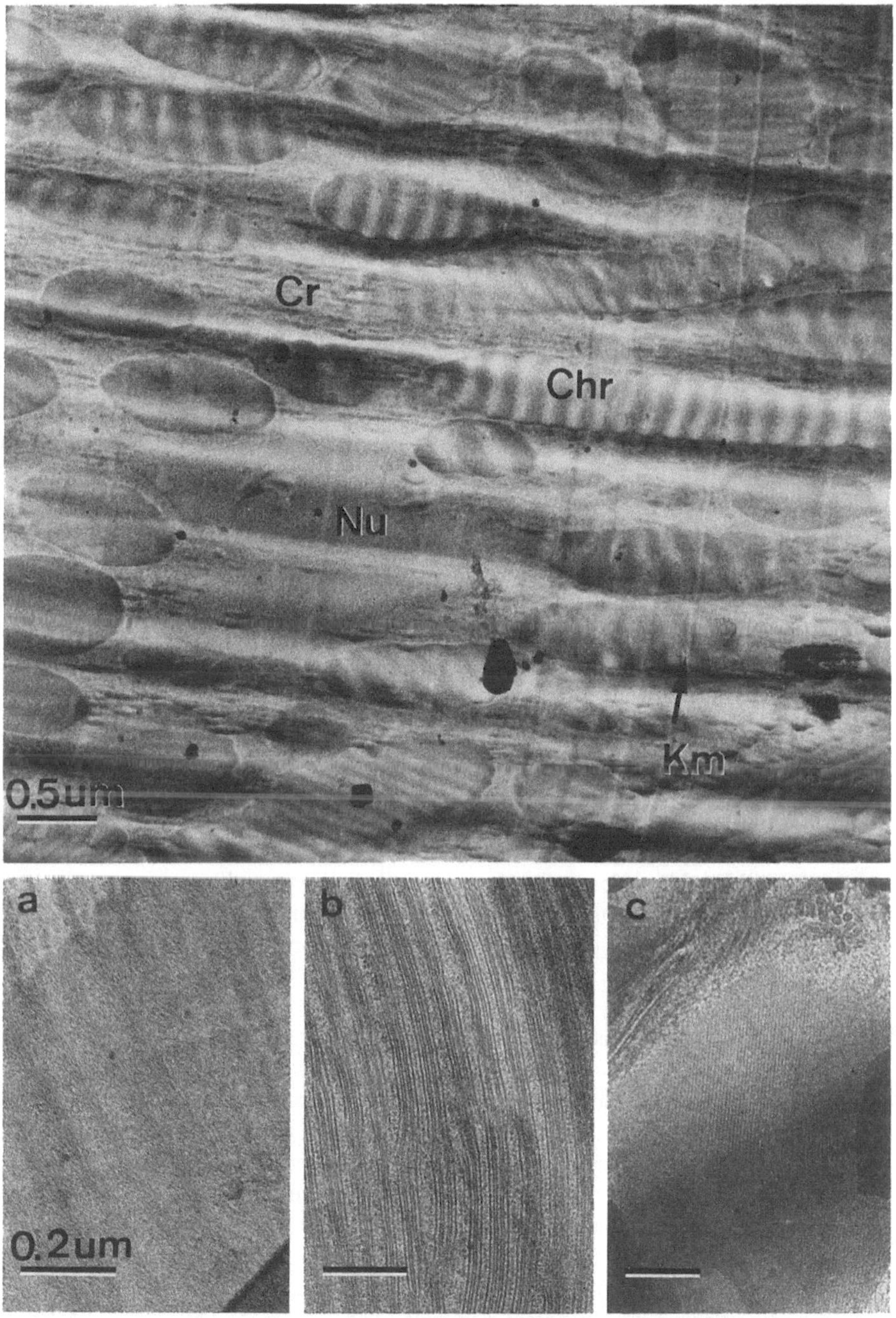

Fig. 1a–c. Thin vitrified section of the nucleus of the dinoflagellate *Porocentrum micans*. The unstained and unfixed sample was vitrified in 20% sucrose and cut and observed in the vitrified state as described previously (McDowall et al. 1983). *Chr* chromosomes; *Nu* nucleolus; *Km* knife marks; *Cr* crevasses. Magnification: 22000 ×.
Insets: Enlarged views of the same section. **a** Part of a chromosome showing the arch pattern formed by the twisted superposition of DNA fragments. **b** Crystalline arrangement of the thylacoid membranes. **c** Fragment of a trichocyst

Vitrified sections seem to suffer much more from the cutting process than plastic sections made at room temperature. In fact, it is not quite so. Plastic sections cut dry at room temperature also show many of the same cutting artefacts as vitrified sections (Griffiths et al. 1984). It seems therefore that the value of the conventional sectioning method owes much to the beneficial effect of floating the section on the water surface. Unfortunately, the search for a liquid which could perform the same role, but at a temperature below 135 K has not been successful and could well remain so in the future.

One of the advantages of vitrification is that vitreous water is much less difficult to cut than hexagonal ice. This is perhaps because the maximum stress which can be put on a hexagonal ice crystal depends strongly on its orientation. At the microscopic level, a polycrystalline ice specimen is thus non-homogeneous and does not cut well. Cubic ice, which is generally observed between vitrified regions and those where ice is in the hexagonal form, can be cut relatively well. Sections of biological samples made in these regions can be useful for structural studies. In general, the structural preservation in the presence of cubic ice crystals seems good. However, one example of structural damage induced by cubic ice formation has been reported: in glycerinated insect flight muscle, cubic ice causes the thick filaments to appear as exaggerated hollow cylinders (McDowall et al. 1984).

The cutting process itself is a combination of plastic deformation and fracture (Dubochet and McDowall 1984b, Frederik et al. 1984; reviewed by Zierold, Chap. 6, this Vol.). Some possible roles of these two mechanisms have been described. Many observations point to the fact that plastic deformation is unusually important in vitrified samples. Indeed, it must be kept in mind that a vitreous sample is a liquid of very high viscosity. Consequently, any deformation can be produced even in a short time provided the force is strong enough. How biological structures can endure this deformation is essentially unknown. The frequent observation that the presence of biological material generally makes the section less prone to crevasses, suggests that it can be deformed even more easily than vitreous water.

3.2 Thin Film of Suspension

The shape and structure of biological particles are well preserved in a thin film of vitrified solution, provided the particles are notably smaller than the thickness of the film. This good preservation has been observed in many specimens and measured quantitatively in several of them. For example, polyheads of the bacteriophage T4 are completely flattened when they are prepared by conventional negative staining. Their cylindrical shape seems perfectly preserved in an unsupported water film of sufficient thickness (Lepault 1985). Nevertheless, the interactions suffered by biological particles at the interface between the solution and the supporting film have been known for a long time to be important (Dubochet et al. 1982c; see also Kellenberger, Chap. 2, this Vol.). The interactions taking place at the liquid-air interface are also well known and sometimes used to advantage in specimen preparation (Kleinschmidt and Zahn 1959). It is therefore not surprising that surface phenomena play an important role when thin films of suspension are prepared. What was surprising, however, was the omni-

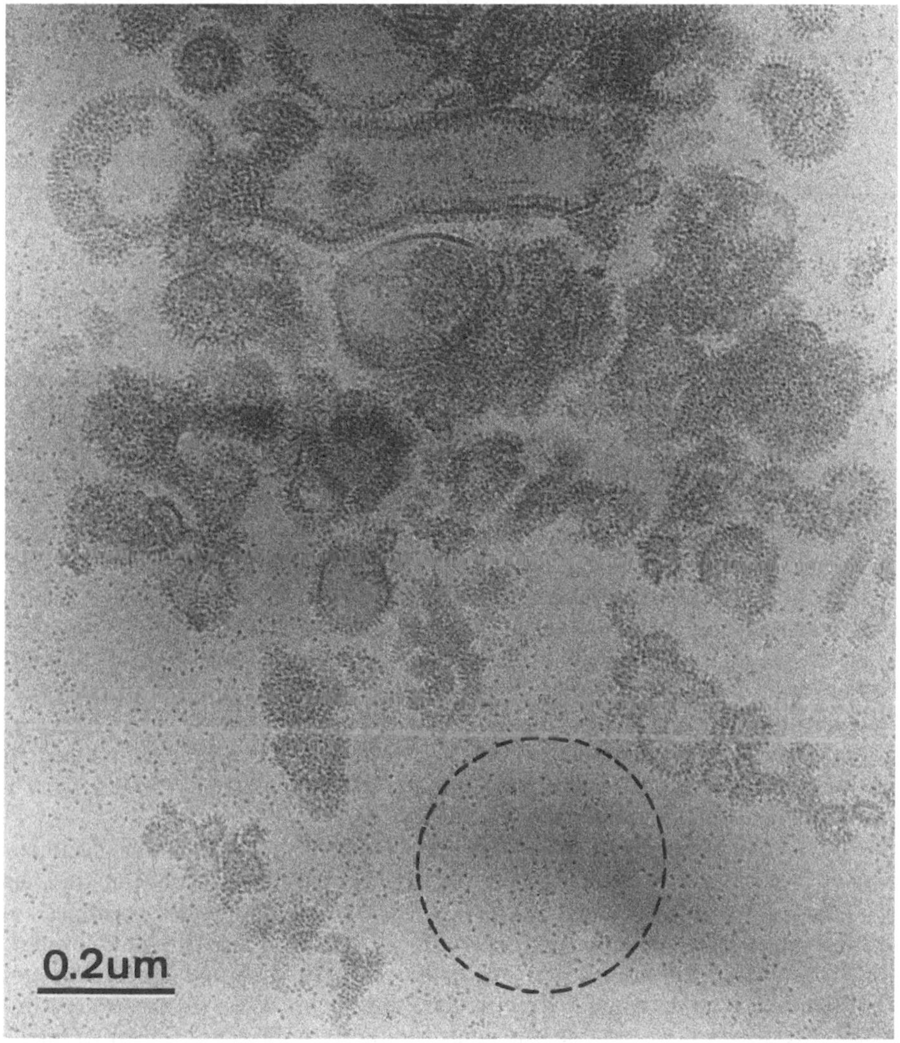

Fig. 2. G protein of the Vesicular Stomatitis Virus, incorporated into the lipid-coated, liquid-air interface (*circled*). The specimen was obtained from a solution of G protein containing artificial lipid vesicles. Surface fusion takes place during preparation by the bare grid method (Adrian et al. 1984), but only the surface which is not blotted away during preparation is kept intact on the thin film

presence and diversity of the surface-induced phenomena. A good example observed, even before vitrification of liquid water was discovered, is provided by an experiment aimed at measuring the density of amorphous solid water obtained by vapour condensation (Dubochet et al. 1982b). We mounted polystyrene spheres on a carbon film which was then cooled in the electron microscope. Water vapour was allowed to condense on the carbon film, but the embedding of the spheres failed because they were floating on top of the growing amorphous layer.

Other examples have been discussed in various articles (see in particular Dubochet et al. 1985). The phenomenon of surface interaction became more interesting when it was found that it was possible to control it by the addition of a minute amount of detergent or lipid to the solution or to the grid itself. Since then we have learned to take advantage of surface effects. In the example shown in Fig. 2, the conditions have been chosen such that some of the viral spike proteins incorporated in artificial vesicles are discharged in the lipid-coated, air-liquid interface.

4 Image Formation

The contrast of unstained biological specimens in water is low. It was even thought to be insufficient for useful observation. Fortunately, Taylor and Glaeser (1976) found that this was not the case and high resolution observations are now made routinely. There are two major reasons for this. The first one is that for judging the achievable resolution, it is not contrast that is the relevant parameter, but signal to noise ratio. A historical example is provided in the study of the purple membrane which gives micrographs of such low contrast that nothing can be seen by direct visual observation, though it allows the highest resolution reconstruction made to date on biological specimens (Unwin and Henderson 1975). For most classical specimens the limiting factor in the signal to noise ratio of biologically relevant information is not noise due to electron statistics, but damage suffered during specimen preparation, irregularities in stain distribution and the actual structure of the supporting film. All of these are nearly suppressed in vitrified thin film preparations thus leaving quantum noise and consequently beam damage, as the major limiting factors.

The second effect that helps record informative images of vitrified specimens is phase contrast. As illustrated in Fig. 3, the phase-contrast transfer function has the value of zero for the largest reciprocal space frequencies and it oscillates between one and zero for smaller values. How to correct for this non-constant transfer function has been known for many years (Erikson and Klug 1970) and the method is frequently applied to periodic specimens. When the specimen comprises a relatively large volume, the contribution of amplitude contrast to image formation becomes noticeable. This effect has sometimes been represented by a non-zero value of the phase-contrast transfer function for large reciprocal space frequencies (Unwin and Henderson 1975). Quantitative determination of the respective role of phase and amplitude contrast is, however, difficult and depends on the specimen. Until recently this has not been very important, because most image reconstruction work has been done on periodic objects where reciprocal space frequencies greater than the period are immaterial. Furthermore, in stained specimens, the contribution of amplitude contrast to the information content of

Fig. 3a–d. Focus series from a thin layer of vitrified Semliki Forest Virus solution and corresponding contrast-transfer function. The specimen was prepared by the bare grid method (Adrian et al. 1984) and the micrographs were underfocussed by 1.5, 3, 6 and 11 μm respectively for **a–d**. (See also text). (After Vogel et al. 1986)

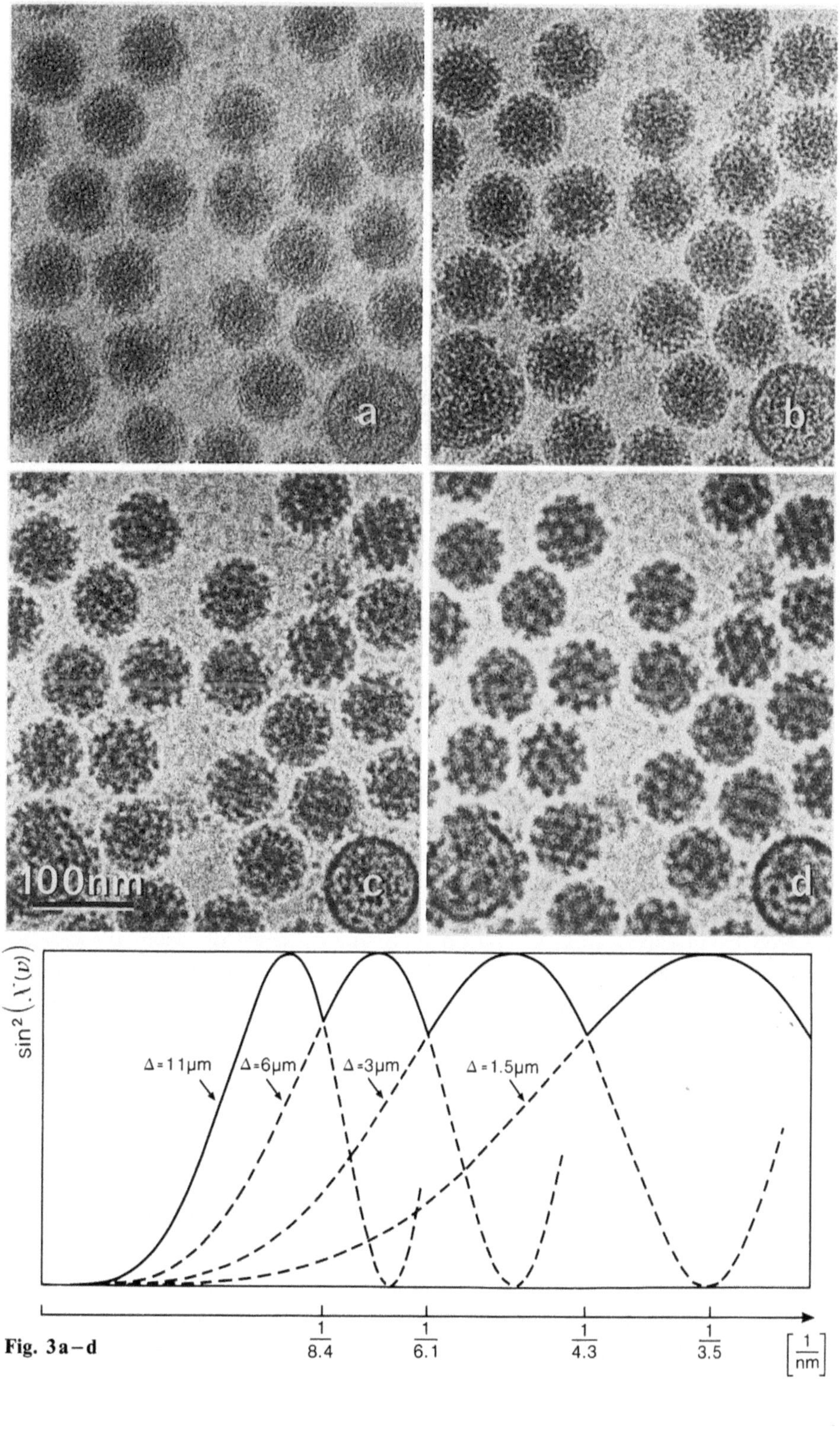

Fig. 3a–d

the image is significant for reciprocal space frequencies above 5 nm. A good complementation between phase and amplitude contrast can therefore be obtained when the first maximum of the transfer function is around 3 nm because this leads to reasonable coverage of the relevant information contained in most biological particles.

The situation is different when imaging unstained vitrified specimens. In most cases the contribution of amplitude contrast is only significant for dimensions much greater than 10 nm. Imaging of any smaller details exclusively rely on phase contrast. Unusually large defocussing values must therefore be used in order to retrieve this middle range information. In practice, experience has shown that for the study of non-periodic specimens, it is convenient to record the micrographs pairwise: the first one underfocussed in such a way as to obtain the best signal from the smallest reciprocal space frequency hoped to be found in the image; the second at a much higher defocussing value in order to best reveal the gross morphology of the specimen. It is also possible to proceed in a more rigorous way. This was done, for example for the study of the Semliki Forest virus (Vogel et al. 1986): the data used for a three-dimensional reconstruction of the virus was a series of four micrographs focussed in such a way as to optimize the information coverage. This was obtained when the first zero of anyone micrograph in the series just corresponded to the first maximum of the preceding one. Such a case is illustrated in Fig. 3, where a portion from each of four micrographs is shown together with their corresponding transfer function. For large reciprocal space frequencies, the contribution of spherical aberration is negligible, and these optimal conditions are simply obtained for series in which the defocus is doubled at each micrograph.

5 Beam Damage

Beam damage is an omnipresent limitation in cryoelectron microscopy of vitrified specimens. Here also, the situation contrasts with conventional electron microscopy where a lot of good work is done without even considering that the electron beam transforms the specimen. In the presence of water, the careless observer will rapidly be brought back to stark reality by the impressive phenomenon of bubbling and, if he persists, by the disappearance of the specimen. This difference is, however, misleading. First of all, beam damage is less severe at low temperatures than in a conventional electron microscope by a factor of three to six (Chiu et al. 1986). Furthermore, beam damage is no more severe in hydrated than in dry specimens at low temperatures (Lepault et al. 1983a; Jeng et al. 1984; Talmon et al. 1986; Talmon, Chap. 3, this Vol.). The only major difference is that in dry, stained specimens beam damage results finally in a "burnt" object which still resembles the original one, whereas beam damage in vitrified specimens results in its total destruction. In both cases, however, it is only the well-preserved specimen which is of interest for the good electron microscopist and it has been amply established that well-preserved specimens can be recorded, provided methods for minimizing the electron exposition are used. The recent three-dimensional reconstruction of clathrin cages on the basis of an 11-micrograph tilt series provides the best illustration that a careful operator can go a long way with beam damage (Vigers et al. 1986a).

Beam damage in vitrified specimens takes three major forms. (1) Destruction of the fine structure of the biological material, (2) bubbling and (3) drift. The first one has been studied quantitatively by measuring the fading of electron diffraction spots of protein crystals in water (Lepault et al. 1983a; Jeng and Chiu 1984). Studies on liquid crystals of filamentous virus or DNA have also led to the same conclusion (J. Lepault, unpubl. results). In general, it is found that the dose D_e required to reduce a given reflection to 1/e of its initial intensity is three to six times larger at low temperature than at room temperature, but it is independent of the specimen being dry or hydrated. An useful, rough empirical relation can be deduced from the above mentioned measurements. It relates the dose D received by the specimen to the best achievable resolution $\Delta = D/1000$, where D is expressed in electrons/nm^2 and Δ in nm. As can be observed in the direct image of vitrified specimens, this relation is also roughly applicable up to a resolution of several nm and doses of several thousand electrons/nm^2. However, when the dose reaches the range of 10000 electrons/nm^2, another phenomenon, bubbling, becomes dominant.

As its name indicates, the term bubbling describes the beam-induced formation of bubbles taking place in organic material in water (Glaeser and Taylor 1978; Dubochet et al. 1982a, b; Talmon 1982). The threshold dose for bubbling depends on the density of the organic material and on the volume of the particles. For example, the head of the T4 bacteriophage (80×100 nm) requires about 7000 electrons/nm^2 before bubbles start to appear in its interior, whereas apoferritin (diameter about 12 nm) resists to more than 50000 electrons/nm^2. The origin of bubbling is still a matter of controversy. For some investigators, it is due to a specific chemical reaction between organic material and water (Talmon 1984). We think it is mainly the sign that the gas produced by beam decomposition of the organic sample is prevented from escaping by the surrounding water layer and that the original damage is basically the same as in dry specimens (Dubochet et al. 1982b). This interpretation is supported by the fact that structural damage is independent of hydration and that mass loss of organic specimens is also the same in dry or in hydrated samples (Talmon et al. 1986). Etching of organic material due to chemical reaction with water also takes place but, as compared to direct decomposition, it becomes important only when the dose is well above the 10000 electrons/nm^2 range.

Drift is a serious beam-induced problem for the cryoelectron microscopist. In this sense, it can also be regarded as a form of beam damage. In contrast to movement of the specimen holder, beam-induced drift varies in direction and magnitude over the field of one micrograph. It originates probably from the electrostatic charges induced by the electron beam on the poorly conducting specimen. This charge distribution causes a field of force which deforms the specimen. Because images must be recorded at low dose, it is not possible to wait until stabilization is reached. Nevertheless, drift can be reduced in several ways. In particular, it is important that illumination is applied in a symmetric way. For example, recording an image just after focussing with a condensed beam at the border of the recorded field, is hopeless. It is also observed that the presence of the objective aperture plays a favourable role. This could be due to electrons ejected from the metal forming a gas neutralizing positive charge on the specimen. The presence of an anticontaminating fork closely sandwiching the specimen also seems favourable in this respect.

The use of very low temperature (4 K) is now generally accepted to provide little additional cryoprotection as compared to the temperature around liquid nitrogen (Chiu et al. 1986). In vitrified specimens the disadvantages of very low temperature can, however, be even more pronounced. In particular, the electrical conductivity decreased rapidly in this temperature range. Experience has shown that the beam-induced drift of hydrated specimens is indeed especially severe. The fact that crystalline ice, which is structurally very resistant around 120 K, is completely destroyed for the small dose of 100 to 500 electrons/nm^2 as soon as the temperature is below 70 K (Lepault et al. 1983b), throws another shred of doubt on the usefulness of very low temperature.

Acknowledgements. The authors express their thanks to Dr. A. Gautier and L.A. Michel-Salamin for the material depicted in Fig. 1, to Drs. K. Metsikkö and K. Simons for that of Fig. 2 and to Christine Barber for faithful help with the preparation of the manuscript.

References

(References marked by an asterisk are review papers or books respectively)

Adrian M, Dubochet J, Lepault J, McDowall AW (1984) Cryo-electron microscopy of viruses. Nature (London) 308:32–36

*Angell CA (1982) Supercooled water. In: Franks F (ed) Water: A comprehensive treatise, vol 7. Plenum, New York London, pp 1–81

Brüggeller P, Mayer E (1980) Complete vitrification in pure liquid water and dilute aqueous solutions. Nature (London) 288:569–571

*Chang J-J, McDowall AW, Lepault J, Freeman R, Walter CA, Dubochet J (1983) Freezing, sectioning and observation artefacts of frozen hydrated sections for electron microscopy. J Microsc (Oxford) 132:109–123

Chanzy H, Guizard C, Vuong R (1977) Electron diffraction on frozen hydrated polysaccharides. J Microsc (Oxford) 111:143–150

*Chiu W (1986) Electron microscopy of frozen hydrated biological specimens. Annu Rev Biophys Chem 15:237–257

Chiu W, Downing KH, Dubochet J, Glaeser RM, Heide HG, Knapek E, Kopf DA, Lamvik MK, Lepault J, Robertson JD, Zeitler E, Zemlin F (International Experimental Study Group) (1986) Cryoprotection in electron microscopy. J Microsc (Oxford) 141:385–391

*Douzou P (1977) Cryobiochemistry: an introduction. Academic Press, London New York

Dubochet J, McDowall AW (1981) Vitrification of pure water for electron microscopy. J Microsc (Oxford) 124:RP3–4

Dubochet J, McDowall AW (1984a) Cryoultramicrotomy: study of ice crystals and freezing damage. In: Csanady A, Röhlich P, Szabo D (eds) Electron microscopy 1984, vol 2. 8th Eur Congr Electron Microsc Budapest, pp 1407–1410

Dubochet J, McDowall AW (1984b) Frozen hydrated sections. In: Revel J-P, Barnard T, Haggis GH (eds) The science of biological specimen preparation. SEM, AMF O'Hare, IL 60666, pp 147–152

Dubochet J, Chang J-J, Freeman R, Lepault J, McDowall AW (1982a) Frozen aqueous suspensions. Ultramicroscopy 10:55–62

*Dubochet J, Lepault J, Freeman R, Berriman JA, Homo J-C (1982b) Electron microscopy of frozen water and aqueous solutions. J Microsc (Oxford) 128:219–237

Dubochet J, Groom M, Müller-Neuteboom S (1982c) The mounting of macromolecules for electron microscopy. In: Cosslett VE, Barer R (eds) Advances in optical and electron microscopy, vol 8. Academic Press, London New York, pp 107–135

Dubochet J, McDowall AW, Menge B, Schmid EN, Lickfeld KG (1983) Electron microscopy of frozen-hydrated bacteria. J Bacteriol 155:381–390

Dubochet J, Adrian M, Lepault J, McDowall AW (1985) Cryo-electron microscopy of vitrified biological specimens. Trends Biochem Sci 10:143–146

Dubochet J, Adrian M, Schultz P, Oudet P (1986) Cryo-electron microscopy of vitrified SV40 minichromosomes. The liquid drop model. EMBO J 5:519–528

*Dubochet J, Adrian M, Chang J-J, Lepault J, McDowall AW (1987) Cryo-electron microscopy of vitrified specimens. Q Rev Biophys (in press)

Echlin P (1971) The examination of biological material at low temperature. In: Johari O, Cerwin J (eds) Proc 4th Annual scanning electron microscope Symp, pt 1. IIT Press, Chicago, pp 227–232

Erikson HP, Klug A (1970) The Fourier transform of an electron micrograph: effects of defocussing and aberrations, and implications for the use of underfocus contrast enhancement. Bunsenges Phys Chem 74:129–1137

Fahy GM, MacFarlane DR, Angell CA, Meryman HT (1984) Vitrification as an approach to cryopreservation. Cryobiology 21:407–426

Fernández-Morán H (1960) Low-temperature preparation techniques for electron microscopy of biological specimens based on rapid freezing with liquid helium II. Ann N Y Acad Sci 85:689–713

Fernández-Morán H (1985) Cryo-electron microscopy and ultramicrotomy: Reminiscence and reflections. In: Advances in electronics and electron physics, suppl 16. Academic Press, London New York, pp 167–223

Franks F (1982) The properties of aqueous solutions at subzero temperatures. In: Franks F (ed) Water: a comprehensive treatise, vol 7. Plenum, New York London, pp 215–338

Frederik PM, Busing WM, Persson A (1984) Surface defects in thin cryosections. Scanning Electron Microsc 1984/I:433–443

Fuller SD (1987) The T = 4 envelope of Sindhis virus is organised by interactions with a complementary T = 3 capsid. Cell 48:923–934

*Glaeser RM, Taylor KA (1978) Radiation damage relative to transmission electron microscopy of biological specimens at low temperature; a review. J Microsc (Oxford) 112:127–138

Griffiths G, McDowall AW, Back R, Dubochet J (1984) On the preparation of cryosections for immunocytochemistry. J Ultrastruct Res 29:65–78

*Hiromi K (1979) Kinetics of fast enzyme reactions. Theory and practice. Kodansha scientific books. John Wiley & Sons, New York

Hutchinson TE, Bacaner M, Brodhurst J, Lilley J (1974) Electron microscopy of frozen biological tissues. Rev Sci Instrum 45:252–255

Hutchinson TE, Johnson DE, MacKenzie AP (1978) Instrumentation for direct observation of frozen hydrated specimens in the electron microscope. Ultramicroscopy 3:315–324

Jaffa JS, Glaeser RM (1984) Preparation of frozen-hydrated specimens for high resolution electron microscopy. Ultramicroscopy 13:373–378

Jeng T-W, Chiu W (1984) Quantitative assessment of radiation damage in a thin protein crystal. J Microsc (Oxford) 136:35–44

Johari GP (1977) On the heat capacity, entropy and "glass transition" of vitreous ice. Philos Mag 35:1077–1090

Kleinschmidt AK, Zahn RK (1959) Über Deoxyribonukleinsäure-Moleküle in Protein-Mischfilmen. Z Naturforsch 14b:770–779

Lepault J (1985) Cryo-electron microscopy of helical particles TMV and T4 polyheads. J Microsc (Oxford) 140:73–80

Lepault J, Leonard K (1985) Three-dimensional structure of unstained frozen-hydrated extended tails of bacteriophage T4. J Mol Biol 182:431–441

Lepault J, Pitt T (1984) Projected structure of unstained, frozen-hydrated T-layer of *Bacillus brevis*. EMBO J 3:101–105

Lepault J, Booy FP, Dubochet J (1983a) Electron microscopy of frozen biological suspensions. J Microsc (Oxford) 129:89–102

Lepault J, Freeman R, Dubochet J (1983 b) Electron beam induced "vitrified ice". J Microsc (Oxford) 132:RP3–RP4

Lepault J, Pattus F, Martin N (1985) Cryo-electron microscopy of artificial biological membranes. Biochim Biophys Acta 820:315–318

Lepault J, Dubochet J, Baschong W, Kellenberger E (1987) Organization of double-stranded DNA in bacteriophages; a study by cryo-electron microscopy of vitrified samples. EMBO J 6:1507–1512

*Luyet BJ, Gehenio PM (1940) Life and death at low temperatures. Biodynamica, Normandy, Miss

*MacKenzie AP (1977) Non equilibrium freezing behaviour of aqueous systems. Philos Trans R Soc London Ser B278:167–189

Mandelkow E-M, Rapp R, Mandelkow E (1986) Microtubule structure studied by quick freezing: cryo-electron microscopy and freeze fracture. J Microsc (Oxford) 141:361–373

Mayer E (1985) Vitrification of pure liquid water. J Microsc (Oxford) 140:3–15

Mazur P (1970) Cryobiology: The freezing of biological systems. Science 168:939–949

Mazur P (1984) Freezing of living cells, mechanisms and implications. Am J Physiol 247, 16:C125–C142

*McDowall AW, Chang J-J, Freeman R, Lepault J, Walter CA, Dubochet J (1983) Electron microscopy on frozen hydrated sections of vitreous ice and vitrified biological samples. J Microsc (Oxford) 131:1–9

McDowall AW, Hofmann W, Lepault J, Adrian M, Dubochet J (1984) Cryo-electron microscopy of vitrified insect flight muscle. J Mol Biol 178:105–111

McDowall AW, Smith JM, Dubochet J (1986) Cryo-electron microscopy of vitrified chromosomes in situ. EMBO J 5:1395–1402

Milligan RA, Flecker PF (1986) Three-dimensional reconstruction of decorated thin filaments in a frozen hydrated state. Biophys J 49:220

Milligan RA, Brisson A, Unwin PNT (1984) Molecular structure determination of crystalline specimens in frozen aqueous solutions. Ultramicroscopy 13:1–10

Moreton RB, Echlin P, Gupta BL, Hall TA, Weis-Fogh T (1974) Preparation of frozen hydrated tissue sections for X-ray microanalysis in the scanning electron microscope. Nature (London) 247:113–115

Parsons DF (1974) Structure of wet specimens in electron microscopy. Science 186:407–414

Rall WF, Fahy GM (1985) Ice-free cryopreservation of mouse embryos at −196 °C by vitrification. Nature (London) 313:573–575

Rasmussen DH (1982) Ice formation in aqueous systems. J Microsc (Oxford) 128:167–174

*Robards AW, Sleytr UB (1985) Low temperature methods in biological electron microscopy. In: Glauert AM (ed) Practical methods in electron microscopy, vol 10. Elsevier, Amsterdam

*Stewart M, Vigers G (1986) Electron microscopy of frozen-hydrated biological material. Nature (London) 319:631–636

Takahashi T, Hirsh A, Erbe EF, Bross JB, Steere RL, Williams RJ (1986) Vitrification of human monocytes. Cryobiology 23:103–115

Talmon Y (1982) Thermal and radiation damage to frozen hydrated specimens. J Microsc (Oxford) 125:227–237

Talmon Y (1984) Radiation damage to organic inclusions in ice. Ultramicroscopy 14:305–316

Talmon Y, Adrian M, Dubochet J (1986) Electron beam damage to organic inclusions in vitreous, cubic and hexagonal ice. J Microsc (Oxford) 141:375–384

Taylor KA, Glaeser RM (1975) Electron diffraction of frozen, hydrated protein crystals. Science 186:1036–1037

Taylor KA, Glaeser RM (1976) Electron microscopy of frozen hydrated biological specimens. J Ultrastruct Res 55:448–456

Tokuyasu KT (1973) A technique for ultracryotomy of cell suspensions and tissues. J Cell Biol 57:551–563

Tokuyasu KT (1980) Immunochemistry on ultrathin frozen sections. Histochem J 12:381–403

Trinnick J, Cooper J, Seymour J, Egelman EH (1986) Cryo-electron microscopy and three-dimensional reconstruction of actin filaments. J Microsc (Oxford) 141:349–360

Unwin PNT, Ennis PD (1984) Two configurations of a channel-forming membrane protein. Nature (London) 307:609–613

Unwin PNT, Henderson R (1975) Molecular structure determination by electron microscopy of unstained crystalline specimens. J Mol Biol 94:425–440
Unwin PNT, Muguruma J (1971) Transmission electron microscopy of ice. J Appl Phys 42:3640–3641
Vigers GPA, Crowther RA, Pearse BMF (1986a) Three dimensional structure of clathrin cages in ice. EMBO J 5:529–534
Vigers GPA, Crowther RA, Pearse BMF (1986b) Location of the 100kD–50kD accessory proteins in clathrin coats. EMBO J 5:2079–2085
Vogel RH, Provencher SW, Bonsdorff von C-H, Adrian M, Dubochet J (1986) Envelope structure of Semliki Forest Virus reconstructed from cryo-electron micrographs. Nature (London) 320:533–535

Chapter 6 Cryoultramicrotomy

KARL ZIEROLD[1]

1 Introduction

The aim of cryoultramicrotomy is to provide a biological specimen suitable for transmission electron microscopy with the ultrastructure and chemical composition representing as close as possible the natural state. Cryosections offer the unique possibility to study cells unaffected by chemical treatment, staining and embedding. The present state of the art of cryoultramicrotomy can fulfill this aim only partly: the section thickness achieved routinely is about 100 nm, thus limiting the ultrastructural analysis. The distribution of water and diffusible substances, however, can be preserved and the chemical properties of biomolecules, in particular antigenicity, can be maintained.

The preparation of cryosections compromises with respect to the preparation procedure, as indicated by the different preparation paths sketched in Fig. 1: im-

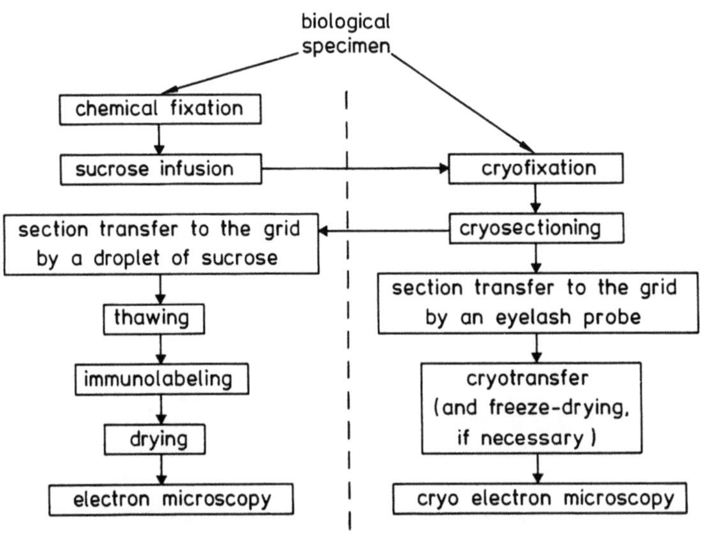

Fig. 1. Preparation paths in cryoultramicrotomy. Wet preparation steps as used for immunocytochemistry are shown on the *left*. Dry preparation steps are shown on the *right*

1 Max-Planck-Institut für Systemphysiologie, Rheinlanddamm 201, D-4600 Dortmund, FRG

Cryotechniques in Biological Electron Microscopy
Edited by R. A. Steinbrecht and K. Zierold

munolabelling of cryosections requires thawing. The cryosection does not tolerate this step without previous chemical fixation and cryoprotection of the biological object by infusion with sucrose or another cryoprotectant. Only then does the section remain stable during immunolabelling. This chapter does not go into details of "wet" cryoultramicrotomy as used for immunolabelling. The reader is referred to Chapter 11 (this Vol.) by Boonstra et al. and to the papers by Tokuyasu (1973, 1986) and Griffiths et al. (1984). The basic protocol of "dry" cryoultramicrotomy is sketched in the right part of Fig. 1. The sections are kept dry until observation in the electron microscope in the frozen-hydrated or freeze-dried state.

As presently the performance of cryoultramicrotomy is much easier than the understanding of the physics, first the method of preparing, handling and studying cryosections will be described. Physical aspects of cryoultramicrotomy and attempts to understand and to improve the cryosectioning process will be treated at the end.

2 Technical Aspects of Cryoultramicrotomy

2.1 Specimen Preparation for Cryosectioning

Cryoultramicrotomy starts by cryofixation of the biological object. General aspects of cryofixation are treated by Bachmann and Mayer (Chap. 1, this Vol.), Sitte et al. (Chap. 4, this Vol.), Plattner and Bachmann (1982) and Robards and Sleytr (1985). Ice-crystal damage in the object turns out to be important for the quality of cryosections. Objects with large ice-crystal damage are brittle, and cutting results in snow and small chips instead of clear and transparent sections. Generally, it can be maintained: the smaller the ice-crystal damage in the specimen, the easier the cryo-sectioning. Consequently, a vitrified object containing amorphous water instead of crystalline ice is the specimen of choice (McDowall et al. 1983; Dubochet and McDowall 1984a). Unfortunately, the vitrification of most biological specimens is very difficult. Thus, cryofixation has to aim for the smallest possible ice-crystal damage, independent of the used freezing method.

Good cryofixation requires an object as small as possible, preferably below 1 mm in size. On the other hand, the frozen object has to be mounted firmly on the cryoultramicrotome holder, which is facilitated by increasing the object size well above 1 mm. This problem can be solved by freezing the wet biological object together with the relatively large object holder, e.g. an aluminum pin. Only the superficial layer of the object will be frozen appropriately for cryoultramicrotomy.

The alternative is to freeze the object without or with any suitable support and to mount the object on the cryoultramicrotome holder by cryoglueing in any desired orientation. This is done usually in the cryoultramicrotome chamber. Figure 2 shows this preparation step schematically. The glue has to be solid at the cutting temperature and should melt at a temperature below the recrystallization temperature of ice, which is about 140 K in pure ice and 190 K in biological cells

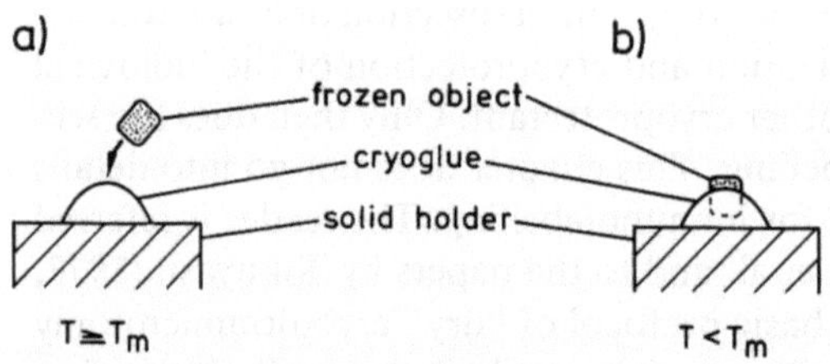

Fig. 2a,b. Principle of cryoglueing. **a** The cryofixed specimen is placed into the droplet of the cryoglue at the temperature immediately above its melting point. **b** The specimen is fastened by lowering the temperature below the melting point of the cryoglue

(Moor 1973). Karp et al. (1982) used toluene with a melting point of 178 K, Roos and Barnard (1985) used n-butylbenzene with a melting point of 185 K, Steinbrecht and Zierold (1984) used n-heptane with a melting point of 183 K. Cryoglueing vitrified objects would require a substance with a melting point below the devitrification temperature of about 140 K. The advantage of heptane in comparison with butylbenzene is that heptane sublimes rapidly in the vacuum and, thus, does not interfere with the biological material during freeze-drying.

Objects have to be small in the sense that the cutting width should be smaller than 200 μm in order to obtain frozen-hydrated, dry-cut sections. Cutting a specimen of broader width ruptures and breaks the sections into small chips. Too broad specimens have to be trimmed until a sufficiently small cutting surface is achieved. Usually, this is not necessary for specimens suitably frozen for cryoultramicrotomy. Almost every biological object requires its special, most appropriate preparation method for cryosectioning. For example, Somlyo et al. (1977) and Wendt-Gallitelli and Wolburg (1984) mounted single muscle fibres on rolled steel mesh holders in order to freeze them suitably for cryosectioning. Schmitz et al. (1985) prepared single cells of *Paramecium* in tiny water droplets on modified Balzers gold planchettes of 1 mm in diameter. Droplets of cell suspensions can be frozen between two 0.1-mm-thick copper sheets, one with a rough plane surface, the other with a smoothly polished concave surface. After freezing, the two plates are separated in the cold, and a solid droplet appropriate for cutting remains on the rough plane copper sheet (Zierold 1986a).

2.2 Preparation of Ultrathin Cryosections

Cryoultramicrotomy is realized technically by combination of an ultramicrotome with a chamber providing a low temperature atmosphere of nitrogen gas. Liquid nitrogen is used as cooling medium. Thus, the temperature range between 110 K and 273 K is at disposal. For an overview of the instrumentation, the reader is referred to Sitte (1982, 1984a), Robards and Sleytr (1985) and Menco (1986).

Originally, sections were cut wet with a flotation liquid at the knife edge. Bernhard and Viron (1971) have used dimethylsulphoxide (DMSO) as trough liquid, Hodson and Williams (1976) used isopentane. However, the use of these and other liquids with low temperature melting points as flotation medium was abandoned, because they are known to affect the ultrastructure and the distribution of diffusible substances. The same holds for the use of chemical fixatives, e.g. aldehydes and cryoprotectants. Therefore, in the following, cryosectioning is defined as cutting natural biological objects with a dry knife. Mostly, glass or diamond knives with a rake angle of 45° are used. Glass knives have the advantage that they are remarkably cheaper than diamond knives and can be replaced easily in routine work when they become dirty, frosted or damaged. Scratch traces occurring sometimes in the sections cut by a glass knife can be avoided by using a diamond knife. Besides that, no distinct difference has been observed so far in the quality of cryosections, whether they are cut by glass or diamond knives. The clearance angle of the glass knife varies between 2° and 10° and does not seem to be a critical parameter. Steel knives were used successfully for cutting about 1-μm-thick sections (Rick et al. 1982), but they have not yet proved feasible in cutting ultrathin sections, i.e. thinner than 200 nm.

The cutting speed reported by different laboratories varies between 0.2 mm s^{-1} and 10 mm s^{-1}. In this range, the cutting speed does not influence the section morphology. A distinctly higher cutting speed (e.g. 50 mm s^{-1}) results in chatter lines (Zierold 1984a) and tremendous deformations.

The temperature during cryosectioning is a definitely more sensitive parameter. Well-frozen homogeneous material, i.e. specimens with vitrified water or very small, visible, ice-crystal damage, can be cut easily at a low temperature down to about 110 K. Inhomogeneous objects (e.g. botanical specimens such as roots, leaves or conifer needles composed of large vacuoles and hard cell wall material) and objects with visible ice-crystal damage larger than about 200 nm crumble immediately to snow and powdery pieces when cut at a temperature below 170 K. Sections can be obtained from these objects by increasing the cutting temperature above 190 K. The problem of the "true" cutting temperature will be treated below in the discussion of physical aspects of cryoultramicrotomy.

Ultrathin cryosections appear on the glass knife as transparent curling sheets with a width between 10 and 200 μm, depending on the particular object and cutting width. Ribbons of sections are obtained very rarely. Special antiroll plates were designed to prevent curling of approximately 1-μm-thick sections at the knife edge immediately after cutting (Saubermann et al. 1981; Rick et al. 1982). However, successful use of these antiroll devices in preparing ultrathin sections has not been reported. Sections are either grey or they show interference colours, depending on their thickness, similar to conventional ultramicrotome sections from plastic material. For immunolabelling, sections are picked up by a wire loop filled with a sucrose solution, as sketched in Fig. 3. For studying cryosections in the frozen-hydrated or freeze-dried state, the sections are transferred to the grid by means of an eyelash probe, as reported by Christensen (1971). Picking up sections is often hampered by static electricity which may be caused by charge separation during cutting. Electrically charged sections are repelled from the approaching eyelash and jump away. In this case, an electrical discharge device generating ions above the sections facilitates the handling (Gupta et al. 1977; McDowall et al. 1983; Zierold 1984a). Usually, sections are mounted on grids coated by Formvar, Pioloform® and/or carbon. Figure 4 illustrates cryosectioning, discharging and picking up of sections schematically. Appleton (1974) has described a method to lift a fresh-cut ribbon of semi-thin cryosections by a vacuum pipette and to lay it on the grid placed on the glass knife near the edge. However, this method is not known to be reproduced for ultrathin sections. Dry cryosections do not

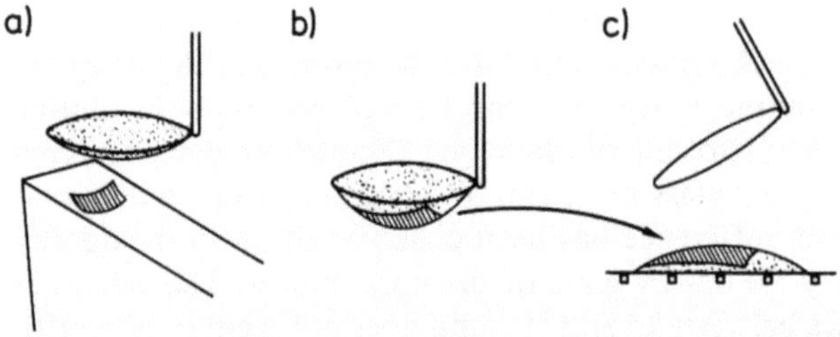

Fig. 3a–c. Section transfer to the grid for immunolabelling. **a** The section is picked up from the knife by means of a sucrose film hanging in a wire loop. **b** The section adhering to the sucrose film is transferred to the grid. **c** The section is deposited with sucrose on the film-coated grid

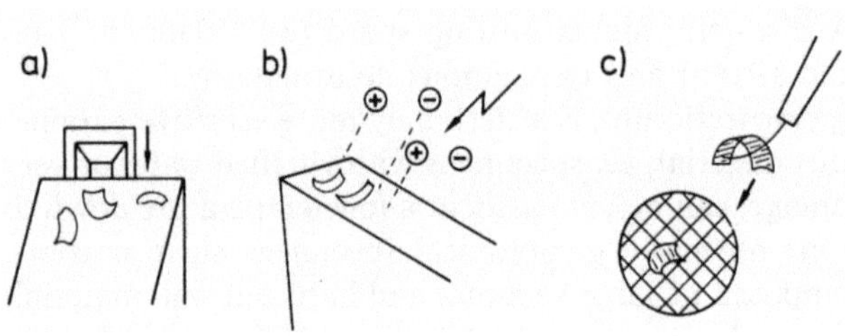

Fig. 4a–c. Dry section transfer to the grid. **a** After cutting, sections on the knife surface are often electrically charged. **b** Discharging of sections by ionization of the atmosphere. **c** Section transfer to the grid by an eyelash probe

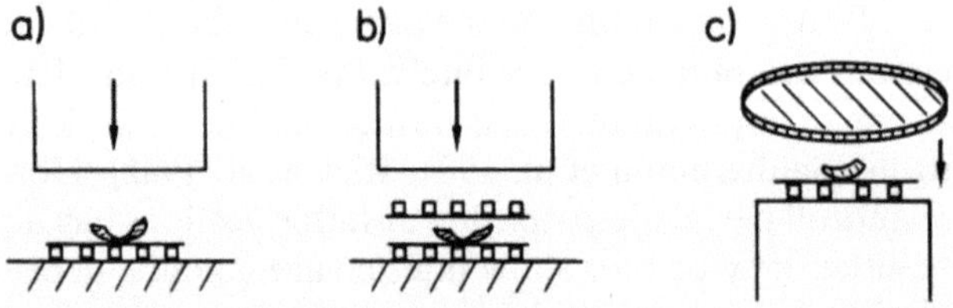

Fig. 5a–c. Mounting sections on the film-coated grid. **a** Sections are pressed by a cold-polished metal rod. Sections may adhere then at the metal rod instead of the grid. **b** Sections are placed between two film-coated grids, and then pressed together by a cold polished metal rod. After defolding, the grid with adhering sections is processed further. **c** A Formvar film held by a metal ring is covered over the section on the grid (Hagler and Buja 1984)

adhere well on the coated grid. Figure 5 shows some techniques to improve the contact between the section and the grid.

The sections mounted on the grid can be stored in a cold nitrogen gas atmosphere until they are processed for electron microscopy. Sublimation of ice does not take place, as shown by Zingsheim (1984). Cryosections mounted on grids can be kept in the frozen-hydrated state at least for several weeks by storing them in plastic capsules filled with cold nitrogen gas. The capsules are closed and placed in liquid nitrogen.

2.3 Frozen-Hydrated Cryosections

In theory, the ultrastructure and the composition of a biological specimen should be preserved best in the frozen-hydrated state. For this purpose cryosections have to be transferred into an electron microscope equipped with a cold stage. This is performed by means of a transportable cold chamber, which can be attached to the cryoultramicrotome and the electron microscope (Ross et al. 1981; Zierold et al. 1981; Biddlecombe et al. 1982; Robards and Sleytr 1985). The cryotransfer is usually done in an atmosphere of cold nitrogen gas in order to protect the specimen against wet air. During transfer and observation in the microscope, the temperature of the specimen should be kept below 140 K. Contamination is prevented by mounting metal surfaces colder than the specimen close to it. Frozen-hydrated sections exhibit very low contrast and are extremely sensitive to radiation damage resulting in mass loss (Zierold 1984a, 1985). Phase-contrast imaging at low electron exposure by conventional transmission electron microscopy is the most promising method to visualize details in frozen-hydrated sections (McDowall et al. 1983; Dubochet et al., Chap. 5, this Vol.). Vitrified sections are interpreted by Dubochet et al. (Chap. 5, this Vol.). Recently, electron energy loss spectroscopic imaging was found to be useful in imaging frozen-hydrated sections (K. Zierold, unpubl.). Contrast may be enhanced also by staining cells before cryofixation with heavy metals (Dubochet et al. 1983; Edelmann 1984). However, staining might interfere with the original ultrastructure and element composition. Scanning transmission electron microscopy (STEM) is not appropriate for imaging about 100-nm-thick frozen-hydrated sections, because of the high portion of multiple-scattered electrons blurring the contrast (Fig. 6). It remains to be proved whether Z-contrast imaging could improve the contrast in STEM (Carlemalm et al. 1985). The high electron exposure required for electron probe microanalysis results in tremendous mass loss or even complete destruction of the irradiated area (as seen in Fig. 7). The mechanism of radiation damage is described by Talmon (Chap. 3, this Vol.). Thus, for microanalytical purposes the section has to be much thicker than the thickness of the specimen layer lost by radiation damage (Zierold 1986a). This condition is met in thick sections or bulk material rather than in ultrathin cryosections. Thus, microanalysis was performed successfully in about 1-μm-thick frozen-hydrated sections (Gupta et al. 1977; Gupta and Hall 1981; see also Zierold and Steinbrecht, Chap. 15, this Vol.).

2.4 Freeze-Dried Cryosections

Freeze-drying is the most common method to overcome the problems of poor contrast and radiation damage in frozen-hydrated cryosections. Freeze-drying is performed by sublimation of ice at a pressure below 10^{-2} Pa (10^{-4} mbar) and a temperature below 190 K in order to prevent recrystallization. Freeze-dried cryosections are extremely hygroscopic. They have to be protected against air contact, because moisture and dirt particles may cause considerable contamination (Zierold 1984a, b). Therefore, they have to be transferred into the electron microscope within a protecting gas atmosphere (Hodson and Williams 1976;

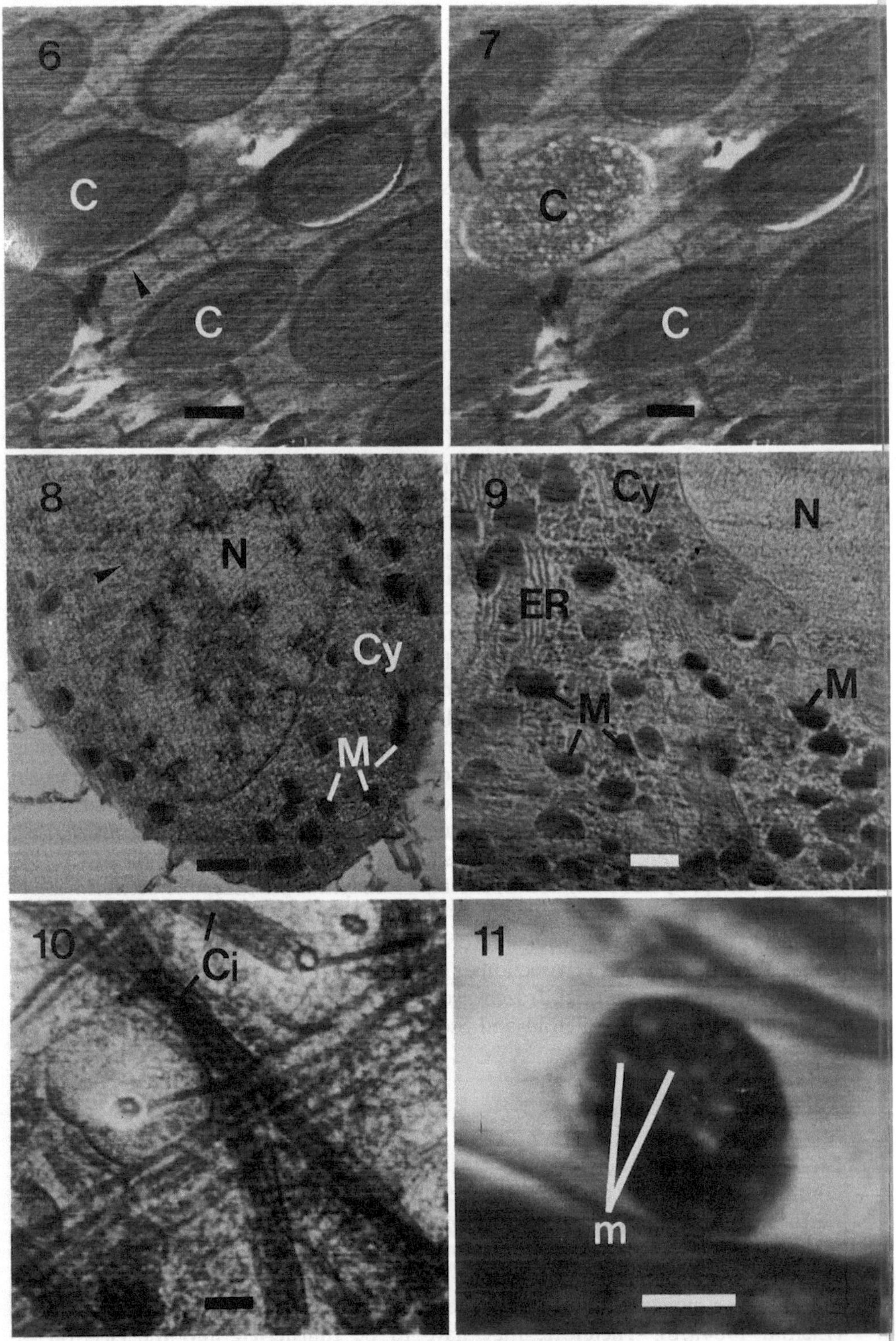
6
C
C
7
C
C
8
N
Cy
M
9
Cy
N
ER
M
M
10
Ci
11
m

Spriggs and Wynne-Evans 1976) or by means of a cryotransfer system. Freeze-dried cryosections can be stabilized by osmium vapour fixation (Frederik and Busing 1981 a, b), however, this might affect the ultrastructure and the distribution of elements as well.

The electron optical contrast in freeze-dried cryosections is much higher than in the frozen-hydrated state, as demonstrated by Figs. 8–11. Completely freeze-dried sections do not undergo observable mass loss by electron irradiation. However, even small amounts of water left in the section may cause radiation damage, resulting in mass loss. Both higher contrast and stability in the electron beam are the main reasons why freeze-dried cryosections are preferably used for electron probe microanalysis (Sevéus 1980; Barnard 1982; Roomans et al. 1982; Somlyo and Shuman 1982; Hagler and Buja 1984; Wendt-Gallitelli and Wolburg 1984; Zierold 1984b; see also Zierold and Steinbrecht, Chap. 15, this Vol.). Cold-stage microscopy of freeze-dried cryosections is found to be advantageous with respect to prevention of contamination (Zierold 1984b, 1986a) and to the protection of the specimen from the possibility of lipid melting by warming up to room temperature (Hagler and Buja 1986).

The main disadvantage in using freeze-dried cryosections is that they have lost the main constituent of cells, water. As a consequence, biomolecules may change their configuration, ions are more or less displaced and cellular structures shrink. By comparing cryosectioned cells in the frozen-hydrated and freeze-dried state, the shrinkage as measured in one lateral dimension is found to be approximately 10–20% (Zierold 1984a). Ions are assumed to be displaced by the sublimation of water up to the diameter of one hole as visible in the resulting ice-crystal segregation pattern. Presumably, the displacement of solutes is remarkably higher in compartments with higher water content such as vacuoles. X-ray microanalytical studies of gradients in the ion distribution, e.g. across cell membranes, indicate that ions obviously are not smeared by cryoultramicrotomy.

Generally, the intracellular matrix appears grainy like a porous mesh, due to ice-crystal damage caused either by cryofixation or by recrystallization of the vitrified object which had to be warmed up for freeze-drying. The size of the holes in the segregation pattern is not the size of the original ice crystals. Accord-

Figs. 6–11. Cryosections in STEM. *Arrowhead* indicates cutting direction. *C* Cell; *Ci* cilia; *Cy* cytoplasm; *ER* endoplasmic reticulum; *M* mitochondrion; *N* nucleus; *m* microtubule

Fig. 6. Frozen-hydrated section of yeast cells. Besides the electron dense cell wall surrounding the cells, no intracellular details can be identified. Bar = 1 μm

Fig. 7. The same section as in Fig. 6. The reduced area in the centre was irradiated by 30000 e nm^{-2}. Bar = 1 μm

Fig. 8. Freeze-dried cryosection of a fibroblast cell in suspension. Bar = 1 μm

Fig. 9. Freeze-dried cryosection of rat liver tissue. Bar = 1 μm

Fig. 10. Freeze-dried cryosection of *Paramecium caudatum* cell. Note the tubular structure of the cilia. Bar = 250 nm

Fig. 11. Freeze-dried cryosection of a cross-sectioned cilium in *Paramecium caudatum*. Two central microtubules are surrounded by nine peripheral pairs of microtubules. Bar = 100 nm ing to Dubochet and McDowall (1984b), the actual ice-crystal size might have been remarkably larger and can be estimated only by electron diffraction in the frozen-hydrated specimen. The parallel lines (Fig. 8) and the oblique shape of the cells (Figs. 6 and 7) will be discussed below as a physical aspect of cryoultramicrotomy. The contrast is caused by differences in the dry-mass density distribution in the intracellular compartments. This means that the contrast in freeze-dried cryosections represents, at least qualitatively, the water content present in the different compartments before ice sublimation (Zierold 1986b). For instance, the nuclear matrix in Figs. 8 and 9 seems to be less dense than the cytoplasm and the condensed chromatin in Fig. 8. Mitochondria are recognized as electron-dense compartments with an outer shape similar to that known from conventional electron microscopy. Cristae are found rarely, in particular in tissue specimens with presumably swollen mitochondria (Zierold 1982). The section thickness is sufficiently thin to identify the endoplasmic reticulum (Fig. 9) and the microtubules of cilia (Figs. 10 and 11). Cytoplasmic and nuclear membranes are visualized only if they are cut perpendicularly, as demonstrated in Figs. 8 and 9. Obliquely cut membranes remain invisible.

From Figs. 8–11 the spatial resolution in freeze-dried cryosections of cells can be estimated to be about 10–20 nm. This approximate limit is determined by the section thickness and by the ice-crystal damage. This approach is a feasible method to study the ultrastructure and the composition of cells in the intermediate magnification range, but it is not sufficient for high resolution studies, e.g. of molecular structures.

3 Physical Aspects of Cryoultramicrotomy

3.1 The Morphology of Cryosections

Cryosections appear in the electron microscope as irregularly shaped sheets. Sections cut by a glass knife often show knife marks in the cutting direction. Cracks are found, predominantly perpendicular to the cutting direction. Cryosections are more or less deformed parallel to the knife edge, as can be seen by the change in shape of the originally round yeast cells (Figs. 6 and 7). Intact cells are round, their sections, however, appear ellipsoid. Additionally, deformation lines parallel to the knife edge (Figs. 8 and 9) can be found described by many authors (Hodson and Williams 1976; Barnard 1982; Frederik et al. 1982, 1984; Chang et al. 1983; Dubochet and McDowall 1984a; Leunissen et al. 1984; Zierold 1984a). In order to reveal the morphology of these deformations, cryosections were cut from frozen droplets consisting of 20% glycerol-gelatine and 80% water as a model substance and studied in STEM and in scanning electron microscopy (SEM). The main results are:

1. All cryosections with visible ice-crystal segregation patterns with holes less than 200 nm in size show parallel, sometimes scaly, deformation lines with a period between 150 and 300 nm. These lines are present on both sides of the sections. The generation of these lines is independent of a cutting temperature

between 115 and 170 K and of a cutting speed between 0.2 mm s^{-1} and 70 mm s^{-1}. These deformation lines are not found in sections with large ice-crystal damage exceeding 200 nm.
2. High cutting speed, above 30 mm s^{-1}, causes additional deformation lines with a period in the μm range. Presumably, they are caused by vibrations between knife and specimen. They vanish when the cutting speed is reduced below 20 mm s^{-1}.
3. The surface of the block from which the sections are cut appears smooth except for some cracks. In particular, it is free of deformation lines as found in the sections.

These results are demonstrated by Figs. 12–18. Any theoretical model explaining the formation of ultrathin cryosections has to fit with these morphological observations.

3.2 The Cryosectioning Process

There is an extended discussion on the nature of the cryosectioning process, particularly on the question whether sections are generated by local melting or by cleaving without any significant increase of the temperature. The answer to this question is not only of academical interest, but would imply important consequences for the concept of cryoultramicrotomy. If sections would transiently melt during cutting, then the risk of redistribution of diffusible substances would increase and recrystallization of ice could affect the ultrastructure. Furthermore, an understanding of the cryosectioning process is expected to help in avoiding or at least reducing the observed deformations (Figs. 6–9 and 12–18).

Thornburg and Mengers (1957) have proposed a model which implies a zone of "micromelting" at the knife edge. Hodson and Marshall (1972) have calculated, by assuming the whole mechanical energy to be transformed into heat during cutting at 180 K or less, that the molten zone would be at maximum 10 nm thick. Saubermann et al. (1977) have measured the force during cutting sections. They find that at 240 K smooth sectioning, consuming little energy, takes place, whereas at a temperature of 190 K or lower fracturing with high energy input to the section prevails. Frederik and Busing (1981 b) have compared ice crystals in cryosections and replicas. They found no indication for cutting-induced melting and refreezing of ice. Karp et al. (1982) have cut solid toluene at a temperature of 1 K below its melting point. The temperature increase caused by cutting was not sufficient to melt the sections. These studies and the experience of cutting even vitrified sections (McDowall et al. 1983) provide evidence that cryoultramicrotomy is not based on melting of ice. However, Chang et al. (1983) reported that some sections show a grainy surface. The authors explained this phenomenon as a cutting-induced melting of a surface layer followed by freezing. According to the phase diagram of water, the lowest achievable melting point is 251 K at 2.1×10^8 Pa (2100 bar). Thus, sectioning pure ice by local melting caused by pressure is impossible at a temperature below 250 K.

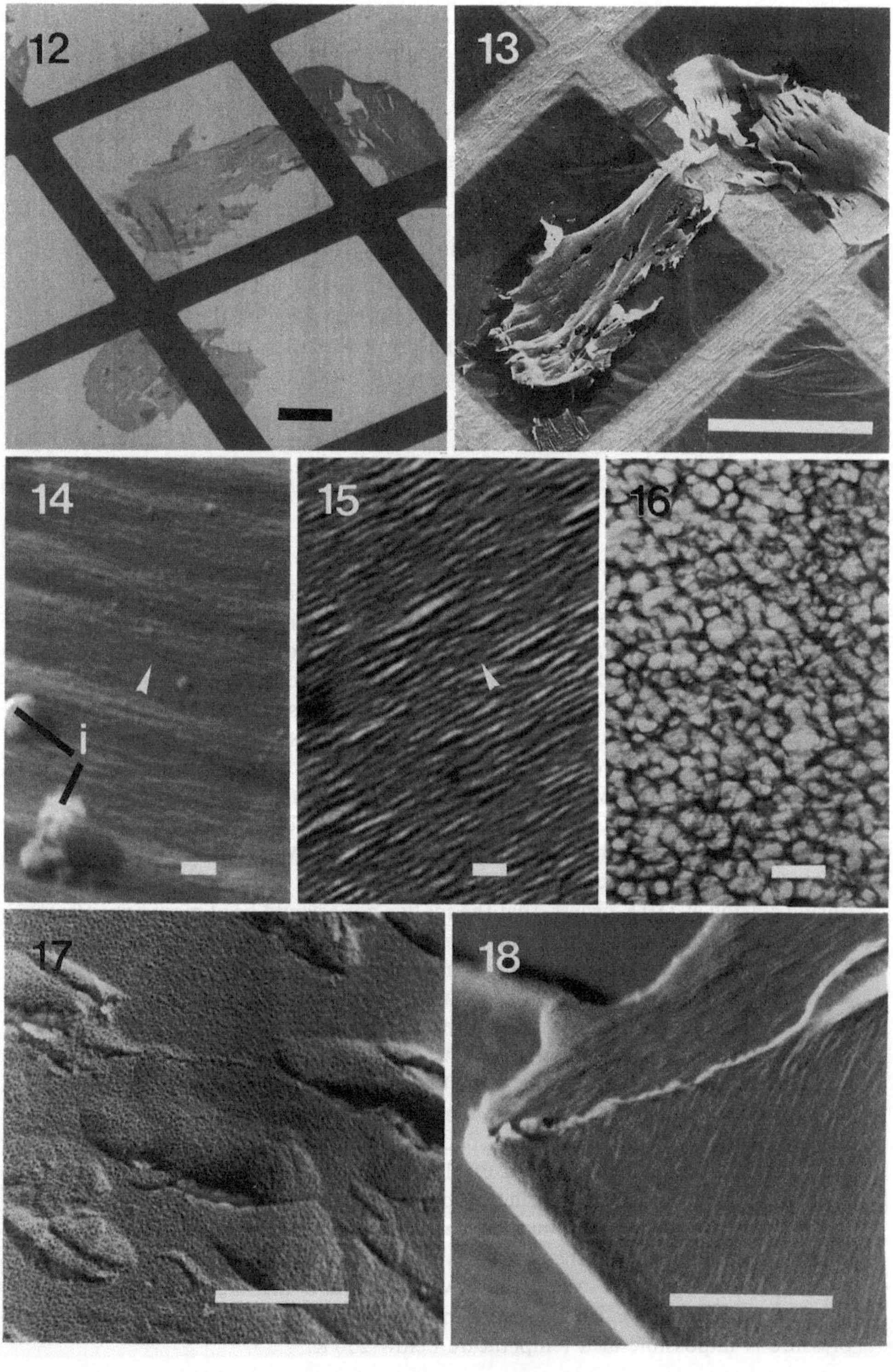
12
13
14
i
15
16
17
18

Studies on chip formation by metal machining (Raupp 1937; Merchant 1945; Degner et al. 1978) show that metal chips exhibit deformation lines caused by pressure due to shearing forces pointing perpendicular to the tool surface. Frederik et al. (1984) compared the periodic deformation lines observed in cryosections and in metal and plastic sections and concluded that they are caused essentially by the same shearing process: during cutting, the sectioned material is periodically collected and shedded away. Sitte (1984b) interpreted the high frequency deformation lines occurring in ultramicrotomy of plastic material by periodic buildup of pressure and friction, alternating with sliding away on the knife surface, similarly to the model proposed by Zierold (1984a). Sitte (1984b) recommended lowering the cutting width, lowering the sum of rake and clearance angle and lowering the cutting speed in order to reduce high frequency deformations.

Figure 19 shows schematically different mechanical models concerning the formation of cryosections. The first row (a and b) illustrates that a real knife has a finite edge curvature, which results in pressure perpendicular to the upper knife surface, rather than "sectioning" in the ordinary sense. The second row (c–f) illustrates chip formation by metal machining according to Merchant (1945). This process would cause an increase of section thickness t_2 compared with the advance setting t_1. The third row (g–j) shows the folding model. It would cause a periodically deformed section without effective alteration of thickness. The fourth row (k–n) illustrates the periodical friction sliding model resulting in a bellow-like section shape. Increase of mean section thickness is possible in this case. Thus, it remains the objective of future studies to determine whether one of these models or combinations could help in elucidating the cryosectioning process. Presumably, the chip formation process (c–f) can be excluded by comparison of dry cryosections with thawed sections picked up by sucrose: the latter exhibit no deformation by compression (Griffiths et al. 1984). This means that the

Figs. 12–18. Studies on the morphology of cryosections of solutions consisting of 20% glycerol-gelatine and 80% water. *Arrowheads* indicate cutting direction

Fig. 12. Cryosections in STEM. Bar = 100 μm

Fig. 13. The same sections as presented in Fig. 12, in SEM. Note that the sections adhere merely partly on the support film. Bar = 100 μm

Fig. 14. Frozen-hydrated cryosection in SEM, cut at 160 K and 50 mm s^{-1} cutting speed. *i* Contaminating ice crystals. Two types of deformation lines with short and long period are present. Bar = 1 μm

Fig. 15. Frozen-hydrated cryosection in SEM, cut at 160 K and 10 mm s^{-1} cutting speed. Note that the long periodic deformation lines are absent, whereas the lines with short period are still present. Bar = 1 μm

Fig. 16. Freeze-dried cryosection with large ice-crystal damage in SEM. No deformation lines occur. Bar = 1 μm

Fig. 17. Freeze-dried block from which cryosections were cut in SEM. No periodic deformation lines are found. Except for some cracks, the block surface appears smooth. Bar = 5 μm

Fig. 18. Folded frozen-hydrated cryosection cut at 160 K and 10 mm s^{-1} in SEM. High frequency deformation lines are present on both sides. Bar = 5 μm

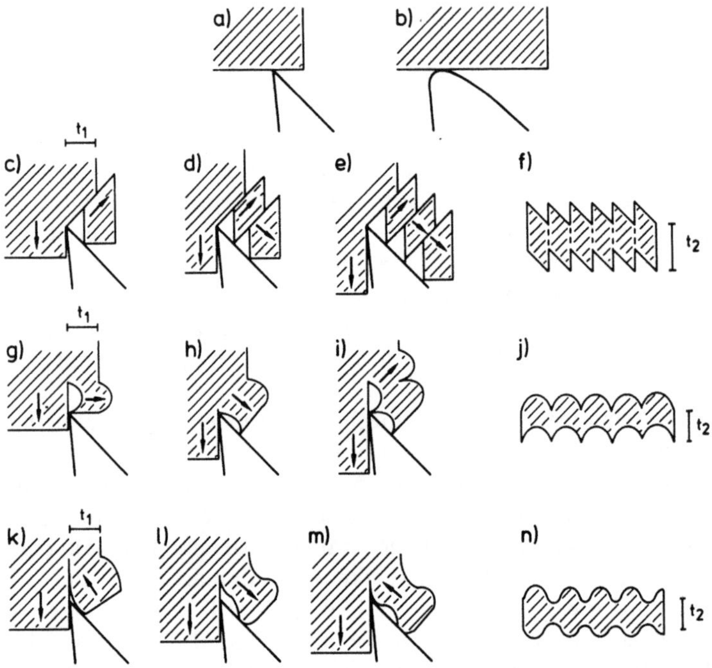

Fig. 19a–n. Theoretical models concerning the cryosectioning process, as seen from the side. The *shaded areas* represent the object to be cut and cryosectioned material. *Arrows* indicate forces. t_1 Advance setting; t_2 effective section thickness. **a** The object has arrived at the knife edge. **b** More realistic shape of the knife edge. Pressure buildup perpendicular to the knife surface becomes evident. **c–f** Chip formation model. The pressure against the knife surface causes successive rupture and dislocation of material resulting in a sawtoothed section profile, as sketched in **f**, with t_2 greater than t_1. **g–j** Folding model. The material pressed against the knife surface is folded and then slips away. The effective section thickness does not increase. **k–n** Friction-sliding model. The friction of the cut material at the knife edge is so high that it causes thickening of the section until relaxation takes place by sliding away. A section with periodically varying thickness results (**n**)

deformation of cryosections can be reversed by thawing and stretching on water. In the chip formation model, material is irreversibly displaced. Restoration to the original structure would be hard to understand, whereas it is compatible with both the folding model and the friction sliding model.

From the point of view of solid-state physics, sectioning of a single ice crystal should take place preferably by shear stress in the basal plane of the crystal, as the hexagonal ice layers can glide over each other like cards over a pack of cards (Hobbs 1974). However, biological samples are quite different from single crystals. They consist either of a vitrified mixture of water with organic molecules and electrolyte ions or of polycrystalline ice within a solidified organic matrix. Most studies on plastic deformation and fracture of polycrystalline ice were performed at temperatures above 200 K. Therefore, extrapolation of data to the lower temperature range of cryoultramicrotomy should be regarded with caution. Studies of plastic deformation and fracture in polycrystalline ice depend in a rather complicated manner on temperature, thermal history, applied stress and stress variation, crystal orientation, grain boundary migration and recrystallization (Hobbs 1974). Nevertheless, it is noteworthy that all parameters concerning the hardness of ice such as the elastic constants, tensile strength and crushing strength increase with decreasing temperature. Also, the elastic region of the stress-strain curve is extended to higher stress values with decreasing temperature. These results are consistent with the experience that cryosectioning of polycrystalline ice becomes harder by lowering the temperature. At present, cryoultramicrotomy of biological objects seems too complicated to be explained by a detailed physical theory.

3.3 Improving Cryoultramicrotomy

Cryoultramicrotomy has to be improved in order to fulfill its potential mentioned in the introduction. This holds, in particular, for the following aspects:

1. Picking up dry cryosections is still a weak link in the whole process. Skill and patience are required to place the sections on the grid by the eyelash probe.
2. Reproducible section thickness is required for many analytical purposes, e.g. mass measurements in subcellular compartments of freeze-dried sections (Zierold 1986b). Wolf and Schwinde (1983) have designed a device to adjust the advance of the ultramicrotome arm reproducibly.
3. Prevention of deformations in the sections is necessary in order to obtain a geometrically undistorted specimen. Hardening the object as suggested by Jesior (1985) for ultramicrotomy of plastic material is very difficult to perform in the case of cryoultramicrotomy. Frozen objects can be hardened by lowering the temperature. Perhaps this idea could be realized by ultramicrotomy at liquid helium temperature, as proposed by Fernández-Morán (1982). Lickfeld (1985) has analyzed the forces caused during ultramicrotomy with the result that deformation of the sections by pressure from the knife surface would be reduced by lowering the rake angle of the knife considerably below 45°. Analysis of the sectioning models sketched in Fig. 19 leads to the same conclusion.
4. Thinner sections are desired in order to improve spatial resolution and to render cryosections useful for electron energy loss spectroscopy (Ottensmeyer 1984). This technique requires that multiple electron scattering in the specimen be excluded. For 100 keV primary electrons, a section thickness of 30 nm is considered to be appropriate. The smallest thickness achieved so far is 42 nm (McDowall et al. 1983). Even this is not yet achievable in routine cryoultramicrotomy. According to Lickfeld (1985), a knife with a considerably reduced rake angle can be expected to cut thinner sections.

4 Conclusions

Cryoultramicrotomy has proved to be an useful tool to study the ultrastructure and composition of biological cells closer to their physiological state than it is possible by conventional techniques. Deriving advantage from a cryosection requires compromises with respect to the preparation procedure: immunocytochemistry necessitates chemical treatment of the biological object and thawing of the section. Frozen-hydrated sections allow one to study subcellular structures in the presence of water, in particular, in the vitrified state. Limits are given by the low contrast and the high mass loss due to radiation damage. Freeze-dried cryosections have proved to be most feasible for microprobe analysis. However, shrinkage and possible redistribution of diffusible substances have to be taken into account in data evaluation. The cutting-induced deformation in the section and the dry transfer of the section to the grid are the main weak points of present cryoultramicrotomy and await improvement.

Acknowledgements. I would like to thank Miss. S. Dongard and Mrs. M. Seiffert for their very careful technical assistance and Dr. M. Schmitz for preparing *Paramecium* cells for cryoultramicrotomy (Figs. 10 and 11).

References

(References marked by an asterisk are review papers or books respectively)

Appleton TC (1974) A cryostat approach to ultrathin 'dry' frozen sections for electron microscopy: a morphological and X-ray analytical study. J Microsc (Oxford) 100:49–74

Barnard T (1982) Thin frozen-dried cryosections and biological X-ray microanalysis. J Microsc (Oxford) 126:317–332

Bernhard W, Viron A (1971) Improved techniques for the preparation of ultrathin frozen sections. J Cell Biol 49:731–746

Biddlecombe WH, McEwan Jenkinson D, McWilliams SA, Nicholson WAP, Elder HY, Dempster DW (1982) Preparation of cryosections with a modified Sorvall MT2B ultramicrotome and cryoattachment. J Microsc (Oxford) 126:63–75

Carlemalm E, Colliex C, Kellenberger E (1985) Contrast formation in electron microscopy of biological material. Adv Electron Electron Phys 63:269–334

Chang J-J, McDowall AW, Lepault J, Freeman R, Walter CA, Dubochet J (1983) Freezing, sectioning and observation artefacts of frozen hydrated sections for electron microscopy. J Microsc (Oxford) 132:109–123

Christensen AK (1971) Frozen thin sections of fresh tissue for electron microscopy, with a description of pancreas and liver. J Cell Biol 51:772–804

Degner W, Lutze H, Smejkal E (1978) Spanende Formung. VEB Verlag Technik, Berlin

Dubochet J, McDowall AW (1984a) Frozen hydrated sections. In: Revel J-P, Barnard T, Haggis GH (eds) The science of biological specimen preparation for microscopy and microanalysis. SEM, AMF O'Hare, IL 60666, pp 147–152

Dubochet J, McDowall AW (1984b) Cryoultramicrotomy: Study of ice crystals and freezing damage. In: Csanády A, Röhlich P, Szabó D (eds) Electron microscopy 1984, vol. 2. Proc 8th Eur Congr Electron Microsc, Budapest, pp 1407–1410

Dubochet J, McDowall AW, Menge B, Schmid EN, Lickfeld KG (1983) Electron microscopy of frozen-hydrated bacteria. J Bacteriol 155:381–390

Edelmann L (1984) Frozen hydrated cryosections of thallium loaded muscle reveal subcellular potassium binding sites. Physiol Chem Phys 16:499–501

Fernández-Morán H (1982) Cryomicroscopy, history and outlook. In: The organizing committee of the 10th Int Congr Electron Microscopy (ed) Electron microscopy 1982, vol. 1. Dtsch Ges Elektronenmikrosk, Frankfurt/M, pp 751–761

Frederik PM, Busing WM (1981a) Ice crystal damage in frozen thin sections: freezing effects and their restoration. J Microsc (Oxford) 121:191–199

Frederik PM, Busing WM (1981 b) Strong evidence against section thawing whilst cutting on the cryo-ultratome. J Microsc (Oxford) 122:217–220
Frederik PM, Busing WM, Persson A (1982) Concerning the nature of the cryosectioning process. J Microsc (Oxford) 125:167–175
Frederik PM, Busing WM, Persson A (1984) Surface defects on thin cryosections. Scanning Electron Microsc 1984/I:433–443
Griffiths G, McDowall A, Back R, Dubochet J (1984) On the preparation of cryosections for immunocytochemistry. J Ultrastruct Res 89:65–78
*Gupta BL, Hall TA (1981) The X-ray microanalysis of frozen-hydrated sections in scanning electron microscopy: An evaluation. Tissue Cell 13:623–643
Gupta BL, Hall TA, Moreton RB (1977) Electron probe X-ray microanalysis. In: Gupta BL, Moreton RB, Oschmann JL, Wall BJ (eds) Transport of ions and water in animals. Academic Press, London New York, pp 83–143
Hagler HK, Buja LM (1984) New techniques for the preparation of thin freeze-dried cryosections for X-ray microanalysis. In: Revel J-P, Barnard T, Haggis GH (eds) The science of biological specimen preparation for microscopy and microanalysis. SEM, AMF O'Hare, IL 60666, pp 161–166
Hagler HK, Buja LM (1986) Effect of specimen preparation and section transfer techniques on the preservation of ultrastructure, lipids and elements in cryosections. J Microsc (Oxford) 141:311–317
*Hobbs PV (1974) Ice physics. Clarendon, Oxford
Hodson S, Marshall J (1972) Evidence against through-section thawing whilst cutting on the ultracryotome. J Microsc (Oxford) 95:459–465
Hodson S, Williams L (1976) Ultracryotomy of biological tissues to preserve membrane structure. J Cell Sci 20:687–698
Jésior J-C (1985) How to avoid compression: A model study of latex sphere grid sections. J Ultrastruct Res 90:135–144
Karp RD, Silcox JC, Somlyo AV (1982) Cryoultramicrotomy: evidence against melting and the use of a low temperature cement for specimen orientation. J Microsc (Oxford) 125:157–165
Leunissen JLM, Elbers PF, Leunissen-Bijvelt JJM, Verkleij AJ (1984) An evaluation of the cryosectioning of fixed and cryoprotected rat liver. Ultramicroscopy 12:345–352
Lickfeld KG (1985) Ein Beitrag zur Frage, welche Kräfte und Faktoren Dünnstschneiden bewirken. J Ultrastruct Res 93:101–115
McDowall AW, Chang J-J, Freeman R, Lepault J, Walter CA, Dubochet J (1983) Electron microscopy of frozen hydrated sections of vitreous ice and vitrified biological samples. J Microsc (Oxford) 131:1–9
*Menco BPM (1986) A survey of ultra-rapid cryofixation methods with particular emphasis on applications to freeze-fracturing, freeze-etching, and freeze-substitution. J Electron Microsc Tech 4:177–240
Merchant ME (1945) Mechanics of the metal cutting process. I. Orthogonal cutting and a type 2 chip. J Appl Phys 16:267–275
Moor H (1973) Cryotechnology for the structural analysis of biological material. In: Benedetti EL, Favard P (eds) Freeze etching techniques and applications. Soc Fr Microsc Electron Paris, pp 11–19
Ottensmeyer FP (1984) Electron spectroscopic imaging: Parallel energy filtering and microanlysis in the fixed-beam electron microscope. J Ultrastruct Res 88:121–134
*Plattner H, Bachmann L (1982) Cryofixation: A tool in biological ultrastructural research. Int Rev Cytol 79:237–304
Raupp A (1937) Über die Spanbildung bei der Metallbearbeitung. Diss, Tech Hochschule Hannover
Rick R, Dörge A, Thurau K (1982) Quantitative analysis of electrolytes in frozen dried sections. J Microsc (Oxford) 125:239–247
*Robards AW, Sleytr UB (1985) Low temperature methods in biological electron microscopy. In: Glauert AM (ed) Practical methods in electron microscopy, vol. 10. Elsevier, Amsterdam
*Roomans GM, Wei X, Sevéus L (1982) Cryoultramicrotomy as a preparative method for X-ray microanalysis in pathology. J Ultrastruct Res 3:65–84
Roos N, Barnard T (1985) A comparison of subcellular element concentrations in frozen-dried, plastic-embedded, dry-cut sections and frozen-dried cryosections. Ultramicroscopy 17:335–344

Ross A, Sumner AT, Ross AR (1981) Preparation and assessment of frozen-hydrated sections of mammalian tissue for electron microscopy and X-ray microprobe analysis. J Microsc (Oxford) 121:261–272
Saubermann AJ, Riley WD, Beeuwkes III R (1977) Cutting work in thick section cryomicrotomy. J Microsc (Oxford) 111:39–49
Saubermann AJ, Echlin P, Peters PD, Beeuwkes III R (1981) Application of scanning electron microscopy to X-ray analysis of frozen-hydrated sections. I. Specimen handling techniques. J Cell Biol 88:257–267
Schmitz M, Meyer R, Zierold K (1985) X-ray microanalysis in cryosections of natively frozen *Paramecium caudatum* with regard to ion distribution in ciliates. Scanning Electron Microsc 1985, I:433–445
Sevéus L (1980) Cryoultramicrotomy as a preparation method for X-ray microanalysis. Scanning Electron Microsc 1980, IV:161–171
*Sitte H (1982) Instrumentation for cryosectioning. In: The organizing committee of the 10th Int Congr Electron Microscopy (ed) Electron microscopy 1982, vol. 1. Dtsch Ges Elektronenmikrosk, Frankfurt/M, pp 9–18
Sitte H (1984a) Instruments for cryofixation, cryoultramicrotomy and cryosubstitution for biomedical TEM. Zeiss Inf MEM Mag Electron Microsc 3:25–31
Sitte H (1984b) Process of ultrathin sectioning. In: Revel J-P, Barnard T, Haggis GH (eds) The science of biological specimen preparation for microscopy and microanalysis. SEM, AMF O'Hare, IL 60666, pp 97–104
Somlyo AP, Shuman H (1982) Electron probe and electron energy loss analysis in biology. Ultramicroscopy 8:219–234
Somlyo AV, Shuman H, Somlyo AP (1977) Elemental distribution in striated muscle and the effects of hypertonicity. Electron probe analysis of cryo sections. J Cell Biol 74:828–857
Spriggs TLB, Wynne-Evans D (1976) Observations on the production of frozen-dried thin sections for electron microscopy using unfixed liver, fast-frozen without cryoprotectants. J Microsc (Oxford) 107:35–46
Steinbrecht RA, Zierold K (1984) A cryoembedding method for cutting ultrathin cryosections from small frozen specimens. J Microsc (Oxford) 136:69–75
Thornburg W, Mengers PE (1957) An analysis of frozen section techniques: I. Sectioning of fresh-frozen tissue. J Histochem Cytochem 5:47–52
Tokuyasu KT (1973) A technique for ultracryotomy of cell suspensions and tissues. J Cell Biol 57:551–565
Tokuyasu KT (1986) Application of cryoultramicrotomy to immunocytochemistry. J Microsc (Oxford) 143:139–149
*Wendt-Gallitelli M-F, Wolburg H (1984) Rapid freezing, cryosectioning, and X-ray microanalysis on cardiac muscle preparations in defined functional states. J Electron Microsc Tech 1:151–174
Wolf B, Schwinde A (1983) Automatischer Ultramikrotom-Antrieb für medizinische Routinehistologie mit dem Elektronenmikroskop. Microsc Acta 87:301–306
Zierold K (1982) Preparation and transfer of ultrathin frozen-hydrated and freeze-dried cryosections for microanalysis in scanning transmission electron microscopy. Scanning Electron Microsc 1982, III:1205–1214
Zierold K (1984a) The morphology of ultrathin cryosections. Ultramicroscopy 14:201–210
Zierold K (1984b) Experiences in X-ray microanalysis of freeze-dried cryosections. Scanning Electron Microsc 1984, IV:1867–1874
Zierold K (1985) Contrast of biological cryosections in scanning transmission electron microscopy. J Microsc (Oxford) 140:56–71
*Zierold K (1986a) Preparation of cryosections for biological microanalysis. In: Müller M, Becker RP, Boyde A, Wolosewick JJ (eds) The science of biological specimen preparation for microscopy and microanalysis 1985. SEM, AMF O'Hare, IL 60666, pp 119–127
Zierold K (1986b) The determination of wet weight concentrations for elements in freeze-dried cryosections from biological cells. Scanning Electron Microsc 1986, II:713–724
Zierold K, König R, Olech K-H, Schäfer D, Lübbers DW, Müller K-H (1981) A cooling chain for studies of cryofixed biological specimens by scanning transmission electron microscopy and X-ray microanalysis. Ultramicroscopy 6:181–186
Zingsheim HP (1984) Sublimation rates of ice in a cryoultramicrotome. J Microsc (Oxford) 133:307–312

Chapter 7 Freeze-Substitution and Freeze-Drying

RUDOLF ALEXANDER STEINBRECHT[1] and MARTIN MÜLLER[2]

1 Introduction

Freeze-substitution (FS) and freeze-drying (FD) are dehydration techniques by which the water is gently removed from a frozen specimen. Both techniques can serve as a link between cryofixation and conventional thin sectioning at room temperature (Fig. 1). They are, therefore, hybrid techniques combining the advantages of the low temperature and the room temperature specimen preparation. With respect to the danger of artefacts, these procedures are much more obscure than "pure" cryotechniques, such as freeze-etching or cryosectioning. Both, FS and FD, are known from light microscopy and have been used in electron microscopy since its early days (for refs. of the older literature see Bullivant 1970; Rebhun 1972; Robards and Sleytr 1985), but only during the last dozen years a breakthrough can be noticed, which is mainly due to improved cryofixation. As for any other cryotechnique in biological electron microscopy, for successful FS and FD the main prerequisite is also good cryofixation with as little freezing damage as possible (see Bachmann and Mayer, Chap. 1; Sitte et al., Chap. 4; Dubochet et al., Chap. 5; Moor, Chap. 8; this Vol.).

Freeze-substitution dissolves the ice in a frozen specimen by an organic solvent, which usually also contains chemical fixatives. It has to be carried out at a temperature low enough to avoid secondary ice-crystal growth. After completion of FS, the temperature can be raised without this risk, since water is now absent. Except for low temperature embedding, infiltration with the embedding resin is done at room temperature followed by heat polymerization. The result is a plastic-embedded specimen which can be sectioned and stained in the routine way.

Freeze-drying eliminates the frozen water by sublimation in a vacuum chamber. Again, the temperature must be sufficiently low to avoid secondary freezing damage. When the specimen is dry, it may be warmed to room temperature for plastic embedding and ultramicrotomy as above.

1 Max-Planck-Institut für Verhaltensphysiologie, D-8131 Seewiesen, FRG
2 Labor für EM I, Institut für Zellbiologie der ETH, CH-9092 Zürich, Switzerland

Cryotechniques in Biological Electron Microscopy
Edited by R.A. Steinbrecht and K. Zierold

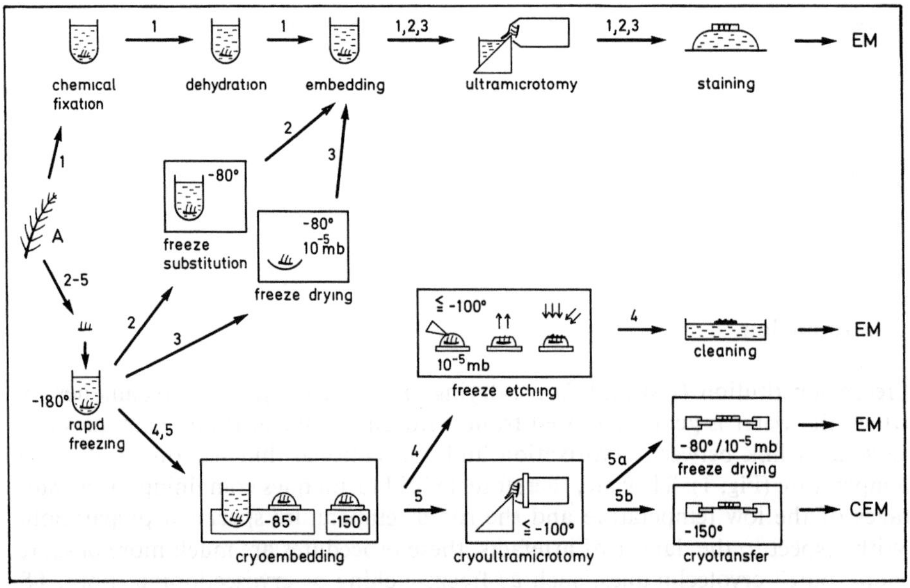

Fig. 1. Flow chart comparing the principal steps of conventional specimen preparation (*1*) and cryofixation with its various follow-up procedures (*2–5*). Rapid freezing is demonstrated by immersion of a small insect antenna (*A*) into liquid propane (for other techniques see Sitte et al., Chap. 4, this Vol.). Freeze-substitution (*2*) and freeze-drying (*3*) link cryofixation with embedding and sectioning at room temperature. Cryoembedding (cryoglueing) with butyl benzene and heptane renders small specimens amenable to freeze-etching (*4*) and cryoultramicrotomy (*5*) respectively. An electron microscope with a cold stage (*CEM*) is essential for the study of frozen-hydrated cryosections (*5b*). Temperatures other than room temperature are given in centigrades, pressure in millibar (Steinbrecht 1984)

Plastic infiltration and thin sectioning are, however, not the only techniques to combine with FS or FD. If the dry specimens are thin enough, they may be directly viewed in the transmission electron microscope after metal shadowing (see Gross, Chap. 10, this Vol.). Freeze-dried bulk specimens can be observed in the scanning electron microscope after metal coating (Sect. 3.4; see also Marshall, Chap. 13, this Vol.). Freeze-drying of ultrathin cryosections is important for microanalysis to increase contrast and stability (see Zierold, Chap. 6, this Vol.).

This chapter provides an overview of the theoretical and experimental data on the methodology of FS and FD, it then describes the most generally used procedures, and closes with a critical evaluation of the present status with selected examples of applications.

2 Methodology: Theoretical and Experimental Data

2.1 Water in the Cell and Its Removal

Both, FS and FD, are essentially dehydration processes. Our current problems in understanding these processes are mainly due to our incomplete knowledge of cellular water. A few points of particular interest should be recalled (for details, see Bachmann and Mayer, Chap. 1; Kellenberger, Chap. 2, this Vol.).

1. Water in the cell is not homogeneous as is water in a dilute solution, but partly exhibits different physicochemical properties. Depending on the type of study performed, this anomalous water is referred to as "bound water", "non-freezable water", "non-solvent water", etc., in contrast to bulk or free water.
2. Most or all of this anomalous water is supposed to be closely associated with surfaces, e.g. of macromolecules or of intracellular membranes (surface-modified water, vicinal water, hydration shell).
3. There is accumulating evidence that this surface-modified water is extremely important for cell metabolism (reviewed by Clegg 1979; Negendank 1986), as well as for the maintenance of the structural integrity of proteins and other cell constituents (reviewed by Tanford 1980).

For the dehydration procedures in biological electron microscopy, two major consequences result from these notions. Firstly, bulk water is more easily removed than water of the hydration shells. Secondly, removing the hydration shells leads to much more serious alterations of the fine structure than removing the bulk water.

This is compatible with, though not proven by, the non-linear shrinkage behaviour of most cells during conventional dehydration with organic solvents at room temperature. Many cells retain their shape and volume until about 70% of their cellular water is replaced by the solvent, but after full dehydration they may have shrunk by 20–70% of the original volume (reviewed by Lee 1984). Freeze-drying below 210 K does not completely remove the water from model (sugar and protein solutions) or biological (yeast, muscle cells) systems, irrespective of drying time and vacuum conditions. The residual water can only be taken away by warming the specimen under vacuum (MacKenzie 1972; Wildhaber et al. 1982; see also Gross, Chap. 10, this Vol.). Again this does not prove, but fits well into the concept of the hydration shells.

MacKenzie (1972) directly observed the collapse phenomena in frozen aqueous model solutions, which happened only, if FD was carried out above a discrete temperature. This "collapse temperature" was found to range from 223 K (8% w/w sucrose – 2% w/w NaCl) to 263 K (10% w/w ovalbumin) and to depend only on the solutes. In the same way collapse temperatures were determined for FS; these were lower than those for FD and depended additionally on the substitution solvent used. Frederik et al. (1984) observed the FD of cryosections directly in the electron microscope: shrinking happened only when the sections were almost completely dry.

Structural alterations during specimen preparation in biological electron microscopy are manifold. In most cases collapse phenomena due to inadequate dehydration cannot be separated from other artefacts (see also Kellenberger, Chap. 2, this Vol.). More systematic studies on this subject are highly desirable. In this context, it is of great help that freeze-fracturing (see Bachmann, Chap. 9, this Vol.) and cryoelectron microscopy of frozen-hydrated specimens (see Dubochet et al., Chap. 5, this Vol.) offer alternatives, in which the water remains in the specimen.

Table 1. Melting point and water capacity of solvents most widely used in freeze-substitution

Solvent	Melting point		% (v) Water dissolved at		
	K	°C	183 K	193 K	203 K
Methanol	179.1	−93.9	29	32	37
Acetone	177.6	−94.4	<2	2.5	3.5
Diethylether	156.8	−116.2	0.06	–	~0.1[a]
Ethanol	155.7	−117.3	13	16	19

[a] 0.14% at 213 K.
Data on water capacity of acetone from Van Harreveld et al. (1965), of diethylether from Humbel (1984), of methanol and ethanol from J.-D. Acetarin and E. Carlemalm (unpubl.).

2.2 Temperature of Freeze-Substitution/Freeze-Drying

The lowest temperature at which FS can be performed is given by the melting point of the solvent used (Table 1). With FD, the limit is set by the extremely low vapour pressure of water at temperatures below 170 K which would lead to impractical drying times (see Sect. 2.3). Therefore, theoretically, the danger of segregation artefacts by secondary ice-crystal growth is imminent with both methods, since they are performed at temperatures which are considerably above the devitrification range of amorphous water (ca. 150 K, see also Bachmann and Mayer, Chap. 1; Dubochet et al., Chap. 5, this Vol.).

It is mainly due to the increasing number of successful studies using FS or FD around 190 K, that these temperatures are now widely believed to be "safe" for cryofixed biological samples. The different behaviour of the "cell sap" as compared to pure water is generally attributed to its complex chemical composition; some components may act as natural cryoprotectants.

Only a few experiments have been performed to determine the temperature at which, for a given specimen, secondary ice-crystal growth actually starts. With human red blood cells this occurred at temperatures as low as 203 K (−70 °C) (Nei 1973). On the other hand, rapidly frozen silk moth antennae could be exposed to temperatures as high as 230 K (−43 °C) for up to 45 min without noticeable ice-crystal growth, whereas at 250 K (−23 °C) damage occurred rapidly (Steinbrecht 1985). Similar high recrystallization temperatures are reported for frost-hardy plant cells (Sakai et al. 1968) and yeast cells (Bank 1973).

2.3 Rate of Freeze-Substitution/Freeze-Drying

In *freeze-substitution*, the dehydration speed depends on the temperature and on the water capacity of the solvent at that temperature (Table 1). On a qualitative basis, this was checked by Zalokar (1966), who observed the extraction of methylene blue from frozen stained agar blocks. A more quantitative approach is the measurement of the ^{3}H-content in the substitution fluid during the FS of

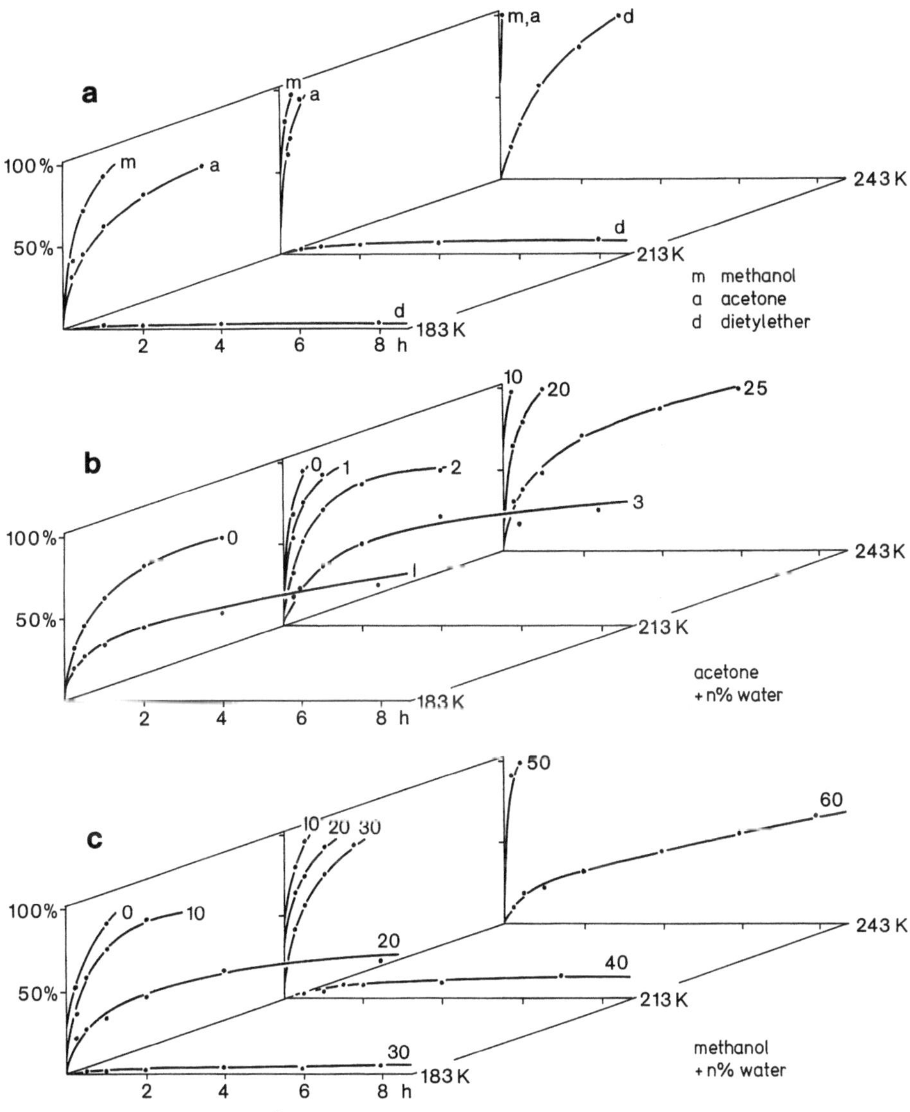

Fig. 2a–c. Substitution of ^{3}H-ice by various solvents as a function of time and temperature. **a** Pure solvents: *m* methanol, *a* acetone, *d* diethylether; **b** acetone containing the indicated (vol %) amounts of water; **c** methanol containing the indicated (vol %) amounts of water. (After Humbel et al. 1983)

known amounts of tritiated ice (Ornberg and Reese 1981; Humbel et al. 1983). Figure 2a shows the substitution velocity of different solvents and its strong temperature dependence. The fastest substituent is methanol, acetone is considerably slower and diethylether by far the slowest medium. At 183 K, the substitution rate of acetone is dramatically reduced by only 1% water (Fig. 2b), but methanol will, even in the presence of 10% water, substitute specimens fairly rapidly (Fig. 2c).

Considering the very slow substitution of acetone below 200 K, it was questioned repeatedly, whether specimens were substituted at all at this temperature or whether this occurred at higher temperatures during the warming-up period. Proof that acetone can substitute the water at 194 K (−79 °C) was given by Steinbrecht (1982) using antennal tissue of moths as a test specimen. Tissue preservation was excellent, if substitution at 194 K was carried out for 7 days, followed by rapid warming (halftime 30 s). On the other hand, substitution time at 194 K could be reduced to 5 min, if the rate of warming was low enough (halftime 3 h). If, however, the substituting acetone contained 5% water, ice crystals had grown during rewarming before the substitution was complete.

The process of *freeze-drying*, at first sight, appears to be only dependent on the vapour pressure of water at the various subzero temperatures. This allows the theoretical calculation of FD times on the basis of the kinetic gas theory (for an extensive compilation, see Umrath 1983). At 150 K it would take 3.5 years until an ice layer of 1 mm thickness is completely removed by FD, even in the best possible vacuum! At 170 K this would require about 10 days and 18 h, but at 191 K only 5 h. Except for very thin cryosections, it is therefore unrealistic to attempt FD at temperatures below 170 K.

Moreover, these data are valid for pure water only and represent the fastest possible drying time. In practice, with a biological specimen, FD will take longer, since (1) drying slows down when the water has to sublimate from the depth of the specimen as compared to its surface; (2) it is also slower if water has to sublimate from a mixed solution of salts and macromolecules as compared to a block of pure ice; and (3) as shown above (Sect. 2.1), a substantial part of the water in a biological specimen may be surface-modified and show a considerably lower evaporation rate (see also Robards and Sleytr 1985). In an elegant experiment this was demonstrated by Wildhaber et al. (1982), who followed with a mass spectrometer the escape of deuterium oxide from a test specimen during FD. A first peak of D_2O appeared between 178 K and 193 K, a second peak was observed only after further heating of the specimen (between 220 K and 235 K). Thus, at 190 K without subsequent warming a biological sample may never become fully dry.

2.4 Chemical Fixatives at Low Temperatures

As pointed out in Section 2.1, complete dehydration may lead to collapse of the ultrastructure of a biological specimen. Therefore, additional crosslinking by chemical fixatives may be desired before the specimen is fully dry. Alternatively, the specimen may be embedded at low temperatures (see Sects. 2.5 and 3.3).

Chemical fixation is applied quite differently in FD and FS. To avoid solvents, with FD only fixing vapours (e.g. OsO_4, formaldehyde) are admitted at a certain time, usually at room temperature after completion of drying. With FS, on the other hand, the fixatives are added to the solvent and are present throughout the substitution process. Therefore, stabilization of the ultrastructure may start at the lowest temperature at which the fixative is still reactive. Unfortunately, these temperatures are largely unknown. White et al. (1976) mentioned that OsO_4 may react with double bonds of unsaturated fatty acids at 200 K. Humbel (1984) found that the addition of uranyl acetate to methanol reduces the extraction of phospholipids from 9% to 2% at 203 K and from 15% to 4% at 243 K. Humbel et al. (1983) studied the crosslinking efficiency of glutaraldehyde using a 2% bovine serum albumin solution as test object. At temperatures of 263 and 243 K, complete crosslinking was observed after 1 and 4 h respectively. At lower temperatures, down to 223 K, crosslinking was still observed, but at a much reduced rate.

2.5 Low Temperature Embedding

Carlemalm and co-workers (1982) found that considerable denaturation and structural decomposition of proteins could be prevented, if chemically fixed specimens were dehydrated and embedded at subzero temperatures (see also Kellenberger, Chap. 2, this Vol.). They developed methacrylates (commercially available under the name Lowicryl) which allow the infiltration of specimens at temperatures down to 190 K and the polymerization by long-wave UV irradiation at these subzero temperatures.

Instead of chemical fixation at room temperature and dehydration at progressively lower temperatures (PLT technique), the low temperature for infiltration and polymerization can be reached from colder regimes after FS or FD. The rationale of this combination is to completely avoid higher temperatures during infiltration and, in particular, polymerization (usually done at $\geqq 330$ K).

Several studies have shown the feasibility of combining FS or FD with low temperature embedding (FS: Humbel et al. 1983; Hunziker et al. 1984; Weibull et al. 1984; Carlemalm et al. 1986; Engfeldt et al. 1986; Humbel and Müller 1986; FD: Wróblewsky et al. 1985; Chiovetti et al. 1986; Edelmann 1986b), but systematic studies as to the effects on fine structure and biological activity are still scarce. A start in this direction is the thesis of Humbel (1984).

3 Procedures and Instrumentation

3.1 Freeze-Substitution

3.1.1 Equipment

Small Eppendorf tubes, scintillation vials or culture tubes (glass tubes with screw caps) may be used as substitution vials. A temperature of 194 K (−79 °C) can be easily produced by submerging the tightly closed vials into dry ice which is mixed with acetone for better thermal contact. Slow rewarming after FS can be achieved in the simplest way by fitting the substitution vials into a large, precooled metal block. In such a device the temperature assymptotically approaches ambient temperature, the halftime depending on the volume of the metal block and its insulation.

More versatile equipment is necessary, if the effects of dehydration, stabilization by fixatives and the interaction with the embedding materials are to be systematically studied at various low temperatures. In this case, free choice of the substitution temperature and of any desired warming schedule is essential; UV polymerization at low temperature should be possible. Suitable equipment has been developed by Müller et al. (1980) and Sitte and co-workers (see Sitte 1984), and is now commercially available from Balzers and Reichert respectively. Both systems are based on liquid nitrogen (LN_2) cooling and can be programmed for largely automatized substitution schedules.

3.1.2 Substitution Media and Procedures

Acetone is now most widely used as a substitution medium for the study of the general morphology of cryofixed tissue. Using 2–4% OsO_4 as a fixing additive, excellent results have been obtained with a great variety of tissues (e.g. Ito and Ichikawa 1982; Akisaka and Shigenaga 1983; Benshalom and Reese 1985; Heath et al. 1985). Substitution is carried out by most workers at dry ice temperature (194 K) for several days, but the minimum time required for a given specimen is generally unknown. After substitution, slow warming (over 0.5–1 day) is recommended to allow enough time for the osmium tetroxide to react with the specimen at elevated subzero temperatures. At about 275 K the OsO_4/acetone is replaced by pure acetone, followed by infiltration with resin and polymerization as usual (for more details see van Harreveld and Crowell 1964; van Harreveld et al. 1965; Hereward and Northcote 1972; Bridgman and Reese 1984; Steinbrecht 1985).

Acetone can be made water-free by adding molecular sieve (e.g. Linde A4), but sometimes components go into solution which reduce the osmium tetroxide. Analytical grade acetone from freshly opened bottles can be used without molecular sieve. OsO_4 dissolves rapidly in acetone, but the solution is not stable at room temperature. It, therefore, should be freshly made and immediately cooled to the substitution temperature. Acrolein (5–10%) has been successfully used as an alternative to OsO_4, in particular, for the purpose of microanalysis or immunocytochemistry (Ornberg and Reese 1981; Bridgman and Reese 1984; Murata et al. 1985). Uranyl acetate or glutaraldehyde are inadequate as fixing additives in acetone, because the former is almost insoluble, while the latter would introduce too much water.

Methanol has the advantage that it can still substitute specimens in the presence of substantial amounts of water (see Sect. 2.3), it also substitutes much faster at low temperatures than acetone (Fig. 2). Zalokar (1966) used absolute methanol containing 20% acrolein. Müller et al. (1980) introduced methanol containing 0.5% uranyl acetate, 1% OsO_4, 3% glutaraldehyde and 3% water (the water being introduced by using a 50% aqueous glutaraldehyde solution). This complex medium is prepared as follows. In a flask (precooled by LN_2) 9 ml of 50% aqueous glutaraldehyde solution are mixed with 60 ml methanol, then 3 ml of a 20% (w/v) solution of uranyl acetate in methanol are added; in a second precooled flask 1.5 g osmium tetroxide are dissolved in 75 ml methanol; both flasks are cooled to about 220 K, their contents poured together and vigorously shaken. The mixture is highly reactive and even at 240 K only stable for a few hours. Storage is possible under LN_2 in a Teflon bottle (CAUTION: Do not trap air bubbles during freezing and do not tightly close the bottle! Danger of explosion due to expanding nitrogen upon rewarming!).

Müller et al. (1980) recommended a stepwise substitution protocol starting at 178 K (−95 °C) and continuing at 213 K (−60 °C) and 243 K (−30 °C) to allow for the reaction of the fixing additives with the specimen. Eight hours for each step proved sufficient in many cases. After 30 min at 273 K (0 °C) the medium is replaced by pure acetone of the same temperature, then the specimens are warmed to 280 K (+7 °C), and infiltrated with araldite/Epon. Polymerization is accomplished at 333 K (+60 °C).

Diethylether has been used as a substitution medium mainly in attempts to retain the original distribution of water-soluble substances, e.g. for the autoradiography of water-soluble sugars (Läuchli et al. 1970). Van Zyl et al. (1976) and Harvey (1982) recommended diethylether containing 20% acrolein for microanalysis of electrolyte elements, mostly in plant material (see also Sect. 4.2). The use of molecular sieve and a very slow rewarming protocol is advisable with these highly apolar solvents.

Ethanol (containing 3.5% glutaraldehyde) was used by Nagele et al. (1985) for cytochemical purposes. Other solvents, such as 1-butanol, 1-hexanol, n-hexane, propylene oxide, tetrahydrofuran or ethylene glycol have only rarely been used, mostly without convincing success (for refs. see Harvey 1982).

Regardless of the solvent used, the volume of the substitution medium should always exceed the specimen volume by at least a factor of 1000, and the vials should be slightly agitated from time to time.

3.2 Freeze-Drying

3.2.1 Equipment

Freeze-drying equipment for biological electron microscopy should fulfill the following demands:

1. Precise control of the specimen temperature, in particular, between 180 and 273 K.
2. An environment around the specimen with a minimal partial pressure of water.
3. Protection from uncontrolled rehydration during the procedures following FD.
4. Versatility of combination with various subsequent procedures, e.g. vapour fixation and vacuum infiltration with resin for thin-sectioning techniques or metal coating for SEM.

Commercial lyophilizers as used in biochemistry usually operate at temperatures above 220 K and, therefore, are not adequate for fine structural work. Most freeze-etching equipment, however, can be easily adapted for FD. A very simple device was used by van Harreveld and Malhotra (1966). Other equipment of various degrees of sophistication has been described, e.g. by Coulter and Terracio

(1977), Pfaller and Rovan (1978), Ingram and Ingram (1984), Lyon et al. (1985) and Linner et al. (1986). Stumpf and Roth (1967) introduced cryosorption pumping to FD, and Edelmann (1978) developed a system which completely avoids rotary and diffusion pumps. A sealed container with a heatable specimen support is immersed in LN_2; molecular sieve inside the container produces a vacuum of about 10^{-3} mbar, but the partial pressure of water is much lower due to the cryosorption effect.

3.2.2 Procedure

Since FD of larger specimens would lead to inappropriately long drying times, usually only the outer zones, which are also those with a minimum of freezing damage, are dried at the lowest temperature ($\leqq$190 K). Then the temperature is progressively raised to accelerate drying of the more interior parts. It is important that the specimen is in good thermal contact with the specimen stage, and that it is protected from radiant heat; otherwise specimen temperature may seriously deviate from the stage temperature. Conservative protocols use drying times of 1–3 days at 190 K (−83 °C), and warm the specimen stepwise or gradually over several days to room temperature (Edelmann 1978, Ingram and Ingram 1984). Occasionally, drying rates are reported in the literature which even exceed the theoretical values for pure water (see Sect. 2.3). In these cases, the specimen must have been warmer than the recorded stage temperature (poor thermal contact) and/or FD had mainly occurred during the warming-up period.

Treatment of the dry specimens with OsO_4 vapours increases the contrast and may help to stabilize the fine structure prior to embedding. This is, however, not always essential for good morphological results (Edelmann 1986b); moreover, osmium might interfere with microanalysis. Rehydration of the dry specimens must be avoided by all means. Therefore, the admission of fixing vapours or resin monomer should be accomplished without venting the vacuum chamber unless dry nitrogen gas is used (Van Harreveld and Malhotra 1966; Edelmann 1978; Pfaller 1979). Spurr's low viscosity epoxy resin is recommended for infiltration by several authors. Although most authors perform resin infiltration under vacuum, Edelmann (1978) reported better preservation of muscle fine structure after infiltration at atmospheric pressure.

3.3 Low Temperature Embedding After Freeze-Substitution and Freeze-Drying

Two low temperature embedding resins have been mostly used in combination with FS and FD: Lowicryl HM20, an apolar resin which remains fairly liquid down to 220 K, and Lowicryl K4M, a polar resin which becomes rather viscous below 240 K. Recently, two more Lowicryls have been introduced, the apolar HM23 and the polar K11M (Carlemalm et al. 1986). Due to the reduced viscosity of these resins, low temperature embedding is now possible down to 190 K.

Equipment for low temperature embedding and the procedure following FS has been described in detail by Humbel and Müller (1986). Warming-up is stopped at the desired infiltration temperature (e.g. 243 K for HM20 and K4M), the substitution medium is exchanged for pure solvent (methanol or acetone) and then for increasing concentrations of Lowicryl in that solvent (30%, 70%, 100%, for at least 2 h each); after a further change of Lowicryl, polymerization is accomplished by indirect UV irradiation (wavelength 360 nm) for 20 h at 243 K and completed at room temperature for 2 more days. The procedure can also be carried out in the commercial FS apparatus of Balzers and of Reichert.

Following FD, Wróblewski et al. (1985) transferred the specimens into Lowicryl HM20 within the vacuum at 243 K; polymerization was initiated in the cold by UV light as described above. Edelmann (1986b) recently succeeded in embedding specimens at 213 K into Lowicryl K11M. After completion of FD, the cold specimens were transferred into precooled resin within the cold chamber of the Reichert FS apparatus at 213 K under a constant stream of cold nitrogen gas, and then polymerized by UV light in the cold.

3.4 Freeze-Drying and Freeze-Substitution of Bulk Specimens for Scanning Electron Microscopy (SEM)

For high resolution SEM the same criteria of good cryofixation are applicable as for conventional TEM; for low resolution SEM, however, the demands are less exacting, the main goal being the preservation of the natural dimensions. Thus, in the case of larger specimens, chemical prefixation may be inevitable, which in terms of overall shrinkage, usually does not produce as severe artefacts as the subsequent dehydration (Lee 1984).

Classical FD from water is described in Section 3.2. FD from an organic solvent such as amyl acetate (Boyde and Wood 1969), tertiary butanol (Wheeler et al. 1975) or Freon 113 (Boyde and Maconnachie 1979) has been proposed for larger specimens. The FD time is considerably reduced due to the high volatility of these solvents.

FS must be followed by FD (see above) or by critical-point drying (CPD) to produce a SEM specimen. Fixing agents, in particular, OsO_4 and $HgCl_2$, not only stabilize the specimen so that shrinkage is reduced during subsequent CPD, but also help to avoid charging in the microscope by metal impregnation. Barlow and Sleigh (1979) systematically tested various FS media for stabilizing the metachronal wave of the cilia of *Paramecium* in the SEM; they recommended a mixture of methanol, ethylene glycol and acrolein [50/40/10% (v)] containing 8% $HgCl_2$ and 3% $CaCl_2$. Bridgman and Reese (1984) optimized FS and CPD of whole tissue culture cells for high voltage electron microscopy. They observed the least shrinkage after FS in acetone containing 10% acrolein and 0.2% tannic acid at 193 K, followed by acetone containing 0.2% OsO_4 at 252 K and finally by a glutaraldehyde/methanol/acetone mixture [10/40/50% (v)] at 273 K. The effects of the various fixatives and stains are discussed in detail.

Rehydration of the dry specimens during transfer to the metal-coating unit or the microscope is an inherent danger and has to be carefully avoided. Specimen preparation for high resolution SEM has been reviewed by Tanaka (1981) and Inoué (1986), metal coating by Peters (1986).

4 Critical Evaluation

Generally, the influence of any preparation technique on the information content extractable from a specimen should be evaluated with respect to:

1. the preservation of structure (shape, dimensions);
2. the preservation of the spatial distribution of diffusible components (ions, small macromolecules);
3. the preservation of biological activity (antigenicity, enzyme reactivity).

It is now common consensus that the most crucial step of *any* cryotechnique is a good cryofixation. Only if the primary artefacts of freezing damage are reduced to a minimum, can the secondary artefacts of FS and FD be evaluated. Criteria of good cryofixation are given by Sitte et al. (Chap. 4, this Vol.) and Dubochet et al. (Chap. 5, this Vol.); in this chapter we shall concentrate on the effects of the follow-up procedures on specimen preservation.

4.1 Structural Effects

Freeze-drying *bulk specimens* for SEM is now widely accepted as the gentlest dehydration technique for delicate objects, preferable to standard drying or critical-point drying (CPD). Shrinkage of glutaraldehyde-fixed liver after CPD may amount to 60% by volume (20% linear), whereas after FD it is reduced to 20% by volume (7% linear) (Boyde et al. 1977, Boyde and Franc 1981). Observation of frozen-hydrated samples (see also Marshall, Chap. 13, this Vol.) in a scanning electron microscope equipped with a cryostage may serve as a reference standard; in addition, the process of FD may be directly observed. However, non-volatile impurities or salts will deposit on the specimen surface, unless FD is performed out of distilled water or pure organic solvents.

Successful application of FS followed by CPD has been shown, for example by Barlow and Sleigh (1979) or Colwell and Kokko (1986). Surface glycoproteins on yeast cells could be demonstrated by SEM only after FD (Käppeli et al. 1984), but not after conventional fixation and CPD (Pringle et al. 1979). Osatake et al. (1985) applied FS with acetone/OsO_4 prior to OsO_4 maceration (Tanaka 1981) for high resolution SEM in cytology. FD and FS have also been successfully used to study the cytoskeleton of whole cultured cells in the high voltage electron microscope (Porter and Anderson 1982; Bridgman and Reese 1984).

The major merit of FS and FD is doubtlessly the applicability of *routine ultramicrotomy* to cryofixed material. There is, at present, no other method which allows oriented serial sectioning with comparable accuracy and reproducibility. When performed correctly, both methods, FS and FD, may produce very similar results (for minor differences, see below), which may, however, be highly different from the outcome of conventional chemical fixation (e.g. van Harreveld et al. 1965, van Harreveld and Malhotra 1966). The specific merits of cryofixation as compared to chemical fixation are discussed elsewhere in this volume (Knoll et al., Chap. 14; Zierold and Steinbrecht, Chap. 15; see also Plattner and Bachmann

1982; Menco 1986). Here, we only recall that FS and FD contributed substantially to the successful application of cryofixation to electron microscopy.

Specimens with rather impermeable walls sometimes cannot at all be well preserved by chemical means, but easily by cryofixation, if they are small enough, e.g. fungus hyphae (Howard and Aist 1979; Hippe 1985), insect antennae (Fig. 3) (Steinbrecht 1980) or spores and other permanent stages of cells (Müller and Moor 1984, Brand and Arnold 1986). FD is usually seriously impeded in such specimens, sometimes even FS (plant material), unless artificial cracks in the cell walls allow access of the solvent.

Good structural correspondence between living and freeze-substituted plant cytoplasm has been reported by McCully and Canny (1985). Wherever freeze-fracturing was carried out in parallel to FS and FD, excellent morphological correspondence has been observed (FS: Dempsey and Bullivant 1976a, b; Steinbrecht 1976, 1980; Browning and Gunning 1977; Franzini-Armstrong et al. 1978; Ebersold et al. 1981; Usukura and Yamada 1981; Hunziker et al. 1984; Nassar et al. 1986. FD: Pfaller et al. 1976; Pfaller and Rovan 1978; Pfaller 1979). Thus, the smooth contours of cell membranes, to mention only one of the most prominent aspects of freeze-substituted tissues, is definitely not a secondary artefact.

Only in regions of major freezing damage were discrepancies between FS and freeze-etch replicas observed. In particular, cytomembranes perforated by ice crystals during cryofixation may be totally lost after FS (Steinbrecht 1980). There are, however, minor differences between FS and FD, and also between different schedules in each procedure, which are too subtle for a straightforward evaluation by referring to freeze etch replicas. Already van Harreveld and Malhotra (1966) mentioned the more frothy appearance of freeze-dried ground cytoplasm as compared to freeze-substituted specimens of rat cerebellum. A white halo (shrinkage zone?) around mitochondria is often observed after FD (Coulter and Terracio 1977; Linner et al. 1986), but apparently can be avoided by prolonged FD (Edelmann 1986b). Myosin and, in particular, actin filaments of skeletal muscle are somewhat perturbed even after optimal FD, but not after FS (Fig. 4) (Edelmann 1986b and personal comm.). Generally, the "nicer picture" after FS is preferred, but we do not know which step in the procedure caused the deterioration in freeze-dried specimens: the dehydration, the action of the chemical fixative, the resin infiltration or the polymerization.

Systematic studies on the effect of different solvents and fixing additives have been carried out by Bridgman and Reese (1984, see also Sect. 3.4), Humbel (1984), Ikeda et al. (1984) and Murata et al. (1985). Considering the highly different specimens and goals of study, an "universal substitution medium" should not be expected. It is noteworthy, however, that all these studies emphasize the importance of chemical fixatives for optimal preservation of fine structure and that in each case optimization could be achieved by systematic experiments. For example, Ikeda et al. (1984) found that 4% OsO_4 in acetone was best for preserving the homogeneous structure of the lipidic inclusion bodies in pneumocytes, while Humbel (1984) showed the advantages of Müller's complex medium (see Sect. 3.1.2) for FS of lymphoma cells in combination with low temperature embedding (Fig. 5).

The convenience of normal ultramicrotomy after FD and FS is an obvious advantage for morphometric work on specimens amenable to cryofixation (e.g.

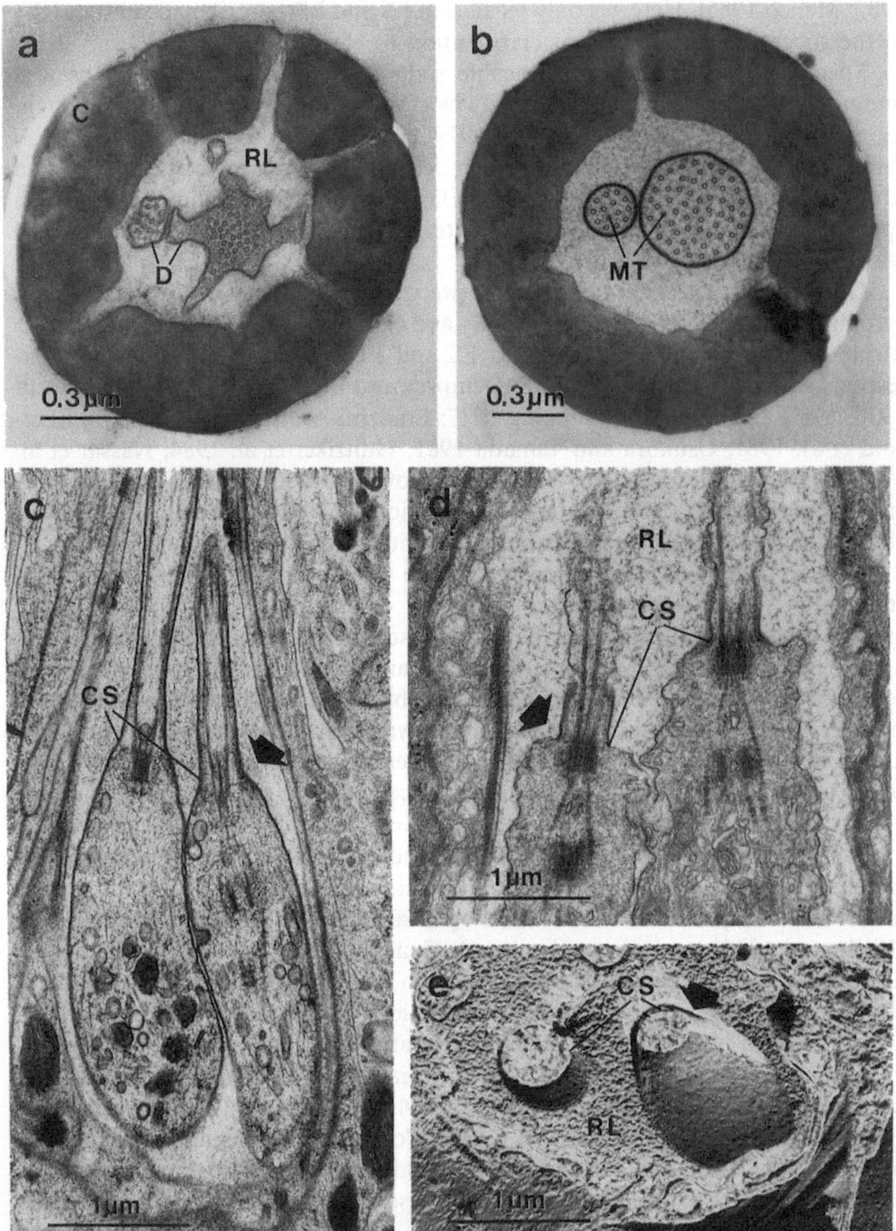

Fig. 3 a–e. Comparison of conventional chemical fixation (**a, d**), freeze-substitution (**b, c**) and freeze-etching (**e**). Insect cuticle is a serious barrier to chemical fixatives; but small sensory hairs (**a, b**) and antennal branches (**c–e**), e.g. of the silk moth *Bombyx mori*, can be easily cryofixed without any chemical pretreatment by rapid immersion into liquid propane (**b, c, e**). Note the different aspect of the sensory dendrites (*D*) and of the ciliary segments (*CS*) of the olfactory receptor cells in **a, b** and **c, d** respectively, but also the good correspondence between freeze-substituted (**c**) and freeze-etched (**e**) ciliary segments (*arrow*). *C* Cuticle; *MT* microtubules; *RL* receptor lymph (**a–d** Steinbrecht 1984; **e** Steinbrecht 1980)

Fig. 4a, b. Comparison between freeze-substituted (**a**) and freeze-dried (**b**) frog sartorius muscle. **a** The specimen was frozen by immersion into liquid propane. FS in acetone (+3% OsO_4) for 2 days at 193 K, warming at a rate of 4 K h^{-1} to room temperature, embedding into Spurr's resin, section stained with uranyl acetate and lead citrate. **b** Rapid freezing by impact on LN_2-cooled Cu block, FD 3 days at 193 K, then 6 days at 213 K as described by Edelmann (1978, 1986b); no chemical fixation, embedding into Spurr's resin and section staining as in **a** (Edelmann 1986b)

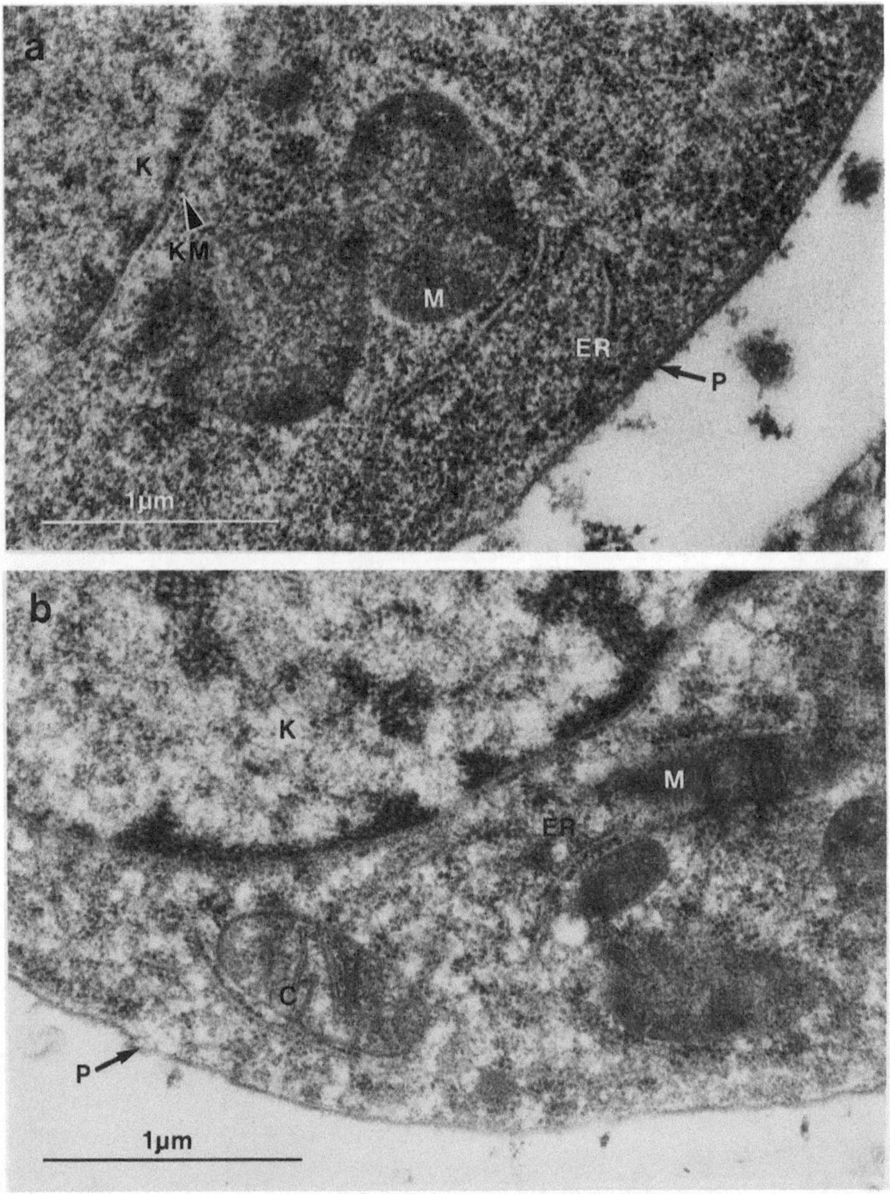

Fig. 5 a, b. Mouse lymphoma cells fixed by propane jet, freeze-substituted and embedded in Lowicryl K4M at 243 K. **a** FS in pure methanol (8 h at 178 K, 8 h at 213 K, 8 h at 243 K); **b** FS in complex medium (Müller et al. 1980), schedule as above. *C* Mitochondrial cristae; *ER* endoplasmatic reticulum; *K* nucleus; *KM* nuclear envelope; *M* mitochondrion; *P* plasmalemma (Humbel 1984)

Gnatzy et al. 1984; Menco 1984). The dimensional changes during the dehydration process, however, have been studied in detail only for FD. Volume shrinkage by 20–30% was measured by different methods (Boyde et al. 1977; Boyde and Franc 1981; Ingram and Ingram 1984). Shrinkage during FS should be less, but Menco (1984) measured the diameter of olfactory cilia after FS to be 20% narrower than after freeze-etching, the morphology being unchanged. On the other hand, microtubules in silk moth olfactory dendrites had the same spacing after FS and freeze-etching (R. A. Steinbrecht, unpubl.). More data of this kind are highly desirable.

Summing up the present state of knowledge, FS appears preferable to FD for pure morphological and morphometric work, because of:

1. easier and simpler performance;
2. the possibility of additional chemical fixation at low temperature;
3. less problematic resin infiltration;
4. usually a better preservation of fine structure.

FD, however, may have advantages over FS, when the retention of soluble elements is concerned (see below).

4.2 Effects on Diffusible Specimen Components

How well can soluble tissue constituents be retained in their natural position during FS and FD? Again the dehydration process itself must not be seen isolated from the following steps of resin embedding and sectioning. FD, for example, avoids organic solvents during dehydration, but the monomer of the embedding resin is an organic solvent as well. There is no liquid water during dehydration, but some resins produce water during polymerization (Marshall 1980). Eventually, water-soluble substances may be lost by sectioning on a water surface (Marshall 1980; Harvey 1982). Therefore, for the retention of water-soluble components careful choice of the resin and sectioning with a dry knife are essential.

Extraction of soluble lipids occurs easily during FS. With methanol as solvent, improvements were observed by adding uranyl acetate (Humbel 1984; see Sect. 2.4) or by using the complex medium of Müller et al. (1980). The latter even enabled the visualization of liposomes, pure lipid vesicles, which cannot be observed after conventional preparation techniques (Müller 1981). Further progress was achieved by combining this FS schedule with low temperature embedding into Lowicryl HM20 (Verkleij et al. 1985). Weibull et al. (1984) found that at 183 K pure methanol extracted much more lipids than pure acetone; 95% of the cellular lipids of the bacterium *Acholeplasma laidlawii* were retained after FS in acetone at 183 K followed by low temperature embedding at 203 K. Lipidic precursors of lung surfactant in pneumocytes could be demonstrated in an autoradiographic study after FS in acetone containing 4% OsO_4 (Ikeda 1985).

Water-soluble components may be identified by X-ray microanalysis in the case of ions or salts (see Zierold and Steinbrecht, Chap. 15, this Vol.), or by autoradiography after radioactive labelling (see Frederik, Chap. 12, this Vol.); by the latter method also sugars, amino acids and other organic compounds can be detected. There is some tradition in botany to use diethylether – sometimes with 20% acrolein added – for FS to prevent extraction or translocalization of such water-soluble components, but comparatively little success has been reported for animal tissues (see reviews by Marshall 1980, and Harvey 1982). Van Zyl et al. (1976) found no significant losses of ^{22}Na, ^{86}Rb and ^{36}Cl after FS in diethylether containing 20% acrolein, in contrast to considerable losses after FS with polar solvents. However, lack of loss of a substance does not ensure lack of displacement within the tissue (see also the round table discussion compiled by Echlin 1978, or the discussion with the reviewers in Marshall 1980, and Ornberg and Reese 1981). Retention of the natural distribution of calcium by FS in acetone is reported for frog skeletal muscle (acetone + 4% OsO_4 + 20 mM oxalic acid, Ornberg and Reese 1980) and for toad retinal rods (acetone + 0.5% OsO_4, Fain and Schröder 1985).

Freeze-drying avoids extraction by the substitution medium, but not that caused by the infiltrating resin monomer. Several workers claim the in situ retention of diffusible substances by FD, mainly in muscle (Ingram and Ingram 1984; Meyer et al. 1985; Wróblewsky et al. 1985; Edelmann 1986a, b). On the other hand, up to 30% extraction of potassium was observed in dry-cut plastic sections of freeze-dried pancreas cells as compared to ultrathin cryosections (Roos and Barnard 1985).

Since even minor changes in the procedure may lead to failure, it is essential that microanalytical data based on FS or FD are verified by results obtained from cryosections. After all, cutting ultrathin cryosections is no longer very difficult with modern equipment, it may even turn out easier than cutting thin dry plastic sections.

4.3 Effects on Antigenicity and Enzyme Activity

In the frozen state, enzyme activity comes to a halt, but the configuration and also the antigenicity of protein macromolecules may be well preserved. If FS or FD is followed by resin embedding, only post-section staining techniques are feasible, which exclude classical enzyme cytochemistry. The FS process per se, however, need not destroy enzyme activity, as has been shown by Klaushofer and von Mayersbach (1979).

Immunocytochemical methods have been successfully applied, usually after FS in pure acetone or methanol, or after FD without vapour fixation with OsO_4. Nagele et al. (1985) localized actin, myosin and tropomyosin in whole glia cells from tissue cultures by immunofluorescence labelling after FS in ethanol containing 3–5% glutaraldehyde at 183 K, followed by rehydration through graded ethanol steps at 277 K. Murata et al. (1985) successfully performed polysaccharide cytochemistry and immunostaining of pepsinogen on Epon sections of rat stomach after FS in acetone containing 5% acrolein. Occasionally, immunolabelling could even be performed after FS in OsO_4/acetone (Inoué et al. 1982).

The theoretical advantages of low temperature embedding after FS or FD have not yet been exploited for immunocytochemistry on a larger scale. Mostly, the PLT technique (dehydration at progressively lower temperatures after "mild"

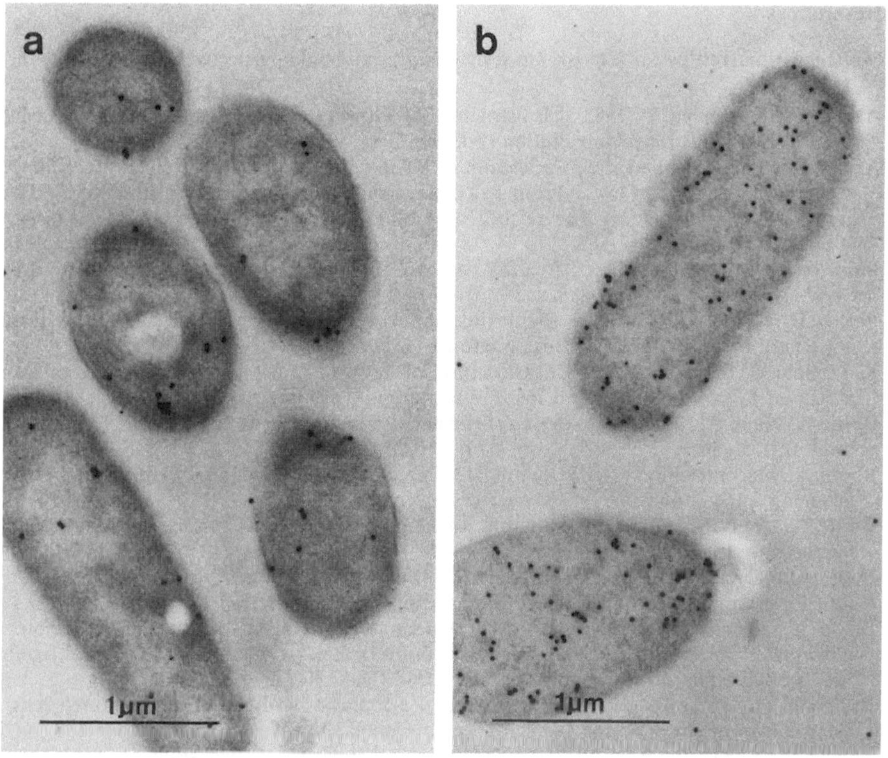

Fig. 6a, b. Comparison of immunocytochemical labelling of the protein groE in *E. coli* cells. **a** After conventional formaldehyde fixation, PLT dehydration and low temperature embedding in Lowicryl K4M at 235 K; **b** after freeze-substitution in acetone containing formaldehyde at 190 K and low temperature embedding at 235 K as above. Note the clearly higher labelling intensity of the cells in **b** (Carlemalm et al. 1986)

glutaraldehyde fixation) is applied (reviewed by Armbruster et al. 1984). However, promising results have been reported by Humbel et al. (1983; see also Humbel 1984): e.g. labelling of M-line protein in muscle cells which were embedded in K4M at 243 K after FS in pure methanol. Carlemalm et al. (1986) clearly demonstrated an increased labelling of cytoplasmatic proteins in *E. coli* cells after FS as compared to PLT preparation (both specimens were embedded in K4M at 235 K; Fig. 6). It is to be hoped that the new resins, Lowicryl K11M and HM23, which have a much reduced viscosity at low temperatures as compared to K4M and HM20, will foster immunocytochemistry of cryofixed specimens after FS and FD.

Acknowledgements. Our sincere thanks are due to Drs. J.-D. Acetarin and E. Carlemalm for allowing us to quote from their unpublished data, to Drs. E. Carlemalm, L. Edelmann and B. Humbel, as well as to their publishers for the permission to reproduce figures to illustrate this chapter. We are grateful to G. Lamprecht, B. Müller and I. S. Rössel for help in preparing the manuscript.

References

(References marked by an asterisk are review papers or books respectively)

Akisaka T, Shigenaga Y (1983) Ultrastructure of growing epiphyseal cartilage processed by rapid freezing and freeze-substitution. J Electron Microsc 32:305–320

*Armbruster BL, Kellenberger E, Carlemalm E, Villiger W, Garavito RM, Hobot JA, Chiovetti R, Acetarin J-D (1984) Lowicryl resins – present and future applications. In: Revel J-P, Barnard T, Haggis GH (eds) The science of biological specimen preparation. SEM, AMF O'Hare, IL 60666, pp 77–81

Bank H (1973) Visualization of freezing damage. II. Structural alterations during warming. Cryobiology 10:157–170

Barlow DI, Sleigh MA (1979) Freeze substitution for preservation of ciliated surfaces for scanning electron microscopy. J Microsc (Oxford) 115:81–95

Benshalom G, Reese TS (1985) Ultrastructural observations on the cytoarchitecture of axons processed by rapid-freezing and freeze-substitution. J Neurocytol 14:943–960

Boyde A, Franc F (1981) Freeze-drying shrinkage of glutaraldehyde fixed liver. J Microsc (Oxford) 122:75–86

Boyde A, Maconnachie E (1979) Freon 113 freeze-drying for scanning electron microscopy. Scanning 2:164–166

Boyde A, Wood C (1969) Preparation of animal tissues for surface-scanning electron microscopy. J Microsc (Oxford) 90:221–249

Boyde A, Bailey E, Jones SJ, Tamarin A (1977) Dimensional changes during specimen preparation for scanning electron microscopy. Scanning Electron Microsc 1977/I:507–518

Brand N, Arnold C-G (1986) Improved structural preservation of the mature zygote of *Chlamydomonas reinhardii* by freeze-substitution fixation compared with chemical fixation with special attention to the mitochondria. Endocyt C Res 3:79–95

Bridgman PC, Reese TS (1984) The structure of cytoplasm in directly frozen cultured cells. I. Filamentous meshworks and the cytoplasmic ground substance. J Cell Biol 99:1655–1668

Browning AJ, Gunning BES (1977) An ultrastructural and cytochemical study of the wall-membrane apparatus of transfer cells using freeze-substitution. Protoplasma 93:7–26

*Bullivant S (1970) Present status of freezing techniques. In: Parsons DF (ed) Some biological techniques in electron microscopy. Academic Press, London New York, pp 101–146

Carlemalm E, Garavito RM, Villiger W (1982) Resin development for electron microscopy and an analysis of embedding at low temperature. J Microsc (Oxford) 126:123–143

*Carlemalm E, Villiger W, Acetarin J-D, Kellenberger E (1986) Low temperature embedding. In: Müller M, Becker RP, Boyde A, Wolosewick JJ (eds) The science of biological specimen preparation 1985. SEM, AMF O'Hare, IL 60666, pp 147–154

Chiovetti R, Little SA, Brass-Dale J, McGuffee LJ (1986) A new approach to low temperature embedding: quick freezing, freeze-drying and direct infiltration in Lowicryl K4M. In: Müller M, Becker RP, Boyde A, Wolosewick JJ (eds) The science of biological specimen preparation 1985. SEM, AMF O'Hare, IL60666, pp 155–164

*Clegg JS (1979) Metabolism and the intracellular environment: the vicinal-water network model. In: Drost-Hansen W, Clegg JS (eds) Cell-associated water. Academic Press, London New York, pp 363–413

Colwell DD, Kokko EG (1986) Preparation of dipteran larvae for scanning electron microscopy using a freeze-substitution technique. Can J Zool 64:797–799

Coulter HD, Terracio L (1977) Preparation of biological tissues for electron microscopy by freeze-drying. Anat Rec 187:477–494

Dempsey GP, Bullivant S (1976a) A copper block method for freezing non-cryoprotected tissue to produce ice-crystal-free regions for electron microscopy. I. Evaluation using freeze-substitution. J Microsc (Oxford) 106:251–260

Dempsey GP, Bullivant S (1976b) A copper block method for freezing non-cryoprotected tissue to produce ice-crystal-free regions for electron microscopy. II. Evaluation using freeze-fracturing with a cryo-ultramicrotome. J Microsc (Oxford) 106:261–271

Ebersold HR, Lüthy P, Cordier JL, Müller M (1981) A freeze-substitution and freeze-fracture study of bacterial spore structures. J Ultrastruct Res 76:71–81

Echlin P (ed) (1978) Specimen preparation techniques (Round table discussion). In: Echlin P, Kaufmann R (eds) Microprobe analysis in biology and medicine. Microsc Acta Suppl 2. Hirzel, Stuttgart, pp 11–45

Edelmann L (1978) A simple freeze-drying technique for preparing biological tissue without chemical fixation for electron microscopy. J Microsc (Oxford) 112:243–248

Edelmann L (1986a) Two opposing theories of the cell: experimental testing by cryomethods and electron microscopy. In: Müller M, Becker RP, Boyde A, Wolosewick JJ (eds) The science of biological specimen preparation 1985. SEM, AMF O'Hare, IL 60666, pp 33–42

*Edelmann L (1986b) Freeze-dried embedded specimens for biological microanalysis. Scanning Electron Microsc 1986/IV:1337–1356

Engfeldt B, Hultenby K, Müller M (1986) Ultrastructure of hyaline cartilage. Acta Pathol Microbiol Immunol Scand Sect A 94:313–323

Fain GL, Schröder WH (1985) Calcium content and calcium exchange in dark-adapted toad rods. J Physiol 368:641–665

Franzini-Armstrong C, Heuser JE, Reese TS, Somlyo AP, Somlyo AV (1978) T-tubule swelling in hypertonic solutions: a freeze substitution study. J Physiol 283:133–140

Frederik PM, Busing WM, Hax WMA (1984) Observations on frozen hydrated and drying thin cryosections. In: Csanády A, Röhlich P, Szabó D (eds) Electron microscopy 1984, vol II. Proc 8th Eur Congr Electron Microsc, Budapest, pp 1411–1412

Gnatzy W, Mohren W, Steinbrecht RA (1984) Pheromone receptors in *Bombyx mori* and *Antheraea pernyi*. II. Morphometric analysis. Cell Tissue Res 235:35–42

*Harvey DMR (1982) Freeze-substitution. J Microsc (Oxford) 127:209–221

Heath B, Rethoret K, Arsenault AL, Ottensmeyer FP (1985) Improved preservation of the form and contents of wall vesicles and the Golgi apparatus in freeze substituted hyphae of *Saprolegnia*. Protoplasma 128:81–93

Hereward FV, Northcote DH (1972) A simple freeze-substitution method for the study of ultrastructure of plant tissues. Exp Cell Res 70:73–80

Hippe S (1985) Ultrastructure of Haustoria of *Erysiphe graminis f. sp. hordei* preserved by freeze-substitution. Protoplasma 129:52–61

Howard RJ, Aist JR (1979) Hyphal tip cell ultrastructure of the fungus Fusarium: improved preservation by freeze-substitution. J Ultrastruct Res 66:224–234

Humbel BM (1984) Gefriersubstitution – Ein Weg zur Verbesserung der morphologischen und zytochemischen Untersuchung biologischer Proben im Elektronenmikroskop. Diss, Eidgen Tech Hochschule Zürich

Humbel BM, Müller M (1986) Freeze substitution and low temperature embedding. In: Müller M, Becker RP, Boyde A, Wolosewick JJ (eds) The science of biological specimen preparation 1985. SEM, AMF O'Hare, IL 60666, pp 175–183

Humbel BM, Marti T, Müller M (1983) Improved structural preservation by combining freeze substitution and low temperature embedding. Beitr Elektronenmikrosk Direktabb Oberfl 16:585–594

Hunziker EB, Herrmann W, Schenk RK, Mueller M, Moor H (1984) Cartilage ultrastructure after high pressure freezing, freeze substitution, and low temperature embedding. I. Chondrocyte ultrastructure – implications for the theories of mineralization and vascular invasion. J Cell Biol 98:267–276

Ikeda H (1985) An ultrastructural analysis of the inclusion body in the type II pneumocyte processed by rapid freezing followed by freeze-substitution – an autoradiographic study. J Electron Microsc 34:398–410

Ikeda H, Ichikawa A, Ichikawa M (1984) The effects of freeze-substitution media on the ultrastructure of inclusion-bodies in type II pneumocytes of mouse lung processed by the cryofixation method. J Electron Microsc 33:242–247

Ingram FD, Ingram MJ (1984) Influences of freeze-drying and plastic embedding on electrolyte distributions. In: Revel J-P, Barnard T, Haggis GH (eds) The science of biological specimen preparation. SEM, AMF O'Hare, IL 60666, pp 167–174

*Inoué T (1986) High resolution scanning electron microscopic cytology – specimen preparation and intracellular structures observed by scanning electron microscopy. In: Müller M, Becker RP, Boyde A, Wolosewick JJ (eds) The science of biological specimen preparation 1985. SEM, AMF O'Hare, IL 60666, pp 245–256

Inoue K, Kurosumi K, Deng ZP (1982) An improvement of the device for rapid freezing by use of liquid propane and the application of immunocytochemistry to the resin section of rapid frozen, substitution-fixed anterior pituitary gland. J Electron Microsc 31:93–97
Ito T, Ichikawa A (1982) An electron microscopic observation of rat peritoneal cells processed by rapid freezing and freeze-substitution fixation. J Electron Microsc 31:235–248
Käppeli O, Walther P, Müller M, Fiechter A (1984) Structure of the cell surface of the yeast *Candida tropicalis* and its relation to hydrocarbon transport. Arch Microbiol 138:279–282
Klaushofer K, Mayersbach H von (1979) Freeze substituted tissue in 5′-nucleotidase histochemistry. J Histochem Cytochem 27:1582–1587
Läuchli A, Spurr AR, Wittkopp RW (1970) Electron probe analysis of freeze-substituted, epoxy resin embedded tissue for ion transport studies in plants. Planta 95:341–350
*Lee RMKW (1984) A critical appraisal of the effects of fixation, dehydration and embedding on cell volume. In: Revel J-P, Barnard T, Haggis GH (eds) The science of biological specimen preparation. SEM, AMF O'Hare, IL 60666, pp 61–70
Linner JG, Bennett SC, Harrison DS, Steiner AL (1986) Cryopreparation of tissue for electron microscopy. In: Müller M, Becker RP, Boyde A, Wolosewick JJ (eds) The science of biological specimen preparation 1985. SEM, AMF O'Hare, IL 60666, pp 165–174
Lyon R, Appleton J, Swindin KF, Abbot JJ, Chesters J (1985) An inexpensive device for freeze drying and plastic embedding tissues at low temperatures. J Microsc (Oxford) 140:81–91
MacKenzie AP (1972) Freezing, freeze-drying, and freeze-substitution. Scanning Electron Microsc 1972/II:273–280
*Marshall AT (1980) Freeze-substitution as a preparation technique for biological X-ray microanalysis. Scanning Electron Microsc 1980/II:395–408
McCully ME, Canny MJ (1985) The stabilization of labile configurations of plant cytoplasm by freeze-substitution. J Microsc (Oxford) 139:27–33
Menco BPM (1984) Ciliated and microvillous structures of rat olfactory and nasal respiratory epithelia. Cell Tissue Res 235:225–241
*Menco BPM (1986) A survey of ultra-rapid cryofixation methods with particular emphasis on applications to freeze-fracturing, freeze-etching, and freeze-substitution. J Electron Microsc Tech 4:177–240
Meyer R, Schmitz M, Zierold K (1985) The influence of different cryopreparations on the distribution of ions in bullfrog myocard cells. Scanning Electron Microsc 1985/I:419–431
Müller M (1981) Demonstration of liposomes by electron microscopy. In: Azzi A, Brodbeck U, Zahler P (eds) Membrane proteins. Springer, Berlin Heidelberg New York, pp 252–256
Müller M, Moor H (1984) Cryofixation of thick specimens by high pressure freezing. In: Revel J-P, Barnard T, Haggis GH (eds) The science of biological specimen preparation. SEM, AMF O'Hare, IL 60666, pp 131–138
Müller M, Marti T, Kriz S (1980) Improved structural preservation by freeze substitution. In: Brederoo P, de Priester W (eds) Electron microscopy 1980, vol II. Proc 7th Eur Congr Electron Microsc, Leiden, pp 720–721
Murata F, Suzuki S, Tsuyama S, Suganuma T, Imada M, Furihata C (1985) Application of rapid freezing followed by freeze-substitution acrolein fixation for cytochemical studies of the rat stomach. Histochem J 17:967–980
Nagele RG, Kosciuk MC, Wang SM, Spero DA, Lee H (1985) A method for preparing quick-frozen, freeze-substituted cells for transmission electron microscopy and immunocytochemistry. J Microsc (Oxford) 139:291–301
Nassar R, Wallace NR, Taylor I, Sommer JR (1986) The quick freezing of single intact skeletal muscle fibers at known time intervals following electrical stimulation. Scanning Electron Microsc 1986/I:309–328
*Negendank W (1986) The state of water in the cell. In: Müller M, Becker RP, Boyde A, Wolosewick JJ (eds) The science of biological specimen preparation 1985. SEM, AMF O'Hare, IL 60666, pp 21–32
Nei T (1973) Growth of ice crystals in frozen specimens. J Microsc (Oxford) 99:227–233
Ornberg RL, Reese TS (1980) A freeze-substitution method for localizing divalent cations: examples from secretory systems. Fed Proc 39:2802–2808
Ornberg RL, Reese TS (1981) Quick freezing and freeze substitution for X-ray microanalysis of calcium. In: Hutchinson TE, Somlyo AP (eds) Microprobe analysis of biological systems. Academic Press, London New York, pp 213–228

Osatake H, Tanaka K, Inoué T (1985) An application of rapid freezing, freeze-substitution for scanning electron microscopy. J Electron Microsc Tech 2:201–208

*Peters K-R (1986) Working at higher magnifications in scanning electron microscopy with secondary and backscattered electrons on metal coated biological specimens and imaging macromolecular cell membrane structures. In: Müller M, Becker RP, Boyde A, Wolosewick JJ (eds) The science of biological specimen preparation 1985. SEM, AMF O'Hare, IL 60666, pp 257–282

Pfaller W (1979) Freeze drying and vacuum embedding of spray-frozen unicellular organisms. Mikroskopie (Wien) 35:37–44

Pfaller W, Rovan E (1978) Preparation of resin embedded unicellular organisms without the use of fixatives and dehydration media. J Microsc (Oxford) 114:339–351

Pfaller W, Rovan E, Mairbäurl H (1976) A comparison of the ultrastructure of spray-frozen and freeze-etched or freeze-dried bull and boar spermatozoa with that after chemical fixation. J Reprod Fertil 48:285–290

*Plattner H, Bachmann L (1982) Cryofixation: A tool in biological ultrastructural research. Int Rev Cytol 79:237–304

Porter KR, Anderson KL (1982) The structure of the cytoplasmic matrix preserved by freeze-drying and freeze-substitution. Europ J Cell Biol 29:83–96

Pringle AT, Forsdyke J, Rose AH (1979) Scanning electron microscope study of *Saccharomyces cerevisiae* spheroplast formation. J Bacteriol 140:289–293

*Rebhun LI (1972) Freeze-substitution and freeze-drying. In: Hayat MA (ed) Principles and techniques of electron microscopy, vol 2. Van Nostrand Reinhold, New York, pp 3–49

*Robards AW, Sleytr UB (1985) Low temperature methods in biological electron microscopy. In: Glauert AM (ed) Practical methods in electron microscopy, vol 10. Elsevier, Amsterdam

Roos N, Barnard T (1985) A comparison of subcellular element concentrations in frozen-dried, plastic-embedded, dry-cut sections and frozen-dried cryosections. Ultramicroscopy 17:335–344

Sakai A, Otsuka K, Yoshida S (1968) Mechanism of survival in plant cells at super-low temperatures by rapid cooling and rewarming. Cryobiology 4:165–173

Sitte H (1984) Instruments for cryofixation, cryoultramicrotomy and cryosubstitution for biomedical TEM. Zeiss Inf MEM Mag Electron Microsc 3:25–31

Steinbrecht RA (1976) Freeze substitution and freeze fracturing of insect sensilla without cryoprotectants. In: Ben-Shaul Y (ed) Electron microscopy 1976, vol 2. Proc 6th Eur Congr Electron Microsc, Isr Soc Electron Microsc, Jerusalem, pp 111–113

Steinbrecht RA (1980) Cryofixation without cryoprotectants. Freeze substitution and freeze etching of an insect olfactory receptor. Tissue Cell 12:73–100

Steinbrecht RA (1982) Experiments on freezing damage with freeze substitution using moth antennae as test objects. J Microsc (Oxford) 125:187–192

Steinbrecht RA (1984) Cryofixation and follow-up techniques in biological electron microscopy – outline and demonstration. Zeiss Inf MEM Mag Electron Microsc 3:9–17

Steinbrecht RA (1985) Recrystallization and ice-crystal growth in a biological specimen, as shown by a simple freeze substitution method. J Microsc (Oxford) 140:41–46

Stumpf WE, Roth LJ (1967) Freeze-drying of small tissue samples and thin frozen sections below −60 °C; a simple method of cryosorption pumping. J Histochem Cytochem 15:243–251

Tanaka K (1981) Demonstration of intracellular structures by high resolution scanning electron microscopy. Scanning Electron Microsc 1981/II:1–8

*Tanford C (1980) The hydrophobic effect: Formation of micelles and biological membranes. 2nd edn. John Wiley & Sons, New York

Umrath W (1983) Berechnung von Gefriertrocknungszeiten für die elektronenmikroskopische Präparation. Mikroskopie (Wien) 40:9–37

Usukura J, Yamada E (1981) Molecular organization of the rod outer segment. A deep-etching study with rapid freezing using unfixed frog retina. Biomed Res 2:177–193

Van Harreveld A, Crowell J (1964) Electron microscopy after rapid freezing on a metal surface and substitution fixation. Anat Rec 149:381–386

Van Harreveld A, Malhotra SK (1966) Demonstration of extracellular space by freeze-drying in the cerebellar molecular layer. J Cell Sci 1:223–228

Van Harreveld A, Crowell J, Malhotra SK (1965) A study of extracellular space in central nervous tissue by freeze-substitution. J Cell Biol 25:117–137

Van Zyl J, Forrest QG, Hocking C, Pallaghy CK (1976) Freeze-substitution of plant and animal tissue for the localization of water-soluble compounds by electron probe microanalysis. Micron 7:213–224

Verkleij AJ, Humbel B, Studer D, Müller M (1985) 'Lipidic particle' systems as visualized by thin-section electron microscopy. Biochim Biophys Acta 812:591–594

Weibull C, Villiger W, Carlemalm E (1984) Extraction of lipids during freeze-substitution of *Acholeplasma laidlawii*-cells for electron microscopy. J Microsc (Oxford) 134:213–216

Wheeler EE, Gavin JB, Seelye RN (1975) Freeze-drying from tertiary butanol in the preparation of endocardium for scanning electron microscopy. Stain Technol 50:331–337

White DL, Andrews SB, Faller JW, Barrnett RJ (1976) The chemical nature of osmium tetroxide fixation and staining of membranes by X-ray photoelectron spectroscopy. Biochim Biophys Acta 436:577–592

Wildhaber I, Gross H, Moor H (1982) The control of freeze-drying with deuterium oxide (D_2O). J Ultrastruct Res 80:367–373

Wróblewski R, Wróblewski J, Anniko M, Edström L (1985) Freeze-drying and related preparation techniques for biological microprobe analysis. Scanning Electron Microsc 1985/I:447–454

Zalokar M (1966) A simple freeze-substitution method for electron microscopy. J Ultrastruct Res 15:469–479

Part III Special Aspects

Chapter 8 Theory and Practice of High Pressure Freezing

HANS MOOR[1]

1 Introduction

Pressure-freezing has often been regarded as a method for highly technical specialists. At the beginning of its development, this may have been true: it was introduced in 1968 by Moor and Riehle at the European conference on electron microscopy in Rome. The interest of the audience was not overwhelming, because everybody thought that this approach is oversophisticated and in principle unnecessary. In the following decade, many technically less pretentious freezing methods have been developed, which work in the absence of pressure. All of them became standardized and their methodology has been described in numerous reviews and textbooks (e.g. Rash 1983; Gilkey and Staehlin 1986; see also Sitte et al., Chap. 4, this Vol.). The compiled experience shows the manifold profits of applying impact-, plunge-, jet- and spray-freezing. In one aspect, however, all of these techniques are inadequate: namely they only enable satisfactory cryofixation of objects or superficial layers, which are not thicker than 10–20 μm. This limitation is caused by the physical properties of aqueous systems and it indicates that thicker specimens can be well cryofixed only if these properties are altered.

One way to accomplish this is the application of highly concentrated "antifreeze" agents. In this case, a chemical fixation has to precede impregnation in order to minimize chemical and physical reactions of the object (Moor 1964). Most of the biological specimens which have been recorded by freeze-etching have been prepared by this means. However, many advantages of cryofixation are thereby lost: we fail to achieve a rapid stabilization of living matter; the properties of membranes are altered, and as a consequence, the compartmentalization of ions is lost and a variety of structural artefacts are obtained (Moor 1971).

Based upon this background, a new interest has arisen for an alternative method, which should enable the vitrification of thick specimens on a purely physical basis and thus avoid all the disadvantages of anti-freeze methods. The method of choice was high pressure freezing. Simultaneously to the development of this new field of interest, the question came up whether a suitable apparatus could be made available commercially. Between 1968 and 1985, a single homemade machine was in use in our laboratory (Moor et al. 1980). But during the last 3 years,

1 Institut für Zellbiologie, Eidgenössische Technische Hochschule, Hönggerberg, CH-8093 Zürich, Switzerland

Cryotechniques in Biological Electron Microscopy
Edited by R.A. Steinbrecht and K. Zierold

BALZERS UNION has developed, together with us, a machine which enables pressure-freezing to be used as a routine laboratory technique.

2 Freezing Under Atmospheric Pressure

In order to make the effect of high pressure comprehensive, the character of freezing under atmospheric pressure shall be summarized. Freezing stabilizes biological specimens, but may kill them at the same time. This killing is caused mainly by the crystallization of water, which destroys essential cellular structures. Stabilization can be achieved by the effect of low temperature. Consequently, a freezing technique, which should conserve both chemistry and structure, has to prevent crystallization!

2.1 The Crystallization of Specimen Water

A first point of consideration is the range in which freezing damage can take place: crystallization is naturally limited to the temperature zone between the melting point and the devitrification temperature of a system. Within this zone, water molecules can be incorporated into ice crystals. When we lower the temperature beyond the melting point, the rate of ice-crystal growth rapidly reaches a maximum, and then slowly drops due to the increasing viscosity of water (see Bachmann and Mayer, Chap. 1, this Vol.). When the devitrification temperature is attained, viscosity reaches a value, which prevents molecule movement and, consequently, ice-crystal growth. In pure water, the zone of crystallization extends from 273 K to about 140 K. This range is related to the water content of biological objects: in physiologically active cells and tissues, the temperature interval may extend from 271 K to 193 K; in frost-hardy cells, where the water content is reduced, the interval extends from about 260 K to 230 K; in dry seeds, spores and pollen grains, there is no water left for crystallization; consequently, neither melting point nor devitrification temperature can be observed (Moor 1964).

Following these lines of observation, one might attempt to suppress crystallization artificially by experimentally reducing the water content of a specimen, e.g. by replacing water by glycerol or sucrose. In practice, this treatment is applied for conservation of sperm or embryos. As we have mentioned already, this way is not acceptable in many cases due to the introduction of artefacts. However, we should bear in mind that different objects may exhibit significantly different freezing properties: certain characteristics may facilitate a vitrification, whereas others may render it very difficult.

2.2 Crystallization Depends on the Cooling Rate

In order to characterize freezing from the point of view of heat extraction, additional parameters must be considered. Primarily the cooling rate is relevant since crystallization is a process which is time-dependent. Therefore, it can be disturb-ed, or even prevented, if the object is cooled so quickly below its devitrification temperature that there is not enough time for crystal growth. In order to describe the freezing process properly, one should distinguish between three different situations, namely slow freezing, rapid freezing and ultrarapid freezing.

2.2.1 Slow Freezing

If the temperature is lowered slowly below the freezing point, supercooling occurs up to the point where impurities start to act as nuclei for crystallization. This type of nucleation is called "heterogeneous". Large ice crystals are produced, since there is only a limited number of nuclei. In addition, the heat of crystallization warms the system up to its melting point. By further cooling, the temperature can only be lowered if water is completely transformed into ice (see Bachmann and Mayer, Chap. 1, this Vol.).

2.2.2 Rapid Freezing

If the temperature is lowered more rapidly, the warming up by the heat of crystallization can be suppressed: the rate of heat extraction must be larger than the rate of heat production by crystallization. As a consequence, the growth of heterogeneous nuclei is slowed down. At 233 K, the system enters the zone of "homogeneous" nucleation; that is, every molecule of water is now potentially a nucleus. Consequently, the number of nuclei is enlarged enormously and only very tiny ice crystals can arise. The transition region from slow to rapid freezing is characterized by a sudden and drastic reduction of the size of segregation compartments produced by the crystallization (Riehle 1968b); the corresponding cooling rate is called "critical" (Moor 1973).

2.2.3 Ultrarapid Freezing

Dubochet et al. (1982) have shown that a layer of water, not thicker than about 100 nm, can be cooled so quickly that no crystallization takes place at all. Such a specimen lacks distinct electron diffraction spots in the microscope, provided it is maintained below 130 K. If warmed up, devitrification takes place at about 145 K. This indicates that a further increase in the cooling rate enlarges the portion of water molecules, which are not incorporated into ice crystals. The ratio can obviously be raised such that water appears to be solidified in the amorphous state!

2.2.4 Vitrification

These findings show that we have to distinguish between three types of frozen specimens: the ice may be macrocrystalline, showing segregation compartments with a diameter of 100 nm or more; it may be microcrystalline, exhibiting segregation compartments with a diameter of about 20 nm; or it is amorphous and does not show any detectable crystallinity. For most electron microscope purposes, macrocrystalline specimens cannot be used due to large-scale segregation phenomena which cause severe structural damage of the object. From a practical point of view, the structure of microcrystalline specimens is in most cases sufficiently preserved. Quite often we are not able to distinguish between the microcrystalline and the amorphous state. From a physical point of view, only an amorphous specimen is vitrified and hence ideally cryofixed.

2.3 Which Cooling Rates can be, and have to be Achieved? (Fig. 1)

From a theoretical point of view, cooling rates are limited by three factors: by the heat capacity and the heat conductivity of water and by the heat flow achieved at the specimen surface. Only the last parameter can be altered. Assuming an infinite value for this heat flow (this is a value which cannot be reached by any improvement of a freezing technique), calculations show that the highest possible rate drops drastically with increasing distance from the cooled surface: at a depth of 0.1 µm a cooling rate of 10^9 K s^{-1} may be attained, while at 10 µm only about 10^5 K s^{-1} and at a distance of 1000 µm from the cooled surface not more than 10 K s^{-1} can be achieved (Riehle 1968a). If a finite value for heat flow is assumed, the situation becomes even more restricted.

Practice has shown, on the one hand, that sufficiently high values can be attained and, on the other hand, that the theoretically predicted limitations are correct: pure water needs more than 10^6 K s^{-1} and has been vitrified in a layer,

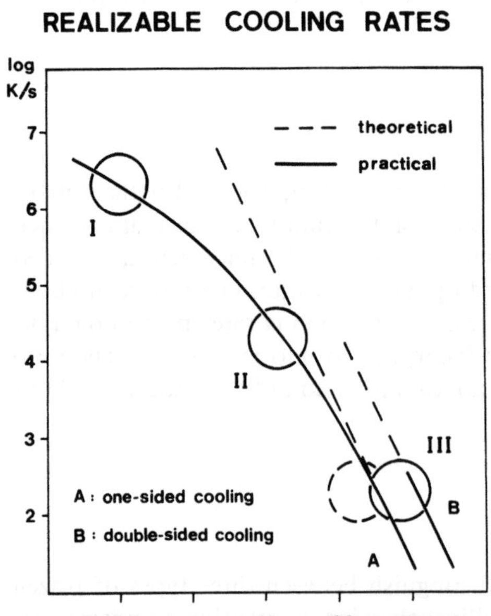

Fig. 1. The *curves* show the dependence of the cooling rate (K s^{-1}) on the distance (µm) of the specimen region from the cooled surface. *I* Range of vitrified pure water. *II* Range of vitrified animal cells and tissues. *III* Range of specimens vitrified under high pressure

which was about 100 nm thick. Cells and tissues need more than 10^4 K s^{-1} and have been satisfactorily cryofixed in superficial layers not much thicker than 10 μm. If such specimens are protected by an anti-freeze agent, they need less than 100 K s^{-1} and can be cryofixed without visible freezing damage up to a thickness of 1 mm. If the use of anti-freeze agents should be avoided, then the alternative is pressure-freezing, which allows the critical cooling rate to be lowered 100-fold. Therefore, tissue blocks, cooled only from one side, can be preserved without visible freezing damage in a superficial zone about 150 μm thick. If cooling of a platelet occurs from both sides, the thickness of well-preserved specimens can be increased to 600 μm, because the freezing of such a specimen improves the cooling rate in the centre of the probe by a factor of four (Riehle 1968 a).

2.4 The Advantage of Double-Sided Cooling is Limited

In this context, a question arises and has to be answered: in the 100 K s^{-1} range, double-sided cooling enlarges the thickness of good cryofixation, and we should expect the same phenomenon in the 10000 K s^{-1} range. That is, instead of 10 to 20 μm, we should be able to vitrify specimen sandwiches about 60 μm thick. Experience shows that this has never been achieved, neither by plunge- nor by jet-freezing. How could this discrepancy be explained? The prerequisite for any improvement with double-sided cooling has to be a sufficient coincidence of the two cooling processes: during the whole procedure, there should be a temperature difference of not much more than 1 K between the two sides of a cooled platelet. This indicates that the two cooling processes should be simultaneous, e.g. within 10^{-4} s, if the cooling rate is 10^4 K s^{-1}, or within 10^{-2} s, if the rate is 10^2 K s^{-1}. It seems likely that deviations of geometry and the turbulance of the streaming of the cryogen prevent sufficiently simultaneous cooling in the 10^4 K s^{-1} range. Therefore, it is only with pressure-freezing that we are able to utilize the advantages of double-sided cooling, due to the reduced demand for precisely simultaneous cooling (Fig. 1).

3 Freezing Under High Pressure

3.1 Pressure Reduces the Critical Freezing Rate

In general, this effect can be exemplified by applying the principle of Le Chatelier: freezing increases the volume of water, and high pressure hinders such expansion, that is, it must hinder crystallization. The effect is made visible by a lowering of the freezing point and by a reduction in the rates of nucleation and ice-crystal growth. Consequently, less heat is produced by crystallization and less heat has to be extracted per unit time by cooling. In other words: the cooling rate necessary for vitrification is reduced.

The optimal range of pressure can be deduced from the phase diagram of water (Kanno et al. 1975; see also Bachmann and Mayer, Chap. 1, this Vol., Fig. 5): the most profitable pressure zone is just above 2045 bar, where the melting point is lowered to a minimum of 251 K. The zone of supercooling is expanded down to 183 K. In addition, water will not crystallize as ice I with a lower density, but as ice II or III with a higher density than liquid water. In this pressure range, heat can be extracted from a specimen in the state of supercooling until the temperature has reached 183 K. If crystallization should occur, it will only proceed very slowly.

The following data are estimates which shall show what could be attained by pressure-freezing: pure water, cooled from one side, may be vitrified in a superficial layer not thicker than 10 μm; double-sided cooling will not improve the situation. Animal tissue slices, e.g. liver (Müller and Moor 1984), cerebral cortex (Moor et al. 1980) and cartilage (Hunziker et al. 1984) can be cryofixed without freezing damage at a thickness up to 600 μm, provided double-sided cooling is applied (see also Fig. 6). Plant tissues, which contain a larger amount of water, must be positioned somewhere between the already mentioned specimens. They can probably be well cryofixed by double-sided cooling in a specimen thickness range of 100–200 μm (Dr. Martin Müller, pers. commun.).

3.2 Reactions of the Biological Specimen on Pressure

The improvement of freezing properties is not the only effect of subjecting living specimens to pressure. High pressure also influences the freezing of the lipid phase in membranes: the transition temperature is raised by about 20 K $kbar^{-1}$ (Macdonald 1984). This indicates that by applying a pressure of more than 2 kbar the membrane lipids can be solidified very quickly, purely by the action of high pressure. In addition, it must be assumed that many membrane-associated processes are altered. This is of great importance when processes such as membrane fusion, vesiculation or the effects of nervous excitation are studied. For instance, a pressure rise of only about 200 bar is sufficient in order to stimulate the firing of crayfish nerves (Kending et al. 1978). This would indicate that pressure-frozen specimens of nerve fibres would all be fixed in the excited state.

The tertiary structure (and also the activity) of proteins should not be significantly affected by high pressure. Irreversible denaturation requires a pressure of about 10000 bar; and reversible denaturation which takes place around 2000 bar requires more time than is available in between pressurizing and freezing the specimen.

Pressure raises the solubility of gases in water. Such uptake should not be harmful, but the formation of gas bubbles during decompression (the so-called Caisson effect) may be critical. However, this effect does not occur after pressure-freezing due to decompression in the solid state.

A pressure-frozen specimen will be solidified in the amorphous state, or it will contain crystallites of ice II or III, which are denser than liquid water. If the pressure is released whilst maintaining the specimen below the devitrification temperature (i.e. well below 145 K), nothing will happen. However, if the temperature is raised, formation of ice I will occur, which causes an expansion of the volume and submits the specimen to tensions. Clefts can be produced which may disturb the representation of the ultrastructure.

4 The Main Practical Problems of Pressure-Freezing

Pressure-freezing comprises a technical challenge which is derived from practical problems which are to be solved, or at least minimized.

4.1 The Specimen may be Destroyed by Applying Pressure

Little detailed information is available about the effect of high pressure on cellular structures. In 1970, Moor and Hoechli showed that the briefer the period of exposure to pressure, the greater is the pressure that can be tolerated: when *Euglena* was exposed to 2000 bar for only 0.1 s, 90% of the cells survived the treatment, while 4 s were sufficient to kill all of them. This indicates that, in practice, freezing has to follow pressurizing instantaneously in order to enable a stabilization of the specimen in the living state. In this respect, instantaneously means not much more than 0.1 s!

4.2 The Specimen may be Frozen too Early

Following the ideas of Riehle (Moor et al. 1980), the specimen is pressurized and cooled with the same medium (liquid nitrogen) simultaneously. The great advantage of this approach consists in a relatively large pressure chamber, in which specimens are exposed directly to the action of the cryogen. The disadvantage is the fact that the pressure rise and temperature fall occur at the same time and the specimen may be frozen before becoming fully pressurized. The problem of premature freezing can be solved by delaying the temperature fall. Such delay must be in the order of the time which is needed for a pressure rise up to 2100 bar! In our hands, the minimal delay time is about 15 ms, which guarantees a most rapid succession of cooling upon pressurizing.

4.3 The Specimen Centre is Cooled Rather Slowly

Pressure-freezing is not very rapid because we take advantage of a reduced critical cooling rate. At the surface of a 0.6-mm-thick specimen, which is shielded with metal platelets against the action of the pressurizing cryogen, a cooling rate of up to 5000 K s^{-1} can be achieved. However, in the centre of the sample, not more than a few hundred K s^{-1} is possible. Higher rates at the surface would not improve the situation.

It is self-evident that pressure has to be maintained until the entire specimen is cooled below the devitrification temperature. If we consider a rate of 500 K s^{-1} for passing through the temperature range between 273 K and 123 K, then the pressure has to be maintained for at least 0.3 s.

4.4 Pressure-Frozen Specimens are Metastable

Amorphous specimens, frozen under atmospheric conditions, or under high pressure, will crystallize if they are warmed up above their devitrification temperature. In addition, crystalline specimens, frozen under high pressure, will recrystallize under atmospheric pressure if warmed up above the same limit. Both phenomena cause an enlargement of the probe volume which may destroy the specimen.

In order to prevent this effect, the period of cooling following pressurizing has to be sufficiently long to enable the removal of the cold specimen (well below 130 K) and its transfer into liquid nitrogen, without provoking devitrification or recrystallization damage.

5 The High Pressure Freezing Machine (Balzers HPM 010)

In the development of a commercial apparatus, two approaches had to be followed: firstly, to find a solution to the problems outlined in chapter 4 (thus establishing a proper physical basis for the whole procedure) and secondly, to solve a number of additional problems in order to guarantee a safe, easy, quick and reproducible handling of the equipment. Two technicians, G. Bellin in my laboratories and M. Wohlwend in the construction bureau of the producer, were engaged in this development (see Moor 1986).

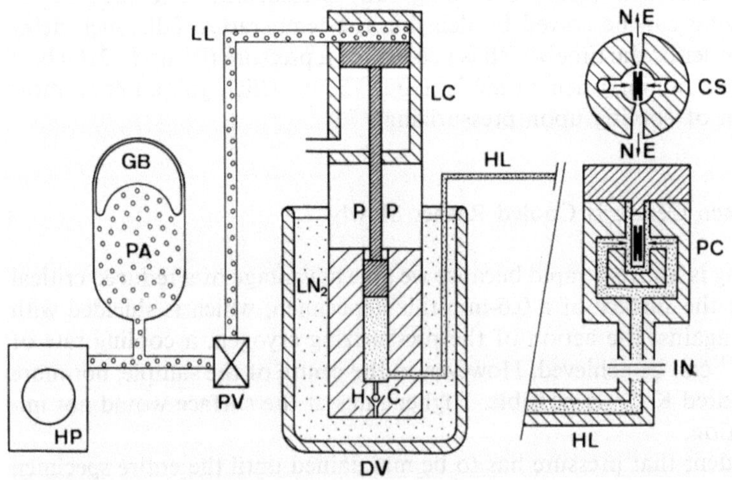

Fig. 2. Diagram of the high pressure freezing apparatus. The operation is explained in the text. Abbreviations: *DV* dewar flask; *GB* gas balloon; *HC* high pressure cylinder; *HL* high pressure line; *HP* hydraulic pump; *IN* position of the injection chamber (see Fig. 3); *LC* low pressure cylinder; *LL* low pressure line; *PA* pressure accumulator; *PC* freezing chamber containing a specimen sandwich; *PP* pressure piston; *PV* pressure valve. *Small circles* oil; *dots* liquid nitrogen (LN_2). Cross-section (*CS*) of the high pressure chamber is shown at the *upper right*. *NE* narrow apertures for gas escape (Moor et al. 1980)

5.1 How Pressure is Generated (Fig. 2)

The technology of pressure generation is the same as that already described in 1980 (Moor et al.). A hydraulic pump forces oil to enter into a pressure accumulator, a 20-litre tank containing a rubber balloon. This balloon is preloaded with 250 bar and compressed with oil up to 300 bar. The pressurized oil can be released from the tank by opening the main pressure valve. It then streams through the low pressure line into the low pressure cylinder and forces the piston to move downwards. In the high pressure cylinder, which is mounted in a Dewar vessel below the low pressure cylinder, the lower part of the piston compresses liquid nitrogen and drives it through the high pressure line into the pressure chamber. The reduction of the piston diameter from 800 to 26 mm increases the pressure nearly tenfold. In the pressure chamber, two jets of liquid nitrogen under a pressure of about 2500 bar are directed from opposite sides onto the surfaces of a sandwiched specimen. The nitrogen leaves the system through two narrow apertures (0.4 mm in diameter). By these means, 100 ml liquid nitrogen pass the chamber in about 0.5 s, pressurize the specimen and cool it down to about 100 K. Incidently, pressurized liquid nitrogen is an extremely effective coolant, because the Leidenfrost phenomenon is suppressed completely.

5.2 How the Liquid Nitrogen is Handled (Fig. 3)

The nitrogen is fed into the Dewar vessel and kept at a constant level automatically. The Dewar vessel is vented through a narrow exhaust pipe. This arrangement enables the Dewar (which is closed with an air-tight cover) to be slightly pressurized and to force the liquid nitrogen to enter, through a check valve, into the lower end of the high pressure cylinder. The total filling of the cylinder is guaranteed by a second check valve situated in the piston. Both valves close when the piston starts to move downwards. The Dewar cover is thermostatically controlled with circulating warm water in order to prevent the low pressure cylinder, which is connected with the cold, high pressure system, from cooling.

5.3 How Premature Cooling is Prevented (Fig. 3)

In order to prevent premature cooling of the specimen, the temperature fall has to be delayed until the pressure has risen above 2100 bar. This is achieved by using a warm pressurized medium, preceding the pressurized liquid nitrogen. For this purpose, isopropyl alcohol is injected into the uppermost part of the pressure line, which is located at the bottom of the pressure chamber. Two ml of alcohol delay the cooling for about 15 ms (see Fig. 4). By releasing a pressure shot, initially pressurized alcohol (at 3 bar) is injected through a check valve into the pressure line and the specimen chamber. A spring-loaded check valve, located at the entry of the pressure line into the chamber, prevents the alcohol from back-streaming into the cold part of the line and its consecutive blockage. After about 0.5 s, the alcohol injection is stopped and the nitrogen is launched: it opens the spring-

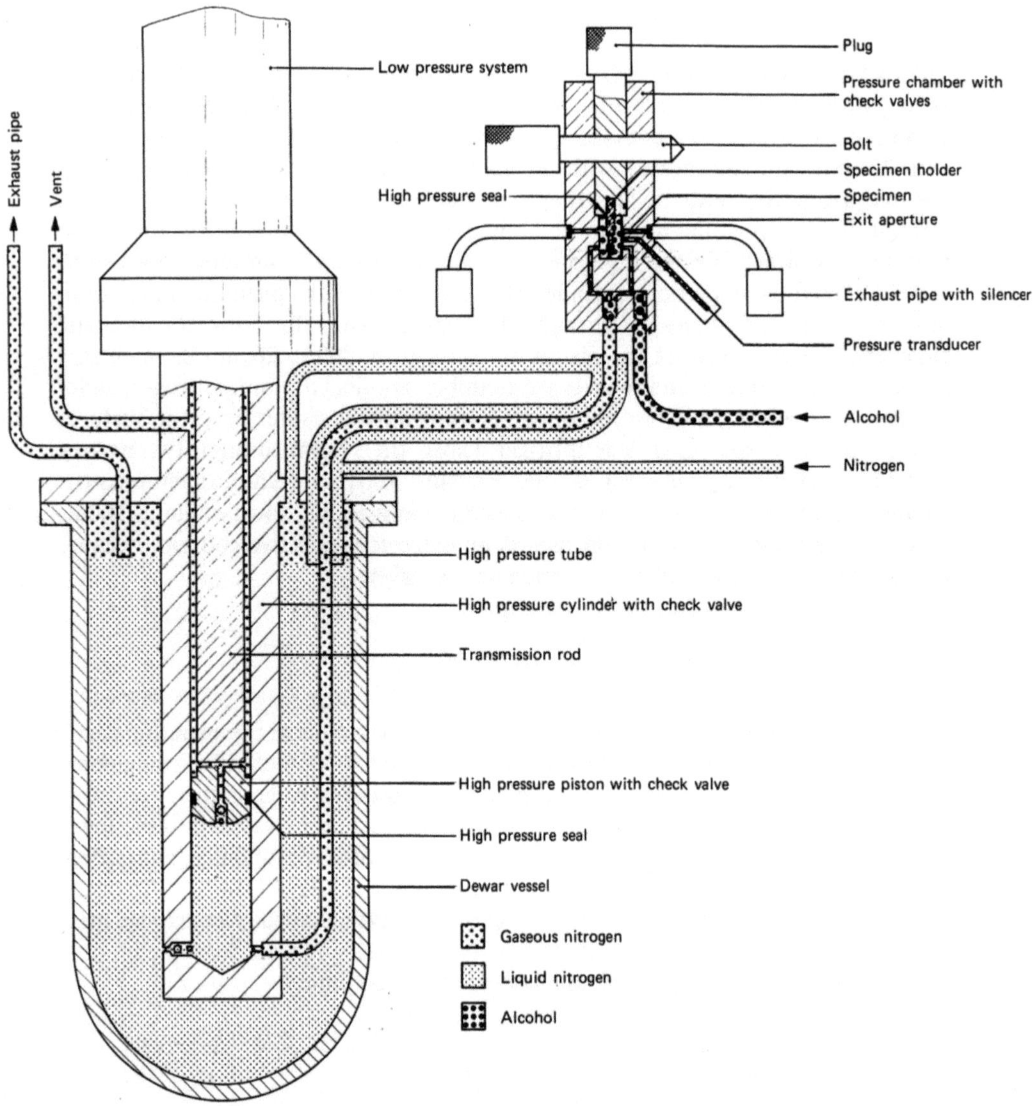

Fig. 3. Diagram of the liquid nitrogen cooled part of the BALZERS HPM 010 apparatus

loaded valve, pressurizes the alcohol in the specimen chamber and then starts to cool the sandwich.

5.4 How the Specimen is Exposed (Fig. 5)

The force, which is borne by the machine, can be compared to that of a gun shot. Therefore, the specimen has to be protected by rather massive metal shields. Good results have been obtained with gold platelets having a 0.3-mm-deep cavity, which

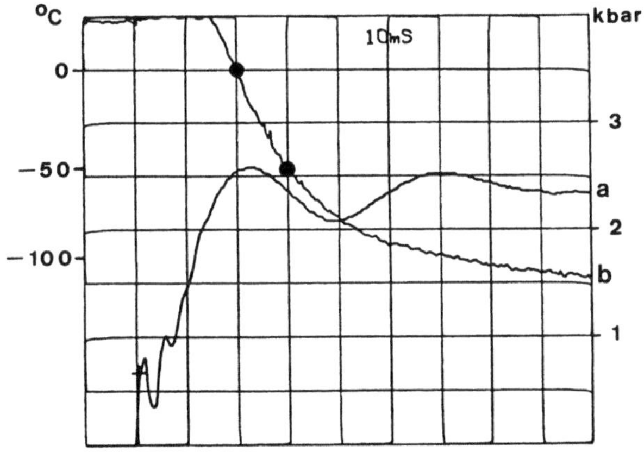

Fig. 4. The time course of pressure (*a*) and temperature (*b*) at the beginning of high pressure freezing, recorded by a digital storage oscilloscope. Distance between grid bars on *abszissa:* 10 ms

enables the uptake of tissue blocks with a diameter of 1 mm and a thickness of 0.6 mm. For all our purposes, we use an asymmetrical sandwich: the upper platelet, which must be coated with egg lecithin, has a dome-shaped cavity, while the lower one contains a cylindrical depression (Moor et al. 1980). After freezing, the upper platelet can be removed by fracturing the sandwich along the coated surface: it leaves the frozen specimen, protruding from the lower platelet, in a fashion which is suitable for cutting with a microtome.

For pressure-freezing, the sandwich is enclosed in a holder which clamps the two platelets firmly together. Only two holes are left open, which enable a direct cooling of the sandwich with the two nitrogen jets (see Moor 1986). The whole block of the pressure chamber is thermostatically controlled via circulating warm water in order to prevent cooling and premature freezing of an introduced specimen.

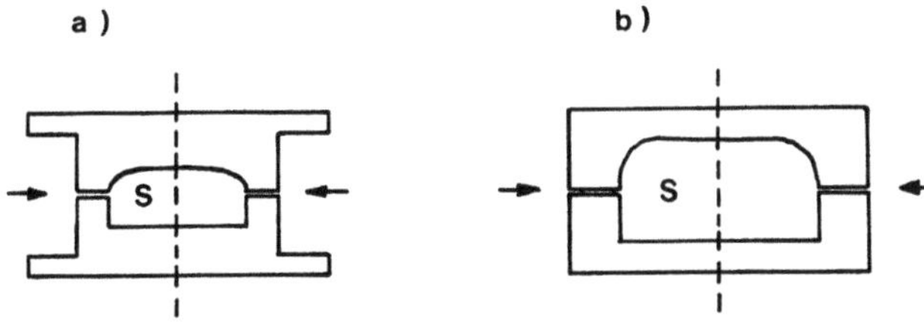

Fig. 5 a, b. Cross-sections through sandwiches, suitable for freeze-fracturing (**a**) and for cryosectioning (**b**). The diameter and thickness of the corresponding specimen space (*S*) for (**a**) is 1.5 mm and 0.6 mm; for (**b**) 2.0 mm and 1.0 mm. Pressure is transduced to the specimen space through the cleft between the sandwich platelets (*arrows*)

5.5 How the Machine is Operated

After 10-min cooling with liquid nitrogen the machine is ready for use. A specimen is inserted, the chamber locked and a shot released. Immediately after collapse of the pressure, the chamber can be opened, the specimen holder withdrawn and plunged into liquid nitrogen. The whole operation can easily be

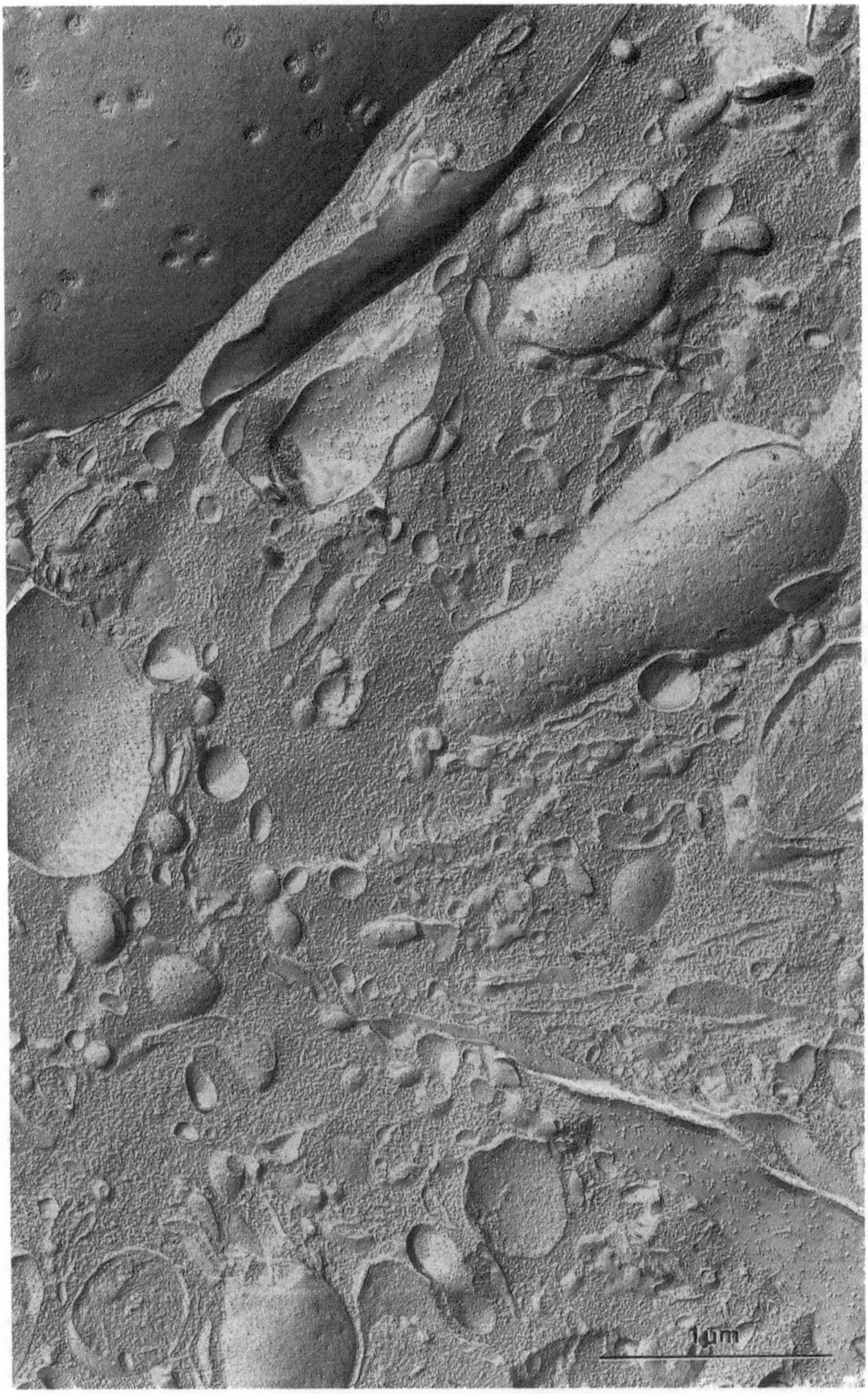

Fig. 6. Pressure-frozen and freeze-fractured fresh liver cell. (Courtesy of M. Müller)

accomplished within 2 to 3 s. Under liquid nitrogen, the holder is opened and the sandwich removed. The machine returns to shot position automatically within less than 2 min.

5.6 How the Machine is Controlled

The whole pressure-freezing process can be controlled by simultaneously measuring the time course of pressure and temperature in the specimen chamber (see Moor 1986). Three essential data can be recorded on an LC display at the keyboard of the machine: the period of temperature fall from 273 K to 223 K (about 10 ms), the period of pressurization above 2100 bar (about 500 ms) and the period of specimen temperature below 173 K (about 7 s). If more detailed information is required, a digital storage oscilloscope can be connected with the display system (see Fig. 4).

6 Some Practical Advice

6.1 The Preparation Before Freezing

It is not the pressure-freezing itself, but the preceding preparation, which is the limiting factor also for this type of cryofixation: it is essential to maintain a specimen under the most natural conditions up to the moment of freezing. If tissues have to be taken from an organism, dissected into pieces, sandwiched, enclosed in the holder and transferred into the pressure-freezer, the whole procedure should be done as quickly as possible in order to hinder physiological and consecutive structural alterations. The specimen should be prevented from drying and must be enclosed as tightly as possible by the metal of the sandwich platelets in order to exclude an excess of surrounding water. Of great advantage is the development of special tools enabling the direct extraction of tissue pieces which tightly fit into the cavity of the sandwich. If a specimen cannot be adapted to the sandwich, it has to be done the other way round: special platelets must be fabricated which fit to the specimen shape. If this cannot be achieved, it might be advantageous to fill the free space in the sandwich cavity with an osmotically inactive hydrophilic substance such as serum albumin, which should be dissolved in a physiological solution (sucrose, NaCl or Ringers salt solution). Such "antifreeze" agents need not impregnate the specimen, but simply have to prevent crystallization which would heat the surrounding medium!

6.2 Adaptation of the Pressure-Freezing Process

The main parameters of pressure-freezing are the cooling rate, the specimen thickness and the period of pressurization!

The cooling rate at the specimen surface is limited by the heat transport in the sandwich metal. The shielding platelets are rather thick (at least 0.3 mm), if the standard gold supports (used for freeze-etching) are applied. With omega-shaped copper supports, which are only 0.1 mm thick, higher rates could be attained, but the reduced stability of the sandwiches has to be taken into account. The advantage of applying higher cooling rates is the greater, the thinner the specimens are; in the case of the 0.6-mm cavity, it is probably negligible.

The specimen thickness is mainly responsible for low cooling rates in the centre of the sandwich. If a pressurized specimen requires a relatively high cooling rate for vitrification (e.g. plant tissues with a high water content), the thickness has to be reduced, i.e. platelets must be used with much flatter depressions (e.g. only about 0.1 mm deep). In this case, the application of a thin-walled copper sandwich might be advantageous.

The pressurization has to last until the specimen centre is cooled down below 140 K. This period can be extended, if the diameter of the apertures, through which the pressurized nitrogen leaves the specimen chamber, is reduced. A reduction from 0.4 to 0.25 mm increases the pressure period from 0.5 to about 2 s. This can only be accomplished at the expense of the cooling rate: a prolonged period of cooling reduces the amount of liquid nitrogen which passes through the pressure chamber per unit time; that is, the rate of heat extraction is also reduced! If specimens, thicker than 0.6 mm should be pressure-frozen, such prolongation must be applied.

6.3 The Preparation After Freezing

The most essential prerequisite of successful preparation after (pressure) freezing consists in maintaining the specimen below its devitrification temperature. With sufficient carefulness, a specimen can be kept below 140 K, if freeze-fracturing is applied; i.e. during the transfer from the freezer to liquid nitrogen, in the storage tank, during the transfer to the freeze-etcher and during evacuation and coating (below 140 K only fracturing, but not etching is possible).

If freeze-substitution with methanol or acetone is to be applied, the specimen has to be warmed up to at least 183 K, and under this condition for many specimens a maintenance of the vitreous state cannot be guaranteed. Therefore, if the efficiency of pressure-freezing is to be controlled, only freeze-fracture replicas or cryosections obtained below 140 K can be conclusive. If such tests indicate devitrification at 183 K, substitution media have to be applied, which are able to dissolve ice at lower temperatures.

6.4 Conclusions

Pressure-freezing facilitates the cryofixation of specimens thicker than 10–20 μm: freezing damage can be reduced considerably; the microcrystalline, or even the amorphous state, may be attained, but wonders should not be expected. The preparator has to adapt both, technique *and* specimen, in order to achieve maximum efficiency. Faults during the preceding preparation and/or after pressure-freezing are the main reasons for failure of this method.

7 Discussion of Results

7.1 The Crystallinity may be Heterogeneous

Experience with tissues of various origins has demonstrated that the whole specimen may be frozen with uniform quality, but quite often differences occur between different compartments or different cells. At least two factors are responsible for this behaviour: one factor is the critical cooling rate, which depends on the actual content of water. Differences in this respect explain why the matrix of mitochondria may be vitrified more easily than the cytoplasmic ground substance and why the same phenomenon can be observed with respect to the karyoplasm and the perinuclear space and concerning the matrix of cartilage and the cytoplasm of chondrocytes. Such differences are most prominent, if the critical cooling rate has been surpassed in only one of two neighbouring compartments (see for instance Hunziker et al. 1984).

The other factor is the supercoolability which depends on the presence and activity of heterogeneous nuclei: tissues, which have attained frost-hardiness, show an increased supercoolability compared to their surrounding media, a phenomenon which disappears as soon as a cell is damaged. This indicates that differences in the crystallinity of similar cells can be expected, if part of them have been injured by the excision procedure, while others were left intact.

7.2 Fracture Faces are Remarkably Etchable

Etchability depends on the amount of non-bound water surrounding a specimen structure. The application of 20 to 30% of an anti-freeze agent increases the amount of bound water drastically and renders a specimen nearly unetchable; the structures can only be made visible by the fracturing process. In this respect, the situation is drastically improved by pressure-freezing an unprotected specimen. We have applied both methods in order to investigate nerve tissue (Moor et al. 1980): in longitudinally fractured axons, the glycerol-treated specimens only showed a regularly fine-grained cytoplasmic ground substance; in pressure-frozen specimens many more details became visible. Most prominent were profiles of the neurotubules, which could not be detected when impregnated with non-etchable glycerol. In pressure-frozen protoplasmic drops of *Physarum polycephalum* (Wolf et al. 1981), freeze-etching enabled the visualization of the microfilament network, which constitutes the cortical area of the cytoplasm. Even surface areas of the cytoplasmic membrane could be observed exhibiting anchoring places of actin filaments at the inner and of mucous filaments at the outer surface.

7.3 Pressure-Sensitive Structures are Preserved

High pressure experiments in molecular biology have shown that the protein tertiary structure can be altered reversibly by the action of 2000 bar. Such alterations most probably cause enzymatic misfunctions, which may kill the pressurized cells after a while (Moor and Hoechli 1970). The structural breakdown of microtubules under high pressure may also be explained in this context. The observation that all these pressure-induced alterations need seconds or minutes, indicates that an interval of less than 0.5 s between pressurizing and freezing is short enough in order to enable the preservation of pressure-sensitive structures. A proof for this assumption has been given by the neurotubuli in pressure-frozen axons (Moor et al. 1980)

7.4 Pressure-Frozen Specimens for Cytochemical Research

Freeze-fracturing, cryosectioning or freeze-substitution are appropriate techniques to follow any cryofixation. They enable cytochemical investigations, e.g. the localization of proteins with antibody labelling or of ions with X-ray analysis. The great advantage of pressure-freezing is to provide satisfactory cryofixation for *thick* specimens. This enables the application of cryotechniques to the investigation of sites far away from the specimen surface. With cryofixation at ambient pressure, in many cases, this surface has to be artificially produced by a dissection procedure, which may destroy the integrity of cells and organelles. In addition, the flattening and compression of surface structures, often produced by impact-freezing, can be avoided. A certain disadvantage of pressure-frozen specimen sites "far away" from the surface consists of the relatively low speed of their stabilization. It reduces the time resolution of dynamic processes by about two orders of magnitude, if compared to the cryostabilization achievable with pressure-less methods (see Knoll et al., Chap. 14, this Vol.).

References

(References marked by an asterisk are review papers or books respectively)

Dubochet J, Lepault J, Freeman R, Berriman JA, Homo JC (1982) Electron microscopy of frozen water and aqueous solutions. J Microsc (Oxford) 128:219–237

*Gilkey JC, Staehlin LA (1986) Advances in ultrarapid freezing for the preservation of cellular ultrastructure. J Electron Microsc Techn 3:177–210

Hunziker EB, Herrmann W, Schenk RK, Müller M, Moor H (1984) Cartilage ultrastructure after high pressure freezing, freeze substitution, and low temperature embedding. 1. Chondrocyte ultrastructure – implications for the theories of mineralization and vascular invasion. J Cell Biol 98:267–276

Kanno H, Speedy RJ, Angell CA (1975) Supercooling of water to −92 °C under pressure. Science 189:880–881

Kending JJ, Schneider TM, Cohen EN (1978) Pressure, temperature and repetitive impulse generation in crustacean axon. J Appl Physiol Respirat Environ Excerc Physiol 45:742–746

Macdonald AG (1984) The effects of pressure on the molecular structure and physiological functions of cell membranes. Philos Trans R Soc London Ser B 304:47–68

Moor H (1964) Die Gefrier-Fixation lebender Zellen und ihre Anwendung in der Elektronenmikroskopie. Z Zellforsch 62:546–580

*Moor H (1971) Recent progress in the freeze-etching technique. Philos Trans R Soc London Ser B:121–131

*Moor H (1973) Cryotechnology for the structural analysis of biological material. In: Benedetti EL, Favard P (eds) Freeze etching, techniques and applications. Soc Fr Microsc Electron, Paris, pp 11–20

Moor H (1986) Recent progress in high pressure freezing. In: Imura H, Maruse S, Suzuki T (eds) Electron microscopy 1986, vol. 3. Proc 11th Int Congr Electron Microsc, Kyoto Jpn Soc Electron Microsc, Tokyo, pp 1961–1964

Moor H, Hoechli M (1970) The influence of high-pressure freezing on living cells. In: Favard P (ed) Microscopie électronique 1970, vol. 1 Proc 7th Int Congr Electron Microsc, Grenoble. Soc Fr Microsc Electron, Paris, pp 445–446

Moor H, Riehle U (1968) Snap-freezing under high pressure: A new fixation technique for freeze-etching. In: Steve Bocciarelli D (ed) Electron microscopy 1968, vol 2. Proc 4th Eur Reg Conf Electron Microsc, Rome, pp 33–34

Moor H, Bellin G, Sandri C, Akert K (1980) The influence of high pressure freezing on mammalian nerve tissue. Cell Tissue Res 209:201–216

Müller M, Moor H (1984) Cryofixation of thick specimens by high pressure freezing. In: Revel J-P, Barnard T, Haggis GH (eds) The science of biological specimen preparation. SEM, AMF O'Hare, IL 60666, pp 131–138

*Rash ER (1983) The rapid-freeze technique in neurobiology. Trends Neurosci 6:208–212

Riehle U (1968a) Schnellgefrieren organischer Präparate für die Elektronen-Mikroskopie. Chem Ing Tech 40, 5:213–218

Riehle U (1968b) Über die Vitrifizierung verdünnter wässeriger Lösungen. Techn Wiss Diss, Eidgen Tech Hochschule Zürich

Wolf KV, Stockem W, Wohlfarth-Bottermann KE, Moor H (1981) Cytoplasmic actomyosin fibrils after preservation with high pressure freezing. Cell Tissue Res 217:479–495

Chapter 9 Freeze-Etching of Dispersions, Emulsions and Macromolecular Solutions of Biological Interest

LUIS BACHMANN[1]

1 Introduction

Dilute dispersed systems which are the subject of this chapter consist mainly of water, their dry weight ranging between less than one per mill and 10%. In previous chapters of this volume, it has been discussed that cryofixation of almost pure water without ice formation and phase separation is a most difficult task. This explains why all attempts to make interactions of colloidal particles ("network formation") visible by cryotechniques have utterly failed unless special freezing procedures were applied. Phase separation in dilute solutions has been demonstrated in the earliest papers on freeze-etching and this problem is still one of the top subjects in biological cryofixation. It is surprising that this discussion has to a large extent been ignored by electron microscopists in the field of non-biological colloid chemistry.

Freeze-etching of liquid specimens is performed to obtain information about the shape, size and number of the dispersed entities. Their spatial distribution and their interrelation in the fluid is undisturbed by interactions with a substrate or the specimen surface. To obtain this various information, different quality standards of cryofixation are required. Size determination, e.g. of latex particles, fat droplets or protein molecules, is less influenced by solute/solvent segregation than the structures of gelled colloidal solutions or multicomponent emulsions (Fig. 1). Freeze-etching and cryoelectron microscopy of frozen fluids are complementary techniques mainly because the information obtainable by them is based on different mechanisms (Bachmann and Talmon 1984).

2 Specific Problems of Specimen Preparation

2.1 Freezing Techniques

The various techniques by which liquids can be frozen have recently been reviewed in detail (Plattner and Bachmann 1982; Robards and Sleytr 1985; Menco 1986;

1 Institut für Technische Chemie, Technische Universität München, D-8000 München, FRG

Cryotechniques in Biological Electron Microscopy
Edited by R. A. Steinbrecht and K. Zierold

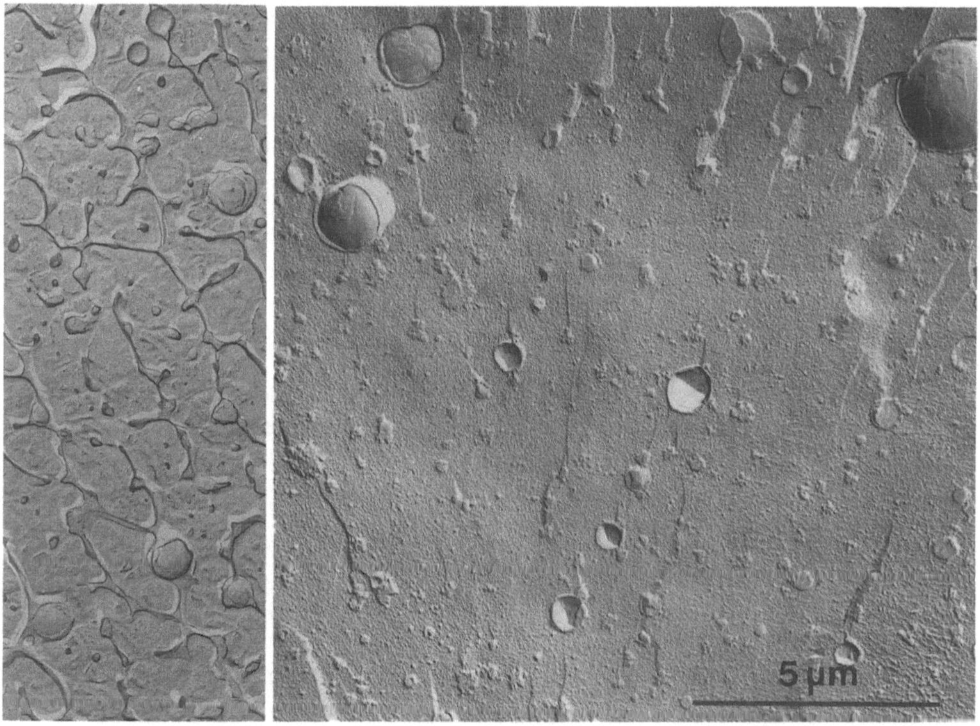

b

Fig. 1a, b. Cow's milk, freeze-etched. **a** Conventionally frozen; **b** spray-frozen (Bachmann and Schmitt-Fumian 1973a)

Sitte et al., Chap. 4, this Vol.), therefore, only some methodical aspects pertinent to the subject of this chapter will be discussed here.

In the course of *spray-freezing* and *specimen jet-freezing*, the sample is either divided into 10–50 µm droplets by an airbrush, or a fine jet stream of the specimen is generated by forcing it through a 10 to 30 µm aperture (Bachmann and Schmitt 1971; Bachmann and Schmitt-Fumian 1973a, b; Ververgaert et al. 1973). The droplets or the jet are injected into the quenching medium with a velocity of some $10\ \mathrm{m\,s^{-1}}$. The specimen is subject to substantial shear stress while the droplets or the jet are being formed and when they hit and penetrate the coolant. No quantitative data on the shear forces occurring in droplets of the discussed size are available (Green 1953; Davies 1972). Their influence on the microstructure of a given specimen must therefore be checked, e.g. by viability tests in the case of living organisms (Plattner et al. 1972) or by comparative light scattering measurements before and immediately after spraying in the case of liposomes (Guiot et al. 1980). In general, the smaller the entities, the less they are affected by shearing. No effect on the molecular weight of polystyrene and dextran molecules has been observed after repeated spraying. During spraying, solvent evaporation in the range of a few percent and a resulting cooling by 5–10 K has been observed. This problem does not arise with the specimen jet. This way solutions can even be processed at 350 K without noticeable loss of solvent (Bachmann et al. 1981).

Specimen spraying and jetting yield excellent cryofixation when the appearance of freeze-etched 5% glycerol is used as a quality criterion. The observed microsegregation in the range of 5 nm has only been matched by high pressure freezing (Riehle and Hoechli 1973). Some of this phase separation might be caused by crystallization or recrystallization during specimen processing and etching. X-ray diffraction and ESR measurements have shown, however, that pure water cannot be truely vitrified in a reliable way by these techniques except when the operating pressure of the specimen jet (∅ ~ 10 μm) is raised to 400 bar (Mayer and Brüggeller 1982).

The various techniques of *impact-freezing* (slamming) (Sitte et al., Chap. 4, this Vol.) exert high shearing forces on the specimen during the impact onto the metal cooling block. The same is true for *splat-freezing* of aerosol droplets, a technique which makes complete vitrification possible (Mayer 1985) and has only recently been applied for freeze-etching.

During *sandwich cryogen-jet freezing* (Moor et al. 1976; Müller et al. 1980; Pscheid et al. 1981) the 50–100 μm-thick specimen layer is protected between two metal foils while the cryomedium is shot at it. Therefore, the specimens experience no more shear stress once the sandwich has been prepared. If wanted they can relax in a controlled atmosphere before being frozen (van Venetie et al. 1980). Because of the necessary handling time, colloidal systems which have a short time of gel formation (seconds) can only be frozen in the relaxed state by the presently used sandwich techniques.

Emulsion-freezing (Buchheim 1972; Buchheim and Welsch 1977) is performed by first emulsifying the specimen in paraffin oil by means of surfactants and a homogenizer followed by conventional freezing. Compared to non-emulsified specimens frozen at similar rates, segregation is substantially reduced (less heterogeneous nucleation). During emulsification, the specimen is subject to shear forces. In addition, it is exposed to the surfactant which is needed as emulsifier. Thus, systems which are sensitive to these substances cannot be investigated. The specimen can mechanically relax before being frozen. Combining emulsification with specimen jet-freezing has made complete vitrifration of water possible (Brüggeller and Mayer 1980; Bachmann and Mayer, Chap. 1, Sect. 3.1, this Vol.).

The *high pressure freezing* technique achieves excellent cryofixation also at "moderate" cooling rates of 10^2 to 10^4 K s^{-1} (Moor, Chap. 8, this Vol.). The effects of mechanical stress connected with pressure-freezing on sensitive specimens has not yet been investigated. The observed influence on the viability of organisms may result from other mechanisms (Riehle and Hoechli 1973).

2.2 Sublimation of Ice

Dispersed particles, in particular, when smaller than 50 nm, are usually exposed by etching before replication. The thickness of the evaporated ice must be known

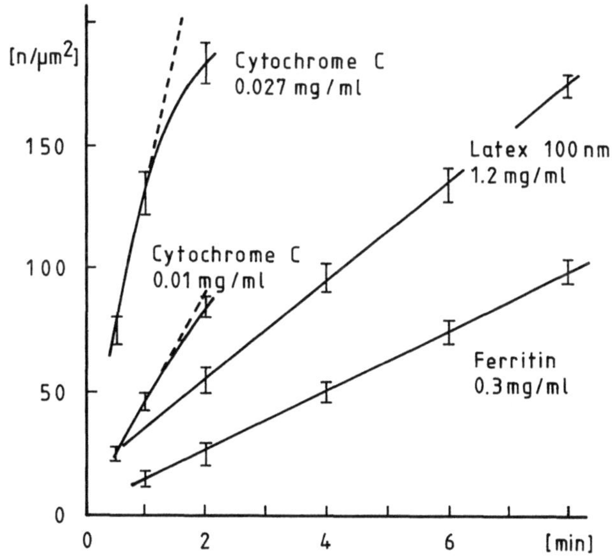

Fig. 2. Particle density versus etching time at 173 K. The particle densities of latex (scale is expanded 100×) and ferritin increase linearly with time, indicating constant sublimation rates. Cytochrome C (MW 12500) is about the smallest molecule which can be reliably counted. Deviation of the curves from linearity after 1 min indicates aggregation of the molecules during etching (Bachmann et al. 1974a)

when absolute particle concentrations are of interest (see Sect. 3.2). The vapour pressure data and rates of sublimation, which are commonly used for estimating the layer thickness are, of course, valid only for pure water. In solutions, the sublimation is progressively slowed down and comes practically to a stop after some time. This slowing down of sublimation is much stronger than the reduction of the vapour pressure in solutions according to the laws of thermodynamics. The reason for this effect is that only water molecules evaporate, while the solute remains on the specimen surface, blocking more and more the escape of water. Unfortunately, there are no salts or buffer systems which are volatile under the conditions of freeze-etching. The better the cryofixation, the sooner sublimation will be impeded because the size of the ice compartments which are exposed by fracturing are smaller. At the grain boundary, however, the segregated solute stops further sublimation.

The presence of low molecular weight substances impairs also the visibility, e.g. of macromolecules. Although not volatile, these substances have a high lateral mobility on the ice surface at the etching temperature. Even small protein molecules tend to clump after prolonged etching (Fig. 2). Miller et al. (1983) have reported on artefacts associated with the "quick-freeze/deep-etch" techniques (cf. Heuser 1981; Heuser 1983): in buffer solutions filament-like structures were observed in the absence of authentic biological filaments in the sample. They may be caused not only by insufficient cryofixation, but also by association of the salts during deep-etching or freeze-drying. If glycerol is used for cryoprotection, no etching occurs and particles can only be observed on the fracture surface (Gulik et al. 1982).

3 Determination of Particle Concentrations and Molecular Weights

The mass of colloidal particles and macromolecules can be determined by measuring a variety of their physicochemical properties, e.g. diffusion, sedimentation velocity or equilibrium, light or X-ray scattering (for a review see e.g. Hiemenz 1984). Each of these methods has advantages and disadvantages and may give very precise data, but all of them require that the particles are of the same size, or that at least their size distribution is known, and that they do not interact. In order to obtain strong signals, measurements frequently have to be made at higher concentrations where the latter condition is not valid so that extrapolations are required. Furthermore, the dispersions have to be stable throughout the measurement. If the above conditions cannot be satisfied, electron microscopy can still give quantitative information on particle weights.

The volume of colloidal objects can be measured by electron microscopy, thus their weight can be calculated if their density is known. The selectivity of this technique cannot be surpassed. The method is hampered by the limited accuracy by which the volume of small shadowed particles can be measured and by the uncertainty of the assumptions concerning density.

3.1 Weight Measurements on Dry Specimens

Williams and Backus (1949) used a more direct method for determining particle weights, which is based solely on the counting of particles in a given volume and which does not need any assumptions on their properties. A virus suspension of known weight concentration was mixed with a dispersion of reference particles (polystyrene-latex) whose concentration (number per volume) was also known. This mixture was sprayed onto mica in a way that discretely bound droplet patterns could be observed on the shadowed replica. When all particles within the boundary of a droplet were counted, its volume could be calculated by the number of reference particles. The weight of the virus particles resulted from their number and concentration relative to that of the reference. This elegant technique has some practical drawbacks: all the particles contained in the drop have to be counted, they are, however, often densely packed after drying; the sample has to be mixed, e.g. with latex, with which it might possibly interact; finally, it is difficult to determine the precise concentration of latex suspensions.

3.2 Weight Measurements by Freeze-Etching

The difficulties of the above procedure can be avoided while their advantages are maintained when the dispersions are freeze-etched instead of dried, thus avoiding particle clumping: the number of particles per unit area is determined; when the thickness of the evaporated layer of the solvent is known, the particle concentration per volume can be calculated.

The calculation of particle concentrations in the fluid specimen from the number of particles per area on a freeze-etched replica is only justified if the ran-

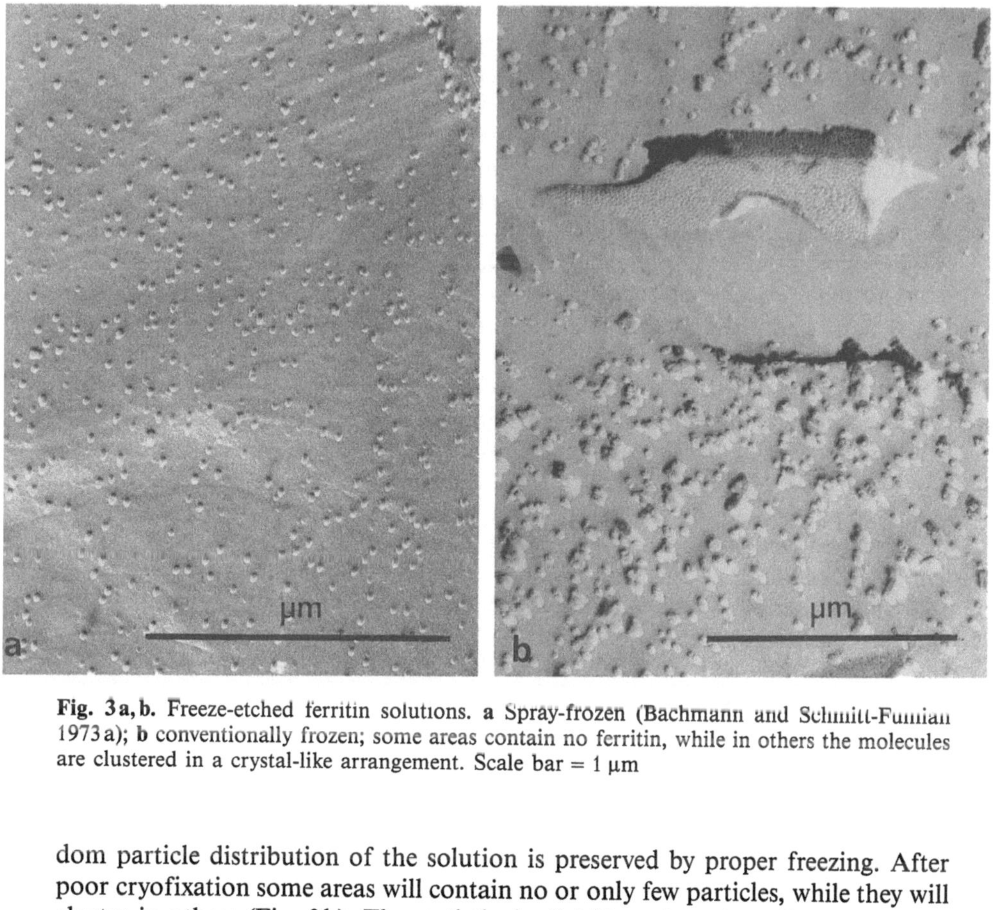

Fig. 3a, b. Freeze-etched ferritin solutions. **a** Spray-frozen (Bachmann and Schmitt-Fumian 1973a); **b** conventionally frozen; some areas contain no ferritin, while in others the molecules are clustered in a crystal-like arrangement. Scale bar = 1 µm

dom particle distribution of the solution is preserved by proper freezing. After poor cryofixation some areas will contain no or only few particles, while they will cluster in others (Fig. 3b). The statistical reliability of random distribution has to be checked by particle counting on different areas or in different droplets of the replica. The obstruction of particle observation by salts and other solutes with low molecular weights has been discussed in Section 2.2. Albumin and fibrinogen molecules can be readily counted in 5×10^{-3} M phosphate buffer, for example.

When the particles to be counted are considerably smaller than the etching depth, the simplification is justified that the observed objects represent the particles which were contained within the evaporated layer of the solvent. The number of particles which are affected or lost by the fracturing process and those only partially exposed by the etching is small compared to the ones which are clearly visible. The magnitude of this systematic error is further reduced when a standard of comparable size is used. Particles with dimensions comparable to or larger than the etching depth require the application of stereological techniques.

The number of particles per area increases linearly with the etching time, at least for a limited period of time after fracturing (Fig. 2). This indicates a constant sublimation rate during that period. Salt concentrations up to 0.3% had no noticeable effect. The inaccuracies involved with determining absolute etching rates have already been discussed (see also Bachmann and Mayer, Chap. 1, this

Vol.). In practical freeze-etching they are believed to be mainly caused by deviations of the specimen temperature and the influence of the solutes. At 170 K a temperature deviation of 1 K changes the sublimation rate by 30%.

All the difficulties connected with the determination of the thickness of the evaporated ice layer can be completely avoided by freeze-etching the solution of a standard molecule with known molecular weight and concentration under exactly the same conditions as the unknown sample, e.g. by using a multiple-specimen stage in the freeze-etching apparatus. The standard solutions may also contain a similar amount and kind of low molecular weight material as the sample. Contrary to the previously mentioned method of Williams and Backus, the standard is not mixed with the sample and the selection of the standard is not restricted to particles which are much larger than those of the sample in order to be distinguishable. Purified solutions of proteins, e.g. bovine serum albumin, have been used as standards; their concentrations can be readily measured in the UV photometer. When spray-freezing is chosen for cryofixation, the two solutions containing the sample and the standard are sprayed one after the other into the same recipient with quenching fluid. When the mixture is freeze-etched, the replica shows side by side spray droplets containing the sample and droplets of the standard solution. The use of a standard makes it unnecessary to calculate the absolute thickness of the evaporated solvent layer. The unknown molecular weight of the sample (M_x) is then given by

$$M_x = M_{st} \cdot \frac{n_{st} \cdot c_x}{n_x \cdot c_{st}} ,$$

where M_{st} is the molecular weight of the standard; n_{st} and n_x the number of particles per area of standard and sample, c_{st} and c_x their weight concentrations. The method is suitable for measuring particle weights ranging from 10^4 to 10^8 (Bachmann et al. 1975; Junger and Bachmann 1977) and is also applicable to micellar solutions (Bachmann et al. 1981).

4 Measurements of Size and Shape

On freeze-etch replicas, macromolecules can be observed even when they have an unusually high degree of hydration. They may not be or may hardly be visible by negative stain or in the cryoelectron microscope because of their low contrast. At present, the exact nature of the surface created by freeze-etching is not known, for example the interface of a hydrated protein molecule with the vacuum. The surface relief which is formed by etching does not represent an equilibrium state with water vapour, but is based on the kinetics of unidirectional evaporation of water onto the cold trap. We cannot yet quantify the influence of the ion distribution close to the protein surface on the sublimation rate of water. We also do not know what kind of surface is finally represented by the replicating material, since we do not know what consequence the local interaction between an impinging "hot" metal atom and the water molecules of the specimen has on the final position of the metal atom.

In the following, size measurements of entities which are smaller or of similar dimension as the etching depth are separately discussed.

4.1 Small Particles (Protein Molecules)

Small particles can be directly measured with a particle size analyzer. The relative influence of shadowing and "hydration layers", on "true" particle size is, of course, most pronounced on objects with dimensions of about 10 nm or less. Preliminary experiments in our laboratory on proteins, colloidal silica and colloidal gold indicate that particles, which were deep-etched from solution and rotary-shadowed may appear up to 2 nm larger in diameter than when measured by other techniques. The thickness of the evaporated film was extrapolated to zero and should therefore not contribute significantly to this discrepancy. Without giving an interpretation we conclude from these studies that size measurements are realistic, at least within these limits. There is increasing evidence that resolution in freeze-etching might not be limited primarily by shadowing (Gross, Chap. 10, this Vol.). For measurements, rotary shadowing with as thin a metal film as possible has definite advantages over unidirectional shadowing. It proved necessary, however, to observe also unidirectionally shadowed replicas for an unambiguous interpretation of the specimen surface. Overshadowing may lead to very impressive micrographs and enhance periodic structures (Heuser 1983), yet this practice should obviously not be used to measure individual small particles.

In spite of the limited accuracy, measurements of freeze-etched macromolecules can be of great value for structural research since both, the volume of the hydrated molecule and the form factor (axial ratio), can be directly obtained. This may be demonstrated for two serum proteins, fibrinogen and albumin. For the fibrinogen molecule, axial ratios have been discussed varying between 1 : 1 and 1 : 10 based on electron microscopy and on physicochemical data (Doolittle 1973). When the molecules are dried on mica and shadowed, they appear 45 nm long and show a trinodular structure (Hall and Slayter 1959). Freeze-etched, the molecules are sausage-shaped, 45×9 nm, with no indication of being composed of three domains (Fig. 4) (Bachmann et al. 1975). The hydration of the molecule, calculated from the freeze-etch dimensions is unusually high, approx. 4 g H_2O g^{-1} protein. The observed data, i.e. the axial ratio *and* the volume, agree, however, very well with measurements of the sedimentation and diffusion coefficients, while they cannot be explained by the electron microscope image of the dry molecule (Doolittle 1973). The water contained in the volume observed after freeze-etching does not need to be bound to the protein; the larger part might just be trapped and have a significantly lower rate of sublimation.

Bovine serum albumin (BSA) is, according to textbooks of biochemistry, an elongated molecule with an axial ratio between 1 : 3 and 1 : 4, and a hydration of 0.34 g H_2O g^{-1} protein. On freeze-etch replicas the molecule appears spherical with a volume corresponding to an unusually high hydration between 2 and 3 g H_2O g^{-1} protein (Fig. 5) (Bachmann et al. 1974b). Again these dimensions conform well with the hydrodynamic data and with the fact that BSA shows no

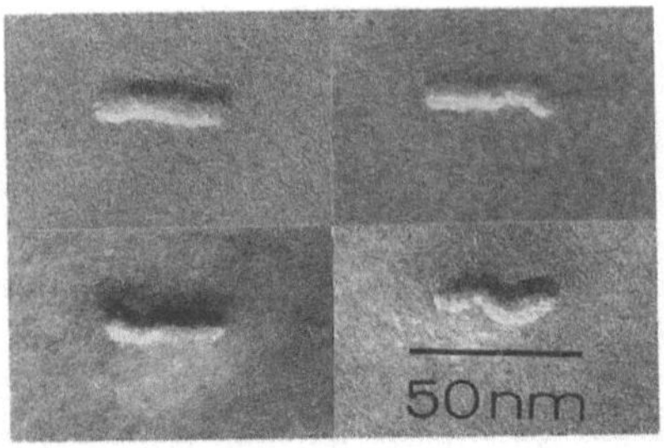

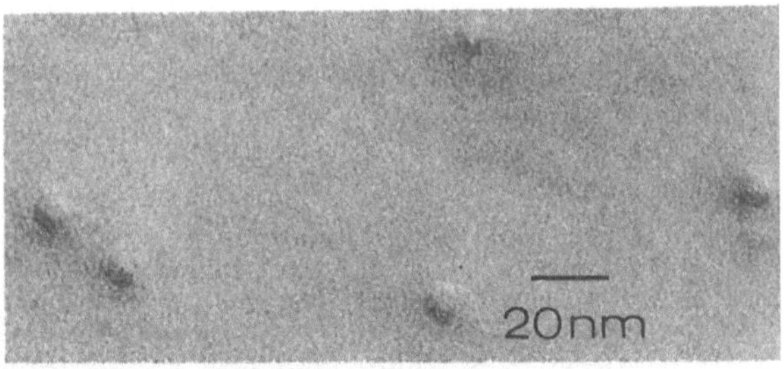

Fig. 4 **Fig. 5**

Fig. 4. Selection of rod-shaped fibrinogen molecules, spray-frozen and freeze-etched (Bachmann et al. 1975)

Fig. 5. Bovine serum albumin molecules, spray-frozen and freeze-etched. No elongated particles were observed (Bachmann et al. 1974b)

contrast in negative stain. The spherical shape of spray-frozen, freeze-etched BSA and the high hydration was confirmed by Schmidt and Buchheim (1982). These examples may show that freeze-etching can be useful for the interpretation of measurements based on physicochemical properties. The latter can be performed with less experimental limits, e.g. on buffer concentration, and they are more sensitive to subtle, conformational changes than electron microscopy. In the case of instable solutions (disintegration or association of particles or molecules), freeze-etching offers the additional advantage that the molecular weight can be checked independently from the apparent size of the particles by their number only.

4.2 Larger Objects (Liposomes, Vesicles, Aggregates)

Measurements on dispersed objects, which are comparable with or are larger than the etching depth, require stereological approaches as they exist for ultrathin sections. The etching depth can to a certain degree be seen in analogy to the section thickness. Zero thickness may be simulated by minimal etching (Gulik et al. 1982) or by measuring structures along the actual fracture zone which can often be distinguished from the etched dimensions. Guiot et al. (1980) calculated the size of liposomes from the profiles using mathematical procedures described by Wicksell (1925) and Bach (1967). In a study of the size of casein micelles, Holt et al. (1978) compared sectioning with freeze-etching using a correction algorithm devised by Goldsmith (1967).

Special techniques can be applied for the analysis of freeze-fractured liposomes which show either convex or concave profiles. Van Venetie et al. (1980) used the technique of Losa et al. (1978) to obtain the real diameter of the liposomes by measuring only the equatorially broken vesicles; those concave-fractured vesicles were selected of which precisely one half of the fracture face is covered with Pt/C, at a shadowing angle of 45 °C. The non-equatorially broken

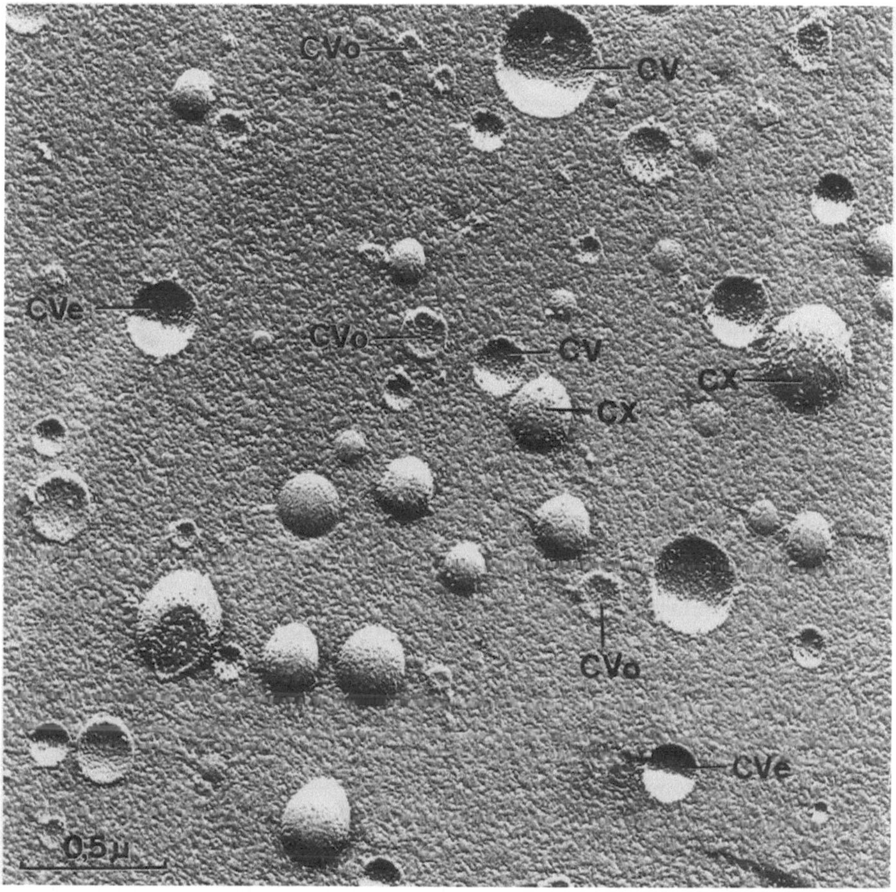

Fig. 6. Overview of freeze-fracture replica of a microsomal fraction. Vesicles of various origin form convex (*CX*) or concave (*CV*) caps. Concave caps without cast shadow (*CVo*) and "equatorial" profiles (*CVe*) with half the cavity covered by cast shadow can be used for analysis (Losa et al. 1978)

vesicles display fracture faces, the surface of which is covered with more than 50% Pt/C (Fig. 6). In a slightly different approach, Weibel et al. (1976) used only those concave fracture faces which were totally covered, with the shadowing material at 45 °C incidence (Fig. 6).

5 Structure of Dispersions in Bulk and at Interfaces

Interaction of biological material with interfaces in the course of specimen preparation is covered by Kellenberger (Chap. 2, Sect. 3, this Vol.). Some aspects in connection with thin frozen specimens are discussed by Dubochet et al. (Chap. 5, this Vol.). Conformational changes on proteins have frequently been observed when the molecules are adsorbed on solid substrates or on the liquid/gas interface

(MacRitchie 1986; Norde 1986). Freeze-etching and cryosectioning are practically the only EM techniques by which entities in the bulk of a liquid dispersion can be observed, undisturbed by phase boundaries. In addition, freeze-etching offers the interesting possibility to check whether adsorption causes structural changes which are on an electron microscope scale e.g. in proteins. Dispersed particles can be observed, for example in bulk and on the surface of spray-frozen droplets. When protein solutions, which contain solid particles, e.g. mica flakes (Heuser 1983), are freeze-etched the replica shows molecules in the adsorbed and in the dissolved state. Since they are seen in the hydrated condition, the effects of adsorption and dehydration in the course of EM-specimen preparation can be separately studied.

References

(References marked by an asterisk are review papers or books respectively)

*Bach G (1967) Kugelgrößenverteilung und Schnittkreisverteilung, ihre wechselseitigen Beziehungen und Verfahren zur Bestimmung der einen aus der anderen. In: Weibel ER, Elias H (eds) Quantitative methods in morphology. Springer, Berlin Heidelberg New York, pp 23–45

Bachmann L, Schmitt WW (1971) Improved cryofixation applicable to freeze-etching. Proc Natl Acad Sci USA 68:2149–2152

Bachmann L, Schmitt-Fumian WW (1973a) Spray-freeze-etching of dissolved macromolecules, emulsions and subcellular components. In: Benedetti EL, Favard P (eds) Freeze-etching: techniques and applications. Soc Franc Microsc Electron, Paris, pp 63–72

Bachmann L, Schmitt-Fumian WW (1973b) Spray-freezing and freeze-etching. In: Benedetti EL, Favard P (eds) Freeze-etching: techniques and applications. Soc Franc Microsc Electron, Paris, pp 73–79

Bachmann L, Talmon Y (1984) Cryomicroscopy of liquid and semiliquid specimens: direct imaging versus replication. Ultramicroscopy 14:211–218

Bachmann L, Fritzmann H, Schmitt-Fumian WW (1974a) Molecular and particle weight determination by counting. In: Sanders JV, Goodchild DJ (eds) Electron microscopy, vol. 2. Proc 8th Int Congr Electron Microsc, Canberra, pp 18–19

Bachmann L, Junger E, Lederer K, Schmitt-Fumian WW (1974b) Size and shape of hydrated protein molecules. In: Sanders JV, Goodchild DJ (eds) Electron microscopy, vol. 2. Proc 8th Int Congr Electron Microsc, Canberra, pp 40–41

Bachmann L, Schmitt-Fumian WW, Hammel R, Lederer K (1975) Size and shape of fibrinogen 1. Electronmicroscopy of the hydrated molecule. Makromol Chem 176:2603–2618

Bachmann L, Dasch W, Kutter P (1981) Elektronenmikroskopische Untersuchungen von Mizellen. Ber Bunsenges Phys Chem 85:883–887

Brüggeller P, Mayer E (1980) Complete vitrification in pure liquid water and dilute aqueous solutions. Nature (London) 288:569–571

Buchheim W (1972) Zur Gefrierfixierung wäßriger Lösungen. Naturwissenschaften 59:121

Buchheim W, Welsch U (1977) Freeze-etching of unglycerinated tissue dispersions by application of the oil emulsion technique. J Microsc (Oxford) 111:339–349

*Davies, JT (1972) Turbulence phenomena. Academic Press, London New York

*Doolittle RF (1973) Structural aspects of the fibrinogen to fibrin conversion. In: Anfisen CB, Edsal JT, Richards FM (eds) Advances in Protein Chemistry, vol 27. Academic Press, London New York, pp 2–102

Goldsmith PL (1967) The calculation of true particle size distributions from the sizes observed in a thin slice. Br J Appl Phys 18:813-830

*Green HL (1953) Atomization of liquids. In: Hermans JJ (ed) Flow properties of disperse systems. Elsevier, North Holland. Biomedical Press, Amsterdam New York, p 299

Guiot P, Baudhuin P, Gotfredsen C (1980) Morphological characterization of liposome suspensions by stereological analysis of freeze-fracture replicas from spray frozen samples. J Microsc (Oxford) 120:159–174

Gulik A, Aggerbeck LP, Dedieu JC, Gulik-Krzywicki T (1982) Freeze-fracture electron microscopic analysis of solutions of biological molecules. J Microsc (Oxford) 125:207–213

Hall CE, Slayter HS (1959) The fibrinogen molecule: its size, shape and mode of polymerization. J Biophys Biochem Cytol 5:11–16

Heuser JE (1981) Quick-freeze, deep-etch preparation of samples for 3-D electron microscopy. Trends Biochem Sci 6:64–68

Heuser JE (1983) Procedure for freeze-drying molecules adsorbed to mica flakes. J Mol Biol 169:155–195

*Hiemenz PC (1984) Polymer chemistry. The basic concepts. Dekker, New York Basel

Holt C, Kimber AM, Brooker B, Prentice JH (1978) Measurements of size of bovine casein micelles by means of electron microscopy and light scattering. J Colloid Interface Sci 65:555–565

Junger E, Bachmann L (1977) Structure of pyruvate dehydrogenase complex comparison between freeze-etching and negative staining. Biochim Biophys Acta 481:364–376

Losa GA, Weibel ER, Bolender RP (1978) Integrated stereological and biochemical studies on hepatocytic membranes III. Relative surface of endosplasmatic reticulum membranes in microsomal fractions, estimated on freeze-fracture preparations. J Cell Biol 78: 289–308

*MacRitchie F (1986) Spread monolayers of proteins. Adv Colloid Interface Sci 25:341–385

*Mayer E (1985) Vitrification of pure liquid water. J Micros (Oxford) 140:3–15

Mayer E, Brüggeller P (1982) Vitrification of pure liquid water by high pressure jet freezing. Nature (London) 298:715–718

*Menco BPhM (1986) A survey of ultra-rapid cryofixation methods with particular emphasis on applications to freeze fracturing, freeze-etching und freeze-subsitution. J Electron Microsc Tech 4:177–240

Miller KR, Prescott CS, Jacobs TL, Lassignal NL (1983) Artifacts associated with quick-freezing and freeze-drying. J Ultrastruct Res 8:123–133

Moor H, Kistler J, Müller M (1976) Freezing in a propane jet. Experientia 32:805

Müller M, Meister N, Moor H (1980) Freezing in a propane jet and its application in freeze-fracturing. Mikroskopie (Wien) 36:129–140

*Norde W (1986) Adsorption of proteins from solution at the solid-liquid interface. Adv Colloid Interface Sci 25:267–340

*Plattner H, Bachmann L (1982) Cryofixation: A tool in biological ultrastructural research. Int Rev Cytol 79:237–304

Plattner H, Fischer WM, Schmitt WW, Bachmann L (1972) Freeze-etching of cells without cryoprotectants. J Cell Biol 53:116–126

Pscheid P, Schudt C, Plattner H (1981) Cryofixation of monolayer cell cultures for freeze-fracturing without chemical pretreatments. J Mixrosc (Oxford) 121:149–167

Riehle U, Hoechli M (1973) The theory and technique of high pressure freezing. In: Benedetti EL, Favard P (eds) Freeze-etching, techniques and applications. Soc Franc Microsc Electron, Paris, pp 31–61

*Robards AW, Sleytr UB (1985) Low temperature methods in biological electron microscopy. In: Glauert AM (ed) Practical methods in electron microscopy, vol. 10. Elsevier, Amsterdam

Schmidt DG, Buchheim W (1982) On the size of small protein particles determined by electron microscopy of unidirectionally shadowed freeze-etched preparations. J Microsc (Oxford) 126:347–351

Venetie R van, Leunissen-Bijvelt J, Verkleij AJ, Ververgaert PHJTh (1980) Size determination of sonicated vesicles by freeze-fracture electron microscopy, using the spray-freezing method. J Microsc (Oxford) 118:401–408

Ververgaert PHJTh, Verkleij AJ, Verhoeven JJ, Elbers PF (1973) Spray-freezing of liposomes. Biochim Biophys Acta 311:651–654

Weibel ER, Losa G, Bolender RP (1976) Stereological method for estimating relative membrane surface area in freeze-fracture preparations of subcellular fractions. J Microsc (Oxford) 107:255–266

Wicksell SD (1925) The corpuscle problem. A mathematical study of a biometric problem. Biometrika 17:84–99

Williams RC, Backus RC (1949) Macromolecular weights determined by direct particle counting. J Am Chem Soc 71:4052–4057

Chapter 10 High Resolution Metal Replication of Freeze-Dried Specimens

HEINZ GROSS[1]

1 Introduction

Freeze-drying followed by heavy-metal shadowing is an established and straightforward approach to routine structural studies of dehydrated biological specimens (Nermut and Frank 1971; Kistler et al. 1977; Gross et al. 1985). Very thin specimens such as isolated membranes or isolated macromolecules can be directly adsorbed on carbon-coated grids. After rapid freezing, the grids are transferred into suitable high-vacuum equipment for freeze-drying and shadowing. After stabilizing the heavy-metal coat with carbon (C-backing), the specimens are warmed to room temperature and transferred via atmospheric conditions into the transmission electron microscope (TEM). In other words, the entire sandwich consisting of C-supporting film, biological specimen, metal coat and C-backing film is examined in the microscope.

In order to improve the structural resolution of the freeze-drying and shadowing approach, at least three lines have to be followed at the same time: the improvement of specimen preparation, the refinement of the shadow-casting technique and the development and application of computer-based image averaging techniques allowing both an image enhancement and an objective comparison of image quality.

Sheet-like structures with a regular lattice are advantageous test specimens to demonstrate methodological progress in freeze-drying as well as in shadowing. Periodic structures allow the application of digital image averaging like Fourier filtration (Crowther and Klug 1975; Unwin and Henderson 1975) or correlation averaging (Frank 1980; Saxton and Baumeister 1982) and surface relief reconstruction (Smith and Kistler 1977; Guckenberger 1985) routinely, with little computer effort.

2 Electron Microscopy and Image Processing

Pt/C and Ta/W films, the most frequently used shadowing materials, are stable enough under electron irradiation so that micrographs can be made under "high-

---

1 Institut für Zellbiologie, ETH-Hönggerberg, CH-8093 Zürich, Switzerland

Cryotechniques in Biological Electron Microscopy
Edited by R. A. Steinbrecht and K. Zierold

dose conditions". This allows focussing and correction of astigmatism on the area of interest.

The micrographs are screened and selected for image processing with a light-optical diffractometer. The resolution of structural detail that is recoverable from the averaged unit cell can be directly evaluated from the highest order spots in the optical transform. In addition, the diffraction pattern allows micrographs to be eliminated which show drift or astigmatism. Finally, the diffraction pattern is used to quantify the defocus. The first zero of the phase-contrast transfer function must lie at a frequency substantially higher than the highest diffraction spots.

The most promising micrographs were digitized and areas of 512×512 picture elements were subjected to digital image processing. Image averaging was performed by Fourier filtration and correlation averaging using the SEMPER system (Saxton et al. 1979). In the case of the specimens studied, which have a nearly perfect crystalline order, both averaging approaches gave exactly the same results. For imperfect crystals or the acquisition of information from crystal fragments, correlation averaging is advantageous.

Unidirectional shadowing has the advantage that three-dimensional information is contained in single micrographs. Based on the a priori assumption that the metal coat deposited by evaporation is continuous and of constant thickness in the direction of metal deposition, the complete surface relief can be reconstructed with relatively little computer effort. To fill the missing information perpendicular to the shadowing direction, one or two additional views with distinctly different shadowing azimuth are necessary. For objects with at least threefold internal symmetry, the complete relief can be reconstructed from a single micrograph (for details see Guckenberger 1985).

In practice, thin metal films are built up of more or less discrete clusters and therefore appear discontinuous when observed at high resolution. If decoration effects can be neglected, the randomly distributed clusters should merge into a continuous film upon averaging. Consequently, modulations of the film thickness should only be related to the geometry of the specimen surface and its orientation with respect to the evaporation source.

That after averaging the model assumptions on which surface relief reconstruction is based are correct, is shown by tilt series reconstruction of Ta/W shadowing films (Wildhaber et al. 1986) and by Monte Carlo simulations of the shadowing process using the relief reconstructed data (Winkler 1986).

3 Characterization of the Test Specimens

The HPI layer (hexagonally packed intermediate layer) is a two-dimensional protein crystal isolated from the cell wall of the bacterium *Deinococcus radiodurans*. Regularly arranged surface proteins ("S-layer") are common amongst prokaryotes for almost all phylogenetic branches (Sleytr and Messner 1983). They must be regarded as multifunctional proteins, which are involved in a wide variety of interactions between the cell and its environment. The HPI layer is a particularly useful test specimen for several reasons. It is an unusually robust structure which can easily be isolated in the form of extensive two-dimensional crystalline sheets and which possesses highly detailed surface structures. In addition, besides shadowing data, independent structural information obtained by negative staining (Baumeister et al. 1986), and unstained freeze-dried and frozen-hydrated preparations (Rachel et al. 1986) is already available.

The light-harvesting chlorophyll a/b protein complex (LHC) plays a key role in the process of photosynthesis occurring in the photosynthetic membranes of higher plants and green algae. The LHC is a multifunctional complex responsible for harvesting solar energy, transferring energy and stacking the thylakoid membranes. A striking property of the LHC that makes it suitable for structural studies is its tendency to form crystalline arrays in vitro. By controlled dialysis against moderate salt concentrations it is possible to obtain extensive, highly ordered two-dimensional crystals (Kühlbrandt et al. 1983). Such crystalline sheets of purified pea LHC have a nearly perfect hexagonal lattice with a lattice constant of 12.5 nm. The 3D-model calculated from tilted views of negatively stained crystalline sheets (Kühlbrandt 1984) shows that the unit cell is composed of two identical triangular units of opposite handedness. The structural units are inserted bidirectionally into the crystals, alternating in orientation perpendicular to the plane of the crystal. Consequently, after shadowing both surfaces reveal basically identical relief data.

4 Controlled Freeze-Drying

Substitution of the specimen water with deuterium oxide (D_2O) allows mass spectrometric separation of subliming specimen water (D_2O) and water vapour (H_2O) originating from elsewhere (Wildhaber et al. 1982). Figure 1 shows qualitatively the course of the sublimation rate of the specimen water (D_2O) during the freeze-drying process. To minimize specimen interactions with residual gases, freeze-drying was performed in the UHV chamber (background vacuum 10^{-9} mbar). The specimens were loaded via the high vacuum lock at 143 K (−130 °C). After warming to 193 K (−80 °C) D_2O appeared, reached a maximum and then slowly decreased and approached an equilibrium at a low sublimation rate after 1−3 h. A further warming to room temperature resulted in a further sublimation of specimen water with at least two characteristic maxima between 223 K (−50 °C) and 243 K (−30 °C) (Fig. 1). Before the specimens reached room temperature the sublimation rate of D_2O was unmeasurably small for both test specimens, indicating that at room temperature practically all specimen water has been removed.

The D_2O-experiments firstly indicate that the stronger bound water, which is removed between 223 K and 243 K, is that of hydration shells, necessary for the structural and mechanical stability of proteins (see Kellenberger, Chap. 2, this Vol.). Secondly, when freeze-drying is performed at specimen temperatures not higher than 193 K, a substantial amount of the hydration shells should be maintained.

This hypothesis is further supported by decoration experiments with silver. After storing thin silver films a few days we observed silver cluster migration and

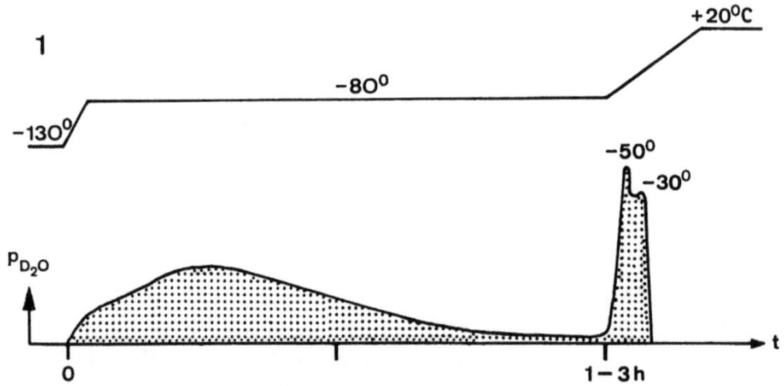

Fig. 1. The course of the subliming specimen water during the freeze-drying procedure. The specimen water was substituted with D_2O allowing mass spectrometric differentiation between specimen water (D_2O) and water vapour originating from elsewhere (H_2O). The specimen table was loaded at 143 K (−130 °C). Under UHV conditions, the temperature was increased to 193 K (−80 °C). After 1−3 h drying an equilibrium at a low sublimation rate was reached. Warming to room temperature resulted in a further sublimation of specimen water. For all tested specimens at least two characteristic desorption peaks between 223 K (−50 °C) and 243 K (−30 °C) were recorded. We believe that this stronger bound water is partly that of hydration shells (see Kellenberger, Chap. 2, this Vol.)

coalescence on both the biological specimen and the C-supporting film, in spite of C-backing. UHV experiments with and without the inlet of pure water vapour (Gross et al. 1978) proved that silver cluster migration occurs exclusively in the presence of surface water. Therefore, the condensation pattern of thin silver films can be used as a sensitive probe to test the existence of "surface water". Since cluster migration and coalescence was observed on both tested protein layers after freeze-drying at 193 K followed by silver coating, the existence of "shell water" is evident after drying at 193 K.

The influence of water shells on the structural preservation of our test specimens was studied by high resolution shadowing with Ta/W at 23 K (−250 °C) (see below). The hydration level was varied by subliming increasing amounts of stronger bound water. For that purpose, after freeze-drying at 193 K the specimen temperature was raised stepwise from 193 K (−80 °C) to 293 K (+20 °C) (Amrein 1985).

For the unusually robust structure of the HPI layer the removal of the stronger bound water results in already moderate structural alterations (Fig. 2a, b). On the micrographs practically no effects are detectable. The corresponding digital averages, however, show that warming to room temperature causes a blurring of structural details. The finest structural features of the inner surface, the small protuberances on the core of the protein are heavily damaged and a deep pore-like opening is formed in the centre of the core (Fig. 2b).

The LHC structure seems to be strongly stabilized by water shells (T. Thaler, H. Winkler, H. Gross, in prep.). A gradual removal of the stronger bound water results therefore in a drastic structural degradation (Fig. 3a−c). After warming to room temperature the surface structure is so dramatically damaged that practically no periodic features can be detected on either the micrograph or the cor-

Figs. 2 and 3. The influence of the hydration shells on the structural preservation of freeze-dried surfaces, revealed by shadowing with Ta/W at 23 K (−250 °C)

Fig. 2a, b. The inner surface of the HPI layer. **a** Freeze-dried at 193 K (−80 °C); **b** freeze-dried at 193 K (−80 °C) and then warmed to 293 K (+20 °C). While the micrographs show only little effect, the corresponding computer averages demonstrate that also the unusually robust structure of the HPI layer is altered by removing the water shells

responding optical diffraction pattern (Fig. 3c). A stepwise variation of the specimen temperature from 193 K to 293 K indicated that the structural alterations are continuous. Figure 3b shows an intermediate situation.

The specimen temperature at which freeze-drying is performed is always a compromise between tolerable length of drying time and structural preservation which is best theoretically at the lowest possible specimen temperature. The maximally tolerable freeze-drying temperature depends very much on the nature of the specimen. Our findings, which are in harmony with many other studies (for a recent review see Robards and Sleytr 1985), show that freeze-drying at 193 K is a good compromise: the time for drying of adsorbed sheet-like structures is tolerable (1 − 3 h) and water shells seem to be substantially maintained.

Collapse phenomena, like structural flattening, reported in connection with freeze-drying, may be caused by removing the water shells by the application of drying temperatures much higher than 193 K. It must be considered that without mechanical clamping of the grids by means of a suitable specimen table, the actual specimen temperature can be much higher than the routinely measured stage temperature. Structural studies with unstained freeze-dried and frozen-hydrated tubular photosynthetic membranes support the above statement. Concerning the flattening of the tubes (diameter 40 nm), there was no significant difference between frozen-hydrated tubes ("fully ice-embedded") and tubes freeze-dried at

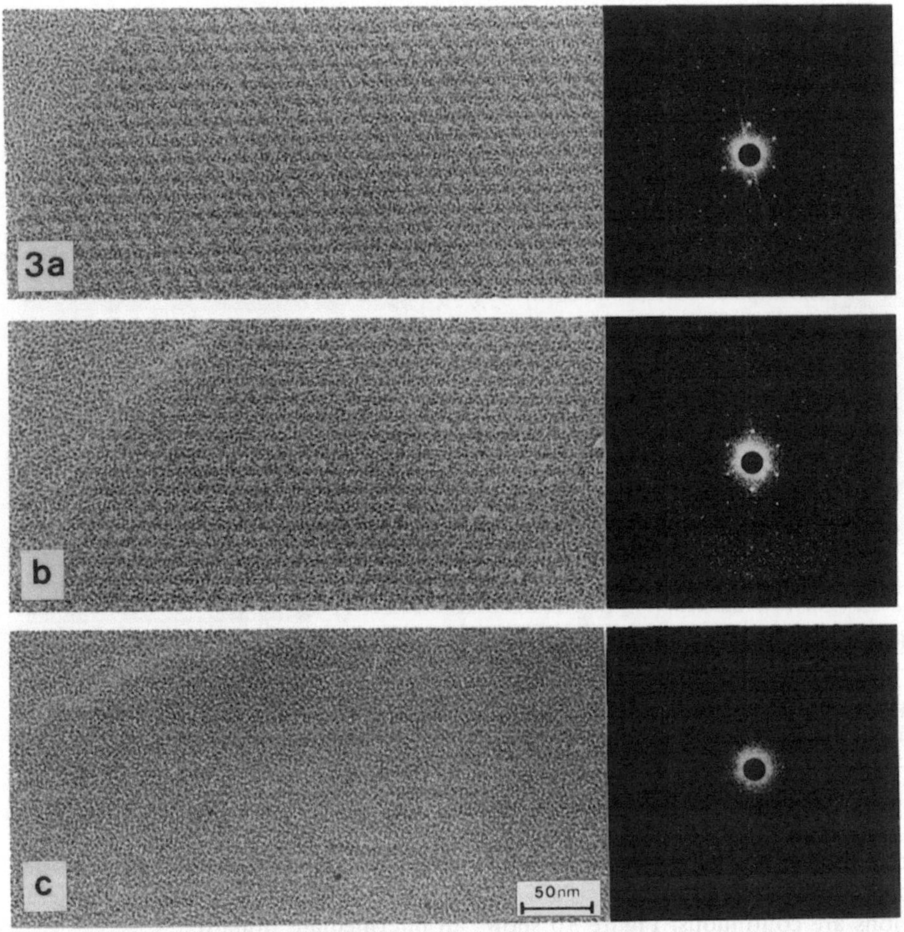

Fig. 3a–c. Crystalline arrays of light-harvesting complexes. **a** Freeze-dried at 193 K (−80 °C); **b** freeze-dried at 193 K (−80 °C) and then warmed to 233 K (−40 °C); **c** freeze-dried at 193 K (−80 °C) and then warmed to 293 K (+20 °C). Both the micrographs and the corresponding optical diffraction patterns prove that subliming of structural water results in a drastic alteration of the surface structure. After warming to room temperature, all periodic features are destroyed (Fig. 2c)

183 K (−90 °C). A succeeding warming to room temperature, however, caused a substantial flattening (Rachel et al. 1986).

5 High Resolution Shadowing

Perfect shadowing can easily be defined: the impinging atoms of the evaporated material must stick to the specimen where they hit. Consequently, a continuous metal film of uniform thickness in the shadowing direction is deposited, modulated only by statistical fluctuations. If viewed from the direction of the evaporation source, no contrast is observed. If the replica is viewed from a different angle, the resulting contrast is strictly related to the surface topography.

In practice, perfect shadowing cannot be attained experimentally. The formation of a thin metal film proceeds via thermal accommodation, surface diffusion, nucleation and crystal growth and crystal coalescence (for details see Maissel and Glang 1970; Venables et al. 1984). The formation of discrete clusters in the condensate is caused by the lateral mobility of trapped atoms and small nuclei of the evaporated material. If the exposed structures cause an unequal distribution of surface forces, nucleation and crystal growth occur preferentially at surface sites with higher binding energy. Therefore, at higher resolution such preferential nucleation (decoration) can be responsible for the revealing of structural details as well as the effect of shadow casting (Gross et al. 1985). Since the information obtained by shadowing and decoration is of a completely different nature, the interpretation of small details (below 5 nm) may be difficult.

In order to approach perfect shadowing, the lateral mobility of the condensing material must be reduced. For that purpose we tested shadowing at very low specimen temperature. At temperatures below 23 K (−250 °C), the saturation vapour pressures of all residual gases (H_2O, CO_2, CO, N_2, O_2) except that for H_2, are below 10^{-10} mbar. Consequently, when cooled to such a low temperature, the specimen surface acts as an effective cryopump for all impinging gas molecules from the residual gas atmosphere or for gas produced during processing (i.e. evaporation). Since the formation of a monomolecular contamination layer at 10^{-6} mbar takes only 1 s, shadowing at extremely low specimen temperature should, in practice, only be done in an adequately low vacuum. This was one of the reasons to develop equipment suitable for shadowing at 23 K (−250 °C) under UHV conditions ($p \leqslant 10^{-9}$ mbar, Gross et al. 1984, 1985) in collaboration with BALZERS (Liechtenstein). The machine (BAF 500 K) is equipped with a high vacuum airlock, enabling fast and contamination-free specimen transfer. UHV is restored within 5 to 10 min after locking.

By shadowing at 23 K improved topographic resolution is achieved with Pt/C and Ta/W, the most frequently used shadowing materials. With Pt/C, the preferred shadowing material for freeze-fracturing, the progress is evident and can be easily explained by the much reduced grain size at 23 K (Gross et al. 1985). With Ta/W the temperature effect is not so conspicuous. The grain size does not seem to be reduced substantially by shadowing at 23 K. Nevertheless, when compared to shadowing at 193 K (−80 °C), the digital averages show significantly sharper and finer structural details (Amrein 1985). The finest "truly shadowed" structures so far, the protuberances on the core of the HPI layer (Fig. 4c′) are not resolved by shadowing at 193 K.

Three factors probably contribute to the increased resolution with lowered specimen temperature: firstly, surface diffusion of the condensing atoms and nuclei of the shadowing material is reduced. Secondly, molecular motion of the specimen itself is reduced. Thirdly, structural alterations caused by heat radiation are reduced. The small granularity observed after shadowing at 23 K may be caused primarily by jumping of the arriving atoms, before they are thermalized. On the other hand, post-shadowing artefacts (warming to room temperature, transfer

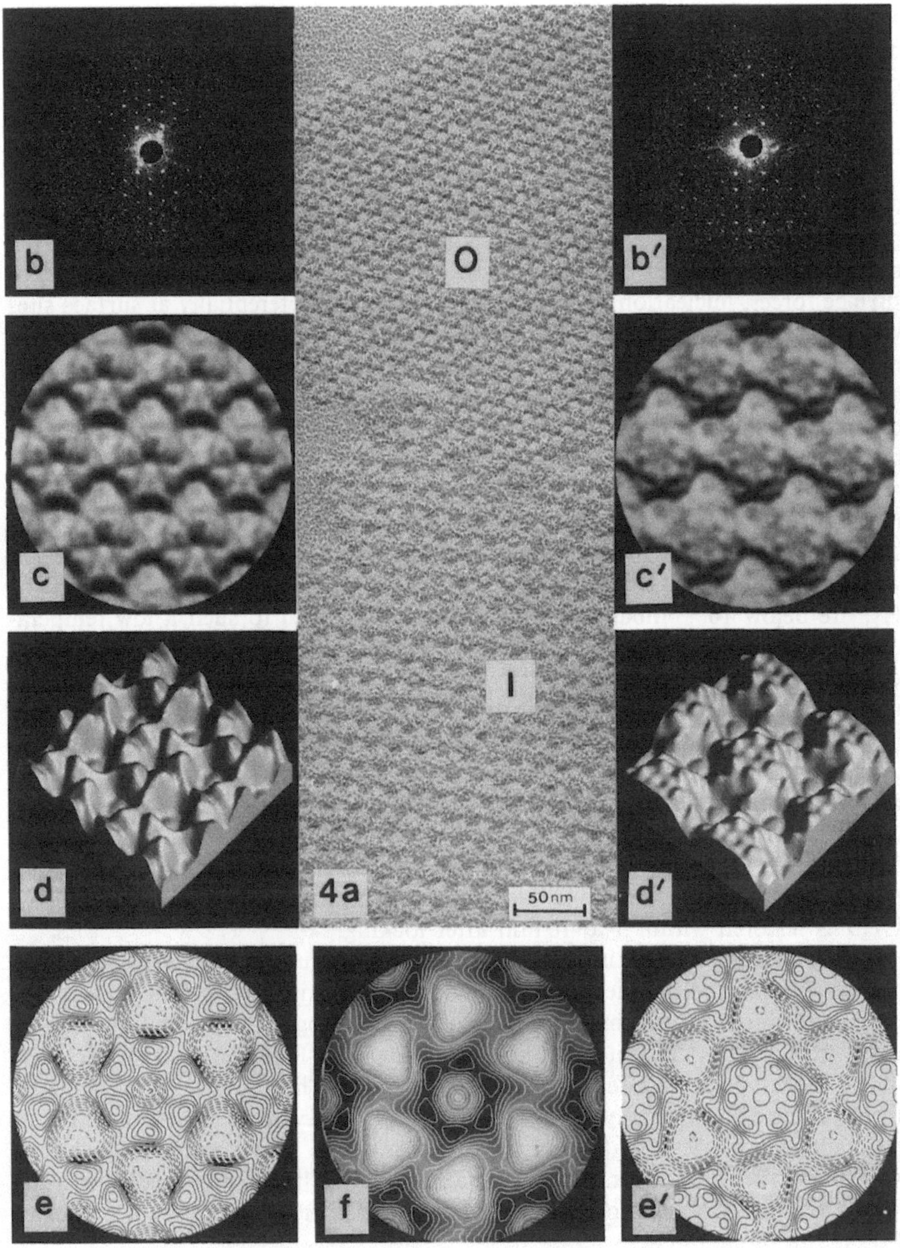

Fig. 4a–f. Surface-relief data of the HPI layer (**a–e**) after freeze-drying at 193 K (−80 °C) and unidirectional shadowing with Ta/W (0.5 nm) at 23 K (−250 °C, elevation angle 45°). The six-fold symmetry of the HPI layer allowed a complete surface reconstruction from one unidirectionally shadowed specimen alone. **a** Micrograph: *I* inside (**b′–e′**); *O* outside (**b–e**); **b, b′** optical diffraction patterns; **c, c′** corresponding computer averages; **d, d′** corresponding reconstructed reliefs, displayed by surface-shading technique ("views"); **e, e′** reconstructed reliefs, displayed as contour-line plots (*continuous lines* represent elevations, *broken lines* valleys). The network-

to the electron microscope via atmospheric conditions) cannot be excluded, in spite of C-backing.

As already discussed, shadowing may be quasi-perfect if the small clusters show a random distribution over the specimen surface. Then averaging results in a continuous film with constant thickness in the direction of metal deposition. The relief reconstruction procedure allows one to separate shadowing from decoration effects (Guckenberger 1985). From such an analysis and from experiments in which decoration was inhibited by the pre-evaporation of a thin carbon film (Winkler et al. 1985), it was concluded that in the case of unidirectional shadowing with Ta/W at 23 K, decoration effects can be neglected. Therefore, averaging enough repeating units allows the extraction of significant structural information down to the average cluster size, which conclusively limits the resolution of the structural record.

In Fig. 4 the best shadowing data obtained from the freeze-dried HPI layer are combined. The micrograph (Fig. 4a) shows that the hexagonal array of morphological units (lattice constant 18 nm) is prominent on both surfaces and that two different surface aspects can be clearly distinguished. The protein complexes consist of massive "cores" composed of two differently sized hexagons and "spokes" interconnecting the complexes and dividing the periphery of the cores into "rosettes" (holes). The spokes emanate from the corner of the larger hexagon. On the inner surface of the smaller hexagon, the already discussed six small protuberances (diameter 1.5 nm) can easily be recognized (Fig. 4c′, d′). The inner surface, as revealed by surface relief reconstruction agrees well with the 3D-model obtained by tilt series reconstruction of the negatively stained HPI layer (Baumeister et al. 1986). For the outer surface, however, this is not true. While the surface of the 3D-model from negative staining data is almost flat, a pronounced relief is obtained after relief reconstruction showing "dumbbell-shaped" morphological subunits (Fig. 4d, e). The flattening of the protein in negatively stained preparations results from interactions with the C-supporting film ("adsorption") probably reinforced by surface tension. It is a distinct advantage of the shadowing approach that freeze-drying can be applied and that we always look at surfaces which are not in contact with the specimen support.

Figure 4f shows the symmetrized average of the unstained, frozen-hydrated HPI layer (Rachel et al. 1986). The projected structure agrees well with the shadowing data (Fig. 4e, e′), revealing the same core structure and the same handedness and location of the spokes. Small differences (amount and position of the mass originating from the "hillocks" of the outside) may be caused by an adsorption artefact, similar to that in negatively stained preparations.

like lattice structure of the HPI layer consists of massive cores and comma-shaped interconnecting spokes. The core is composed of two differentially sized hexagons and reveal highly differentiated and characteristic surface reliefs. The finest resolved structural details are below 2 nm. **f** Projected structure of the unstained, "frozen-hydrated" HPI layer (sixfold, rotationally symmetrized average with contour lines superimposed). The good agreement of the reconstructed external morphology (**e**, **e′**) with the ice-embedded projected structure (**f**), reveals that after freeze-drying at 193 K (−80 °C) the structure is well preserved (**f** after Rachel et al. 1986)

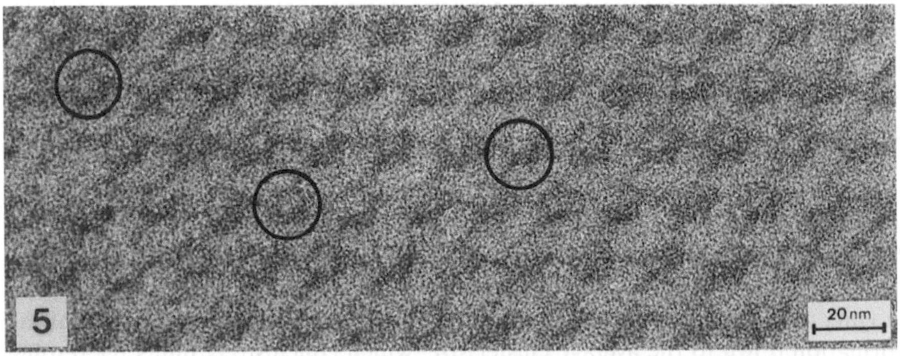

Fig. 5. High magnification (500000×) of the inner surface of the HPI layer. Optimal shadowing clearly reveals the structural heterogeneity of the single repeating units, indicating that the currently achieved averaged structural resolution might not be limited by shadowing

The good agreement of the external morphology (Fig. 4e, e′) with the projected structure obtained from the unstained, frozen-hydrated HPI layer (Fig. 4f) indicates that an excellent structural preservation can be achieved with carefully controlled freeze-drying ("maintaining hydration shells"). This is supported by a close agreement of the structural data obtained from the unstained, frozen-hydrated ("fully ice-embedded") and unstained, freeze-dried HPI layer (Rachel et al. 1986). It should be mentioned, however, that this is true for relatively rigid structures only which withstand the transfer into distilled water without serious changes in morphology. The specimens must be (briefly) rinsed with distilled water before freezing, in order to reveal uncontaminated surfaces after drying.

For specimens suitable for freeze-drying, however, the combination of optimized freeze-drying followed by high resolution shadowing at 23 K is a powerful tool to extract relief data at a high resolution level. This is particularly interesting for very thin structures like fibrous proteins, which do not exclude a sufficient amount of stain to be visualized routinely with negative staining, and for freeze-fracturing where shadowing is the sole contrasting method.

The high contrast in combination with the improved shadowing quality at 23 K allowed the detection of the unexpectedly high structural heterogeneity of the repeating morphological units of the HPI layer (Fig. 5). Intrasubunit variations ("short-range disorder") can cause a blurring of the averaged features and therefore limit the structural record. In other words, the currently achieved structural resolution does not seem to be limited by the shadowing technique, but by the specimen itself, the specimen preservation technique (isolation, adsorption, dehydration) or by post-shadowing artefacts (warming to room temperature, specimen transfer via atmospheric conditions), in spite of C-backing.

References

(References marked by an asterisk are review papers or books respectively)

Amrein M (1985) Der Einfluß von Restwasser nach Gefriertrocknen auf die Strukturerhaltung eines zweidimensionalen Proteingitters – untersucht mittels Schwermetallbeschattung bei extrem tiefer Objekttemperatur. Diplomarbeit, Eidgen Tech Hochschule Zürich

Baumeister W, Bart M, Hegerl R, Guckenberger R, Hahn M, Saxton WO (1986) Three-dimensional structure of the regular surface layer (HPI) of *Deinococcus radiodurans*. J Mol Biol 187:241–253

Crowther RA, Klug A (1975) Structural analysis of macromolecular assemblies by image reconstruction from electron micrographs. Annu Rev Biochem 44:161–182

*Frank J (1980) The role of correlation methods in computer image processing. In: Hawkes PW (ed) Topics in physics, vol 13. Springer, Berlin Heidelberg New York, pp 187–222

Gross H, Kuebler O, Bas E, Moor H (1978) Decoration of specific sites on freeze-fractured membranes. J Cell Biol 79:646–656

Gross H, Müller T, Wildhaber I, Winkler H, Moor H (1984) Freeze-fracturing and replication at −260 °C. In: Proc 42nd Annu EMSA Meet, Detroit MI. San Francisco Press, San Francisco, pp 12–15

Gross H, Müller T, Wildhaber I, Winkler H (1985) High resolution metal replication, quantified by image processing of periodic test specimens. Ultramicroscopy 16:287–304

Guckenberger R (1985) Surface reliefs derived from heavy-metal-shadowed specimens – fourier space techniques applied to periodic objects. Ultramicroscopy 16:357–370

Kistler J, Aebi Y, Kellenberger E (1977) Freeze-drying and shadowing a two-dimensional periodic specimen. J Ultrastruct Res 59:76–85

Kühlbrandt W (1984) Three-dimensional structure of the light-harvesting chlorophyll a/b-protein complex. Nature (London) 307:478–480

Kühlbrandt W, Thaler Th, Wehrli E (1983) The structure of membrane crystals of the light-harvesting chlorophyll a/b protein complex. J Cell Biol 96:1414–1424

*Maissel LI, Glang R (eds) (1970) Handbook of thin film technology. McGraw-Hill, New York

Nermut MV, Frank H (1971) Fine structure of influenza A2 (Singapore) as revealed by negative staining, freeze-drying and freeze-etching. J Gen Virol 10:37–51

Rachel R, Jakubowski U, Baumeister W (1986) Electron microscopy of unstained, freeze-dried macromolecular assemblies. J Microsc (Oxford) 141:179–191

*Robards AW, Sleytr UB (1985) Low temperature methods in biological electron microscopy. In: Glauert AM (ed) Practical methods in electron microscopy, vol 10. Elsevier, Amsterdam

Saxton WO, Baumeister W (1982) The correlation averaging of a regularly arranged bacterial cell envelope protein. J Microsc (Oxford) 127:127–138

Saxton WO, Pitt TJ, Horner M (1979) Digital image processing: the Semper system. Ultramicroscopy 4:343–354

*Sleytr UB, Messner P (1983) Crystalline surface layers on bacteria. Annu Rev Microbiol 37:311–339

Smith PR, Kistler J (1977) Surface reliefs computed from electron micrographs of heavy metal shadowed specimens. J Ultrastruct Res 61:124–133

Unwin PNT, Henderson R (1975) Molecular structure determination by electron microscopy of unstained crystalline specimens. J Mol Biol 94:425–440

*Venables JA, Spiller GDT, Hanbücken M (1984) Nucleation and growth of thin films. Rep Prog Phys 47:399–459

Wildhaber I, Gross H, Moor H (1982) The control of freeze-drying with deuterium oxide. J Ultrastruct Res 80:367–373

Wildhaber I, Hegerl R, Barth M, Gross H, Baumeister W (1986) Three dimensional reconstruction of a freeze-dried and metal shadowed bacterial surface layer. Ultramicroscopy 19:57–68

Winkler H (1986) Computerunterstützte Interpretation elektronenmikroskopischer Bilder von dekorierten und beschatteten biologischen Oberflächen. PhD, Diss (Nr 8186) Eidgen Tech Hochschule Zürich

Winkler H, Wildhaber I, Gross H (1985) Decoration effects on the surface of a regular layer protein. Ultramicroscopy 16:331–339

Chapter 11 Immunogold Labelling of Cryosections and Cryofractures

JOHANNES BOONSTRA[1], PIM VAN MAURIK[2], and ARIE J. VERKLEIJ[1]

1 Introduction

In cell biological research the localization of particular molecules at the microscopical level has been an important aspect, leading to the development of a wide variety of cytochemical methods. These methods are based upon the property that specific markers can be coupled chemically to a wide variety of molecules, such as fluorescent markers for light microscopy and electron dense markers for electron microscopy. An important feature in cytochemistry concerns the specificity of the labelling and in this respect, the use of poly- and monoclonal antibodies caused a considerable improvement. The antibodies were bound to the various markers (both fluorescent and electron dense) and the molecules of interest were visualized via direct or two- or three-step labelling methods. Alternative methods use radioactively labelled compounds that can be visualized by autoradiography (Hayat 1972; Williams 1977; Wischnitzer 1981).

The application of antibodies in labelling protocols, so-called immunocytochemistry, implies that the labelling has to be performed in an aqueous environment in order to ensure detectable antigenicity. This necessity requires alternative conditions for tissue preparation, since the classical methods to prepare biological material for electron microscope examination, i.e. fixation, dehydration and plastic embedding, results in most cases in a complete loss of antigenicity. Furthermore, it has been recognized that the classical electron microscope methods have an enormous impact on molecular interactions so that it is difficult to draw conclusions about the morphology in the native state. These considerations have led to the development of cryofixation, and at present, cryofixed biological material can be used for ultrathin sectioning and a number of freeze-fracture techniques. More recently, the latter methods have been combined with immunocytochemistry and, in particular, the use of colloidal gold has been proven to represent an attractive marker system. The properties of enzyme-gold and protein A-gold cytochemistry have been reviewed extensively (Bendayan 1984a, b) and therefore these aspects will not be considered here.

1 Department of Molecular Cell Biology, University of Utrecht, Padualaan 8, 3584 CH Utrecht, The Netherlands
2 Hubrecht Laboratory, Uppsalalaan 8, 3584 CT Utrecht, The Netherlands

Cryotechniques in Biological Electron Microscopy
Edited by R. A. Steinbrecht and K. Zierold

In this chapter we will discuss the combination of immunogold labelling with cryoultramicrotomy and two freeze-fracture replica methods, i.e. freeze-etching and label-fracture. The main advantage of cryoultramicrotomy is that it enables post-sectioning labelling, thus providing complete accessibility of all cellular antigens, located both in the cytoplasm and on the cell surface. The replica methods have the advantages of enabling a detailed analysis of the lateral distribution of antigens located on the cell surface. We will discuss these methods with special emphasis on the visualization of the epidermal growth factor (EGF) receptors in cultured epidermoid carcinoma A431 cells. Special attention will be given to label efficiency, a parameter which has an enormous impact on the reliability of the immunolabelling methods.

2 Cryoultramicrotomy

Immunocytochemical labelling, using colloidal gold as electron dense marker on ultrathin cryosections, has been developed by Tokuyasu (1973, 1980). Since then, it has been applied in a wide variety of biological systems to visualize a wide variety of antigens. Immunolabelling on biological material for examination at the ultrastructural level has to obey two conflicting requirements, i.e. maintenance of full antigenicity, on the one hand, and preservation of ultrastructural detail, on the other. The conditions for immunogold labelling on ultrathin cryosections, as developed by Tokuyasu, have been demonstrated to fulfill these criteria. The main advantage of this method above others is that it enables post-sectioning labelling, and thus all cytoplasmic and cell surface-located antigens are, in principle, fully accessible to the antibodies and gold labels. Moreover, the availability of gold particles of differently defined sizes allows for the labelling of various antigens simultaneously (Geuze et al. 1983a). The applications and problems of cryoultramicrotomy have been described in detail before (Sitte 1984; Zierold, Chap. 6, this Vol.).

2.1 Preparation of Cryosections

A431 cells were grown in Dulbecco's modified Eagle's medium (DMEM) supplemented with 10% foetal calf serum in a 7% CO_2 humidified atmosphere. The cells were washed twice with phosphate buffered saline (PBS, pH 7.4) and fixed with 2% formaldehyde, made freshly from paraformaldehyde, containing 0.1% acrolein in a sodium-cacodylate-HCl buffer (pH 7.4) for 1 day at 4 °C. This chemical fixation is required for ultrastructural definition and preservation of ultrastructure after thawing of the cryosections. The cells were subsequently washed thrice with PBS containing 50 mM glycine (pH 7.4) and scraped from the culture dish with a rubber policeman. The cells were pelleted and embedded in 10% gelatine in PBS, followed by fixation for 30 min in 2% formaldehyde solution. Gelatine blocks (1 mm^3) were incubated in 1 M sucrose for 60 min at room temperature, followed by 2.3 M sucrose for 24 h at 297 K. The gelatine blocks were then mounted on a specimen stub, and these were frozen in liquid nitrogen. The specimen stubs, containing frozen cells, were transferred to a Reichert OMU-4/

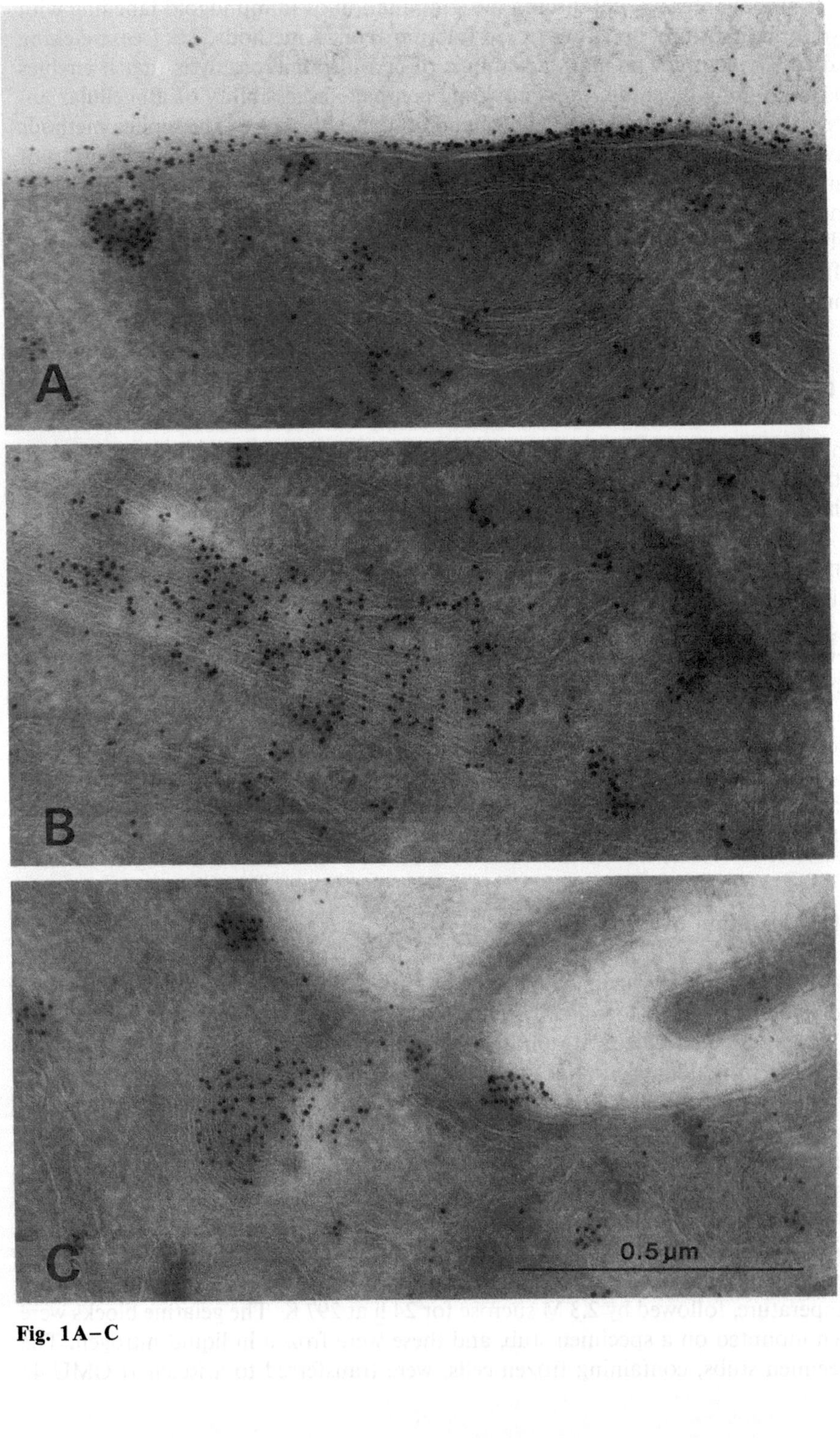

Fig. 1A–C

FC4 cryoultramicrotome and sectioned at 203 K. The sectioning speed was 1 mm s^{-1}, the section thickness control set at 0.1 µm, resulting in an 80% section yield. Sectioning was performed with a dry glass knife with a scoring angle of 45° and a clearance angle of 6°.

Cryosections were picked up from the knife edge with a 2.3 M sucrose droplet according to Tokuyasu (1980). The sections were thawed and attached to grids covered with carbon-coated parlodion film and the grids were placed upside down on droplets for the consecutive incubations.

2.2 Immunogold Labelling

Grids, containing thawed cryosections, were washed once in PBS containing 50 mM glycine (pH 7.4) and thrice in PBS containing 0.2% gelatine and 0.5% bovine serum albumin (PBG, pH 7.4). The grids were incubated for 1 h at room temperature with the various antibodies and the protein A-gold complex, each incubation followed by six washes in PBG over 30 min.

After the protein A-gold complex incubation, the grids were washed six times in PBG and seven times with PBS followed by incubation on aq. dest. for 10 min. The sections were stained by incubation for 10 min in 2% neutralized uranyl acetate (pH 7.0) and 5 min in 2% uranyl acetate (pH 4.0). The sections were subsequently embedded in 1.1% methylcellulose containing 0.1% uranyl acetate at 297 K for 1 min and dried at room temperature.

2.3 Results

We have applied cryoultramicrotomy in combination with immunogold labelling to visualize cell surface and intracellularly located EGF receptors in A431 cells (Boonstra et al. 1985a, b). Immunogold labelling was performed using monoclonal anti-EGF receptor antibodies, designated 2E9 and 2D11, raised and characterized as described (Defize et al. 1986). It was shown that 2E9 recognizes an amino acid sequence located at or close to the EGF-binding domain of the receptor, while 2D11 was shown to be directed against a sugar moiety of the receptor (Defize et al. 1986).

Ultrathin cryosections were consecutively labelled with 2E9 or 2D11, rabbit antimouse antibody (RAM) and protein A-gold as described under Section 2.2. As shown in Fig. 1A, a high number of gold particles was observed on the cell surface. The majority of the gold particles were located at the external side of the plasma membrane, consistent with the notion that the anti-EGF receptor anti-

Fig. 1A–C. Visualization and localization of EGF receptors in A431 cells by cryoultramicrotomy and immunogold labelling. The cells were fixed and processed for cryoultramicrotomy as described in Section 2.1. The cryosections were thawed and consecutively labelled with anti-EGF receptor antibody, rabbit antimouse antibody and protein A-gold conjugate at room temperature as described in detail in Section 2.2. **A** Surface labelling of the plasma membrane; **B** intracellular labelling of Golgi membranes; **C** intracellular labelling of a lysosome. The gold particles had an average diameter of 8 nm bodies bind to the external domain of the EGF receptor. The varying distance of the gold particles to the membrane, as well as some apparent labelling at the cytoplasmic side of the membrane, are most likely due to "plane of section" artefacts (McKanna et al. 1979). In addition, also intracellularly located EGF receptors were visualized, associated to both the biosynthetic and endocytotic pathway. Thus, a dense labelling was observed on membranes of the Golgi complex (Fig. 1 B) and endoplasmic reticulum and, furthermore, in lysosomes (Fig. 1 C), multivesicular bodies (MVB) and a network of vesicular and tubular membranes, probably representing the compartment of uncoupling receptor and ligand, called CURL (Geuze et al. 1983 a). In some cases gold label was found in pits at the plasma membrane, although the number of these, most likely coated, pits, was very low (Boonstra et al. 1985 a). Recently, the EGF-induced redistribution of receptors of A431 cells was demonstrated by Miller et al. (1986), using the same method. From pulse-chase experiments these authors demonstrated an EGF-induced redistribution of EGF receptors from the plasma membrane to the peripheral endosome compartment and then to the pericentriolar compartment and lysosomes.

The advantages of cryoultramicrotomy in combination with immunogold labelling in the study of intracellular pathways of ligands and receptors have been demonstrated so far most elegantly by Geuze and co-workers (1983 b). Using double-labelling, it was demonstrated that uncoupling of the asialoglycoprotein receptor and its ligand occurred in an organelle called CURL (Geuze et al. 1983 b, 1984). The main advantage of the method is the post-sectioning labelling, allowing the immediate fixation of live cells without preceding antibody incubation, as well as the full accessibility of intracellular antigenic sites. Today, the combination of cryoultramicrotomy and immunogold labelling has been applied to visualize a wide variety of proteins in eukaryotic cells, amongst them lysosomal hydrolases (van Dongen et al. 1984), mannose receptor (Rossi and Himmelhoch 1985), sialyltransferase (Roth et al. 1985), serum albumin (Christensen et al. 1985), main intrinsic peptide (Fitzgerald et al. 1983) and a large number of others. This incomplete summary demonstrates already that immunogold labelling in combination with cryoultramicrotomy provides an universal method, which can be applied to visualize all cellular proteins at the ultrastructural level, depending on the availability of well-defined specific antibodies.

However, it should be realized that the ultrastructural delineation in cryosections is not as detailed as routinely obtained in conventionally prepared ultrathin plastic sections. This feature can be improved by treatment of cryosections with osmium tetroxide after immunolabelling, followed by dehydration and embedding in plastic (Keller et al. 1984). Using this procedure, the ultrastructural preservation of cryosections appears as good as that of conventional plastic sections, while the advantages of the immunolabelling are retained.

3 Cryofractures

As described above, cryoultramicrotomy provides information on the presence of antigenic sites located in the plasma membrane, but it does not provide appropriate information for the analysis of the lateral distribution of these antigens. In particular in hormone and growth factor research, the lateral distribution of the respective receptors is important in the biological response of the ligands. Thus, for example, various studies have demonstrated that EGF binding causes a clustering of EGF receptors in coated pits (Haigler et al. 1978, 1979; McKanna et al. 1979; Willingham et al. 1983), followed by internalization via receptor-mediated endocytosis (RME). The coated pits-associated macroclustering is probably preceded by microclustering, as concluded from studies on the lateral mobility of EGF receptor by fluorescence photobleaching recovery (Zidovetski et al. 1981). Binding of EGF to EGF receptors caused a rapid decrease of lateral mobility in both whole cells and isolated plasma membranes, indicating a clustering, probably involving five to ten EGF receptors per cluster (Zidovetski et al. 1981, 1986). The microaggregation of receptors appears to play an important role in EGF action, as indicated by studies using Fab fragments of a bivalent monoclonal anti-EGF receptor antibody (Schreiber et al. 1983).

In order to provide reliable information about the planar distribution of cell surface-located proteins at the electron microscope level, a number of methods have been developed, amongst them cryofractures. Immunolabelling of cryofractured cells was pioneered by Pinto da Silva and co-workers (1973, 1981, 1984) and, at present, several applications in combination with immunogold labelling can be used, such as freeze-etching and label-fracture.

3.1 Immunogold Labelling

A431 cells were grown on thermanox coverslips in DMEM as described in Section 2.1. Usually, the cells were fixed using 2% formaldehyde + 0.1% acrolein solution (see Sect. 2.1). With these procedures chemical fixation is not required for the preservation of ultrastructure, but for the immobilization of the cell surface-located antigens. After fixation the cells were washed once in PBS + 50 mM glycine (pH 7.4) and thrice in PBG (pH 7.4). Subsequently, the cells were incubated in the presence of various antibodies and protein A-gold complex, essentially as described in Section 2.2.

3.2 Preparation of Cryofractures

Following immunolabelling, the coverslips containing the cells were mounted upon a specimen holder and frozen in solid nitrogen without using cryoprotectants for freeze-etching and after incubation with glycerol for label-fracture. The frozen samples were transferred to a cryofract (Reichert-Jung), fractured and etched at 173 K in a vacuum of better than 10^{-4} mbar for 1 h for freeze-etching, or fractured at 123 K in a vacuum of better than 10^{-7} mbar without etching for label-fracture. The etched and fractured samples respectively, were replicated with Pt/C at 153 K. For freeze-etching the replicas were washed in 15% sodium hypochlorite for 1 h, followed by three washes on water and then collected on a 700-mesh copper grid. For label-fracture the replicas were washed in PBS and mounted on a parlodion covered grid.

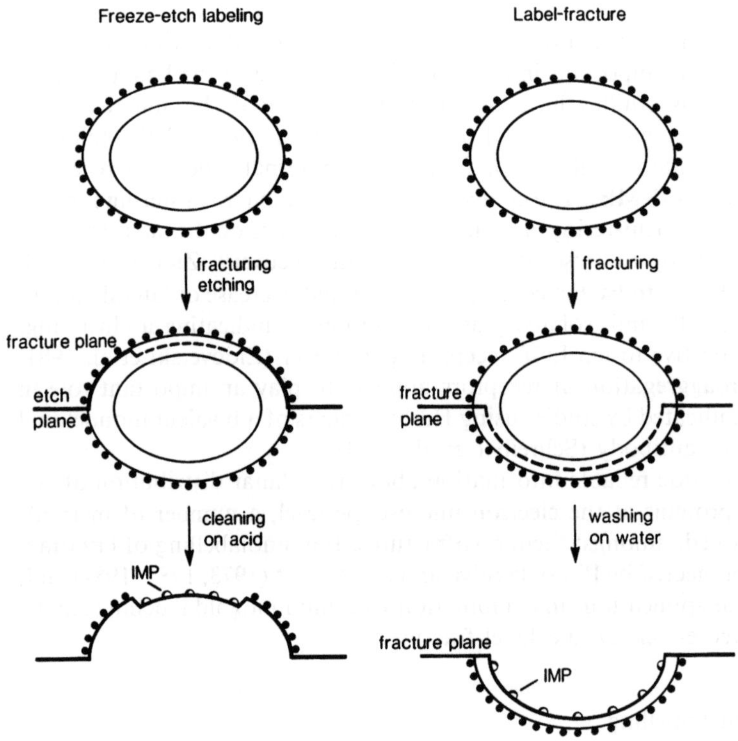

Fig. 2. Schematic representation of immunogold labelling in combination with freeze-etching and label-fracturing

3.3 Freeze-Etching

Theoretically, freeze-etching is the simplest technique to visualize plasma membrane-associated proteins, if these proteins are labelled at the external domain. After immunolabelling, the ice is etched away and the cell surface becomes visible, including the labelled proteins, as schematically shown in Fig. 2. We have used this method to visualize the cell surface-located EGF receptors on A431 cells and to analyze the lateral distributions of these receptors (Boonstra et al. 1985b).

As is shown in Fig. 3, essentially all gold particles are replicated showing shadow cones, demonstrating that the gold particles are trapped into the replica. The fact that no shadow cones without gold particles are observed indicates that no gold particles were lost during the cleaning procedure of the replicas.

Immunogold labelling combined with freeze-etching has been used to demonstrate a variety of cell surface-located antigens, such as PhoE pore proteins in *Escherichia coli* (Voorhout et al. 1986), virus structures and viral antigens on virus-infected cell cultures (Hohenberg et al. 1986), and glycolipids on erythrocyte membranes (Tillack et al. 1983).

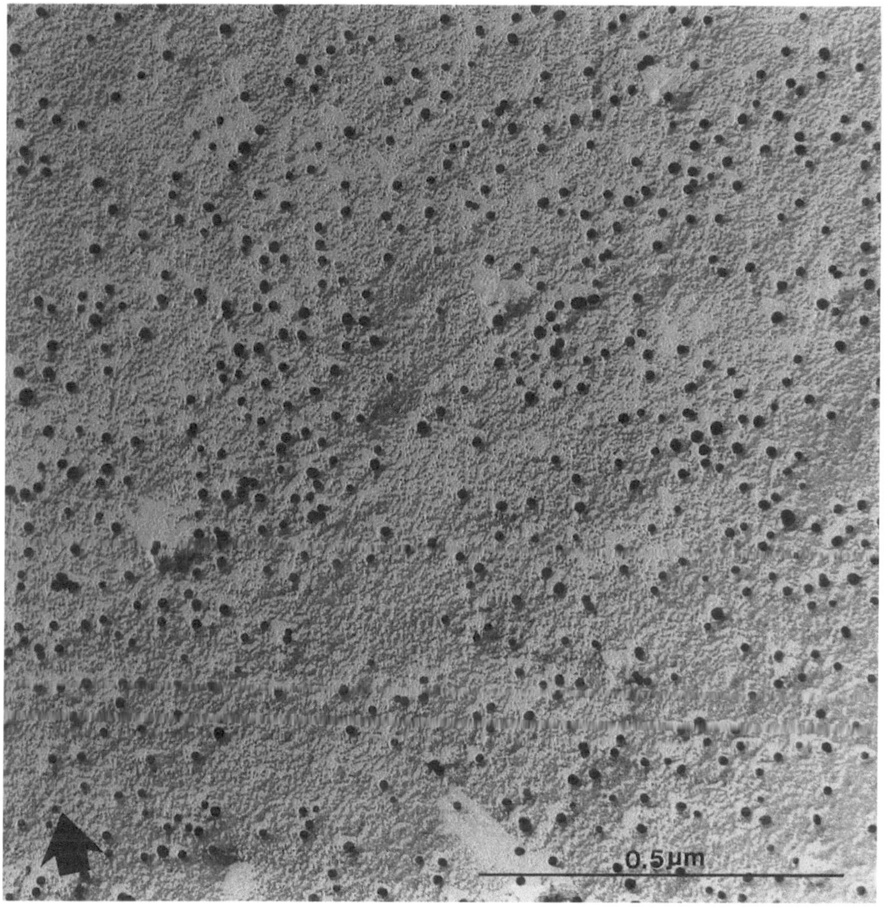

Fig. 3. Cell surface distribution of EGF receptors in A431 cells visualized by freeze-etching. The cells were fixed and immunogold labelled as described in Section 3.1. After freezing, the cells were processed as described in Section 3.2. The gold particles had an average diameter of 12 nm. *Arrow* in this and the following figures denotes the direction of shadowing

3.4 Label-Fracture

An alternative method to freeze-etching for the analysis of the lateral distribution of cell surface-located antigens is provided by the label-fracture method developed by Pinto da Silva and Kan (1984). According to this method, the cells are essentially treated as described above for freeze-etching, but instead of etching, the cells are fractured without etching. The fracture faces are subsequently replicated according to standard methods as shown schematically in Fig. 2. The clue of the method is that the replicas are not washed on detergent, but on PBS, so that the biological material is not removed. Due to the hydrophobic nature of the replica, the outer monolayer sticks then to the replica. As shown in Fig. 4, the gold label is visualized in projection by the high-resolution image of the exo-

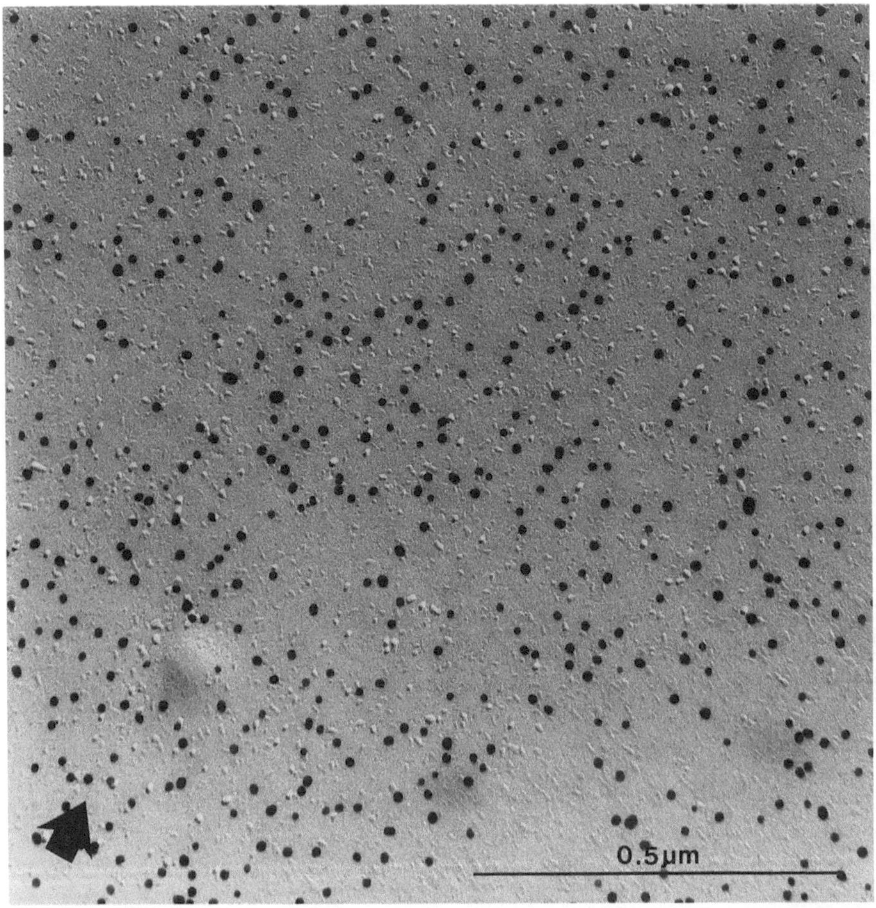

Fig. 4. Cell-surface distribution of EGF receptors in A431 cells as visualized by label-fracturing. The cells were fixed and immunolabelled essentially as described in Section 3.1. The cells were frozen and processed for label-fracture using glycerol as a cryoprotectant. The average diameter of the gold particles was 12 nm

plasmic fracture face (EF face). The lateral distribution of EGF receptors can be analyzed relatively simple from images as shown in Fig. 4. However, the label-fracture method faces a number of disadvantages as well, related in particular to the lateral distribution analysis. When using extremely flattened cells, the fracture plane might cross the cytoplasm and continue through the lower cell membrane, thus yielding the image of an apparently gold labelled PF face. This phenomenon has been described in detail previously (Boonstra et al. 1985b). In addition, a number of other artefacts might arise due to the presence of membrane ruffles or microvilli. As shown in Fig. 5, some cells contain numerous microvilli which are heavily labelled with EGF receptors; since the replicas are washed on water or buffer only, in addition to the outer monolayer, the microvilli also stick to the replica. A similar phenomenon can be observed when the cells contain membrane

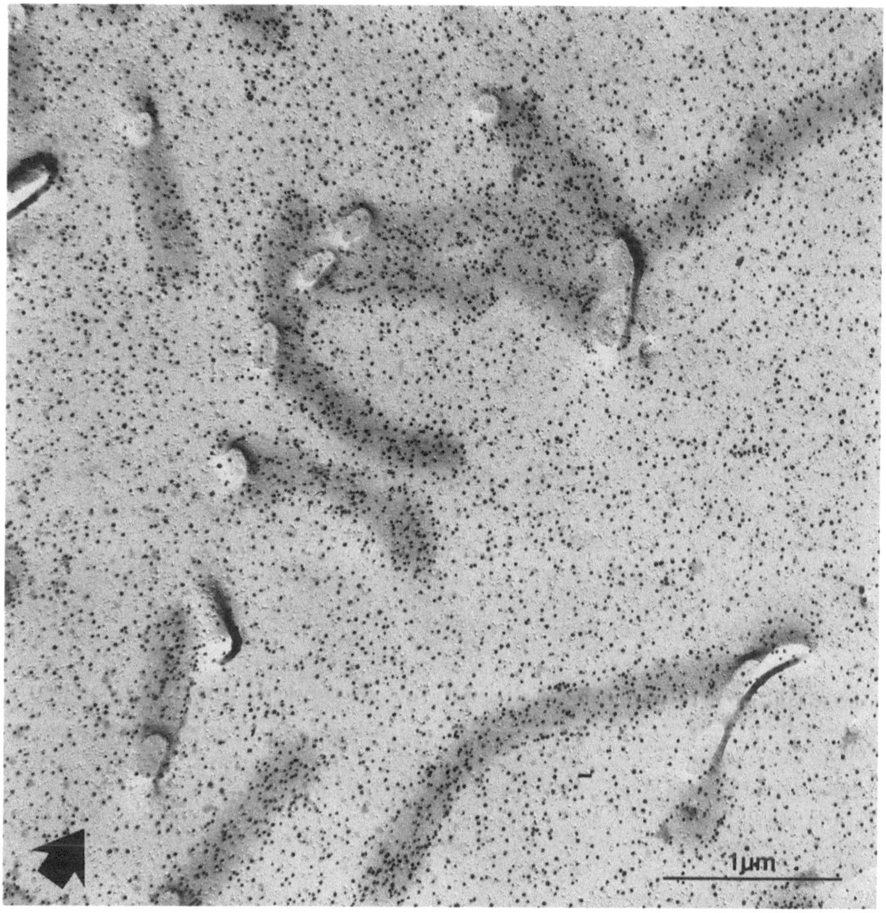

Fig. 5. Cell-surface distribution of EGF receptors in A431 cells as visualized by label-fracturing. The cells were treated as described in Fig. 4. The microvilli on the cell surface are immunogold-labelled and cause a clear artefactual labelling on the EF face

ruffles. These artefacts might easily cause misinterpretations in the analysis of the lateral distribution of receptors.

4 Label Efficiency

An important aspect in all immunological labelling studies concerns the label efficiency, particularly when a study is performed on a number of antigenic sites in a given biological specimen. The label efficiency may be affected by a number of parameters, such as the effect of fixatives, steric hindrance, diffusion and penetration phenomena, concentrations and affinities of the compounds involved. We will discuss a few of these phenomena below.

4.1 Fixatives

Chemical fixatives are, in principle, not necessary when fixation is performed by fast freezing. However, for cryoultramicrotomy chemical fixatives are required for the preservation of ultrastructure after thawing, while the surface-labelling methods often require fixatives to immobilize the proteins under study. Several studies have demonstrated that various fixatives, even when used at moderate concentrations, decrease antigenicity, e.g. by changes in the tertiary structure of the antigenic molecule. Thus, it has been shown for five cell-surface antigens on mouse T-lymphocytes that fixatives, such as osmium tetroxide or acrolein, decreased antibody binding significantly, whereas low concentrations of formaldehyde or glutaraldehyde did not (Van Ewijk et al. 1984). In cryoultramicrotomy, glutaraldehyde combined with formaldehyde has been used most frequently as fixative (Tokuyasu 1973). From a number of studies, reviewed by Bullock (1984), it has become clear that glutaraldehyde and formaldehyde provide the best results, with regard to both preservation of ultrastructure and antigenicity. We have measured the effects of various combinations of fixatives on antigenicity of the EGF receptor in A431 cells by studying the effects on binding of radioactively labelled EGF and anti-EGF receptor antibodies (Boonstra et al. 1985a, b). It has been shown that glutaraldehyde caused a severe inhibition of the binding properties of the EGF receptor for both EGF and anti-EGF receptor antibody. In contrast, acrolein caused only a small decrease in binding properties and formaldehyde alone no decrease at all (Boonstra et al. 1985a, b). These results clearly demonstrated that application of cryoultramicrotomy and immunogold labelling should be accompanied by studies on the effect of fixatives on antigenicity. These effects might be different in various cell types and for various antibody-antigen combinations.

4.2 Penetration Phenomena

An important feature in immunogold labelling comprises the accessibility of antigenic sites on and in the cryosection. In principle, all antigenic sites on the surface of the cryosection are freely accessible to the antibody, as clearly demonstrated in a stereoimage (Fig. 6). However, inside the cryosection the penetration of antibodies and gold label might be differently hindered by cytoplasmic structures, while the accessibility of antigenic sites on the cell surface might be decreased in a cryosection by the embedding material, usually gelatine. As shown in a stereoimage, in areas where the gelatine is retracted from the cell surface (arrow in Fig. 6), all cell surface-located EGF receptors are freely accessible and hence labelled. Similar phenomena can be achieved inside the cell. These aspects might clearly influence the apparent distribution of antigenic sites in a cryosection. Another important aspect of the accessibility concerns the size of the gold label and antibody-gold or protein A-gold conjugates. Thus, it has been demonstrated that the efficiency of labelling decreased with increasing particle size (De Mey 1983; Horisberger 1983; Slot and Geuze 1984; Vernooy-Gerritsen et al. 1984; van Bergen en Henegouwen and Leunissen 1986). In particular, a reduced labelling ef-

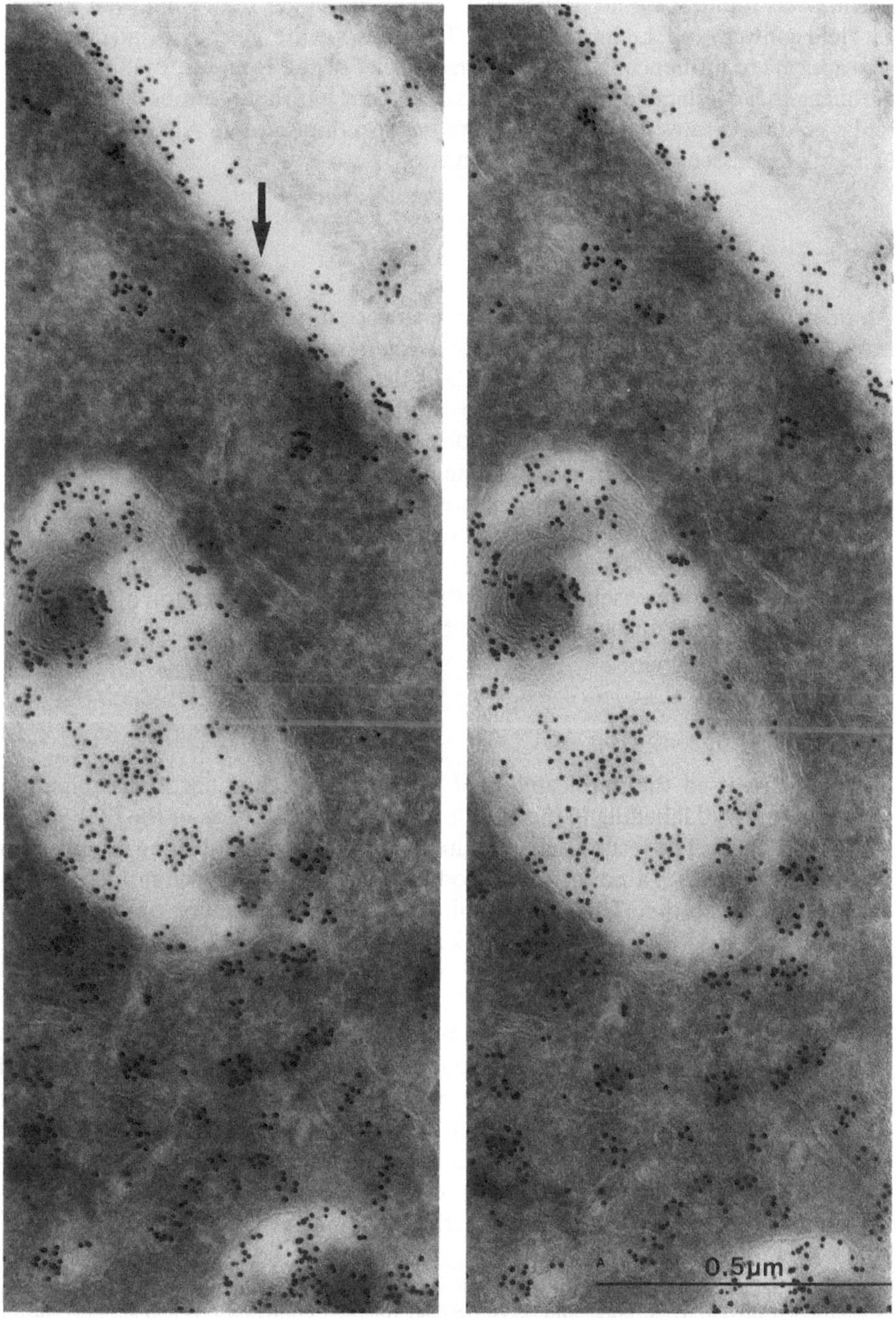

Fig. 6. Stereopair of EGF-receptor labelling in a cryosection of A431 cells. The cells were treated essentially as described in the Legend of Fig. 1. The stereomicrographs were taken at tilt angles of 9°. *Arrow* shows heavy labelling at locations where cell surfaces were exposed due to gaps in the embedding gelatine

ficiency was found for gold probes with a diameter larger than 8 nm (Van Bergen en Henegouwen and Leunissen 1986). In addition, the gold-associated protein also appears to influence the penetration capacity of the complex. Preliminary experiments have indicated that a conjugate of goat antirabbit antibodies and colloidal gold penetrates deeper into the cryosection than protein A-gold conjugates (J. L. M. Leunissen, unpublished observation).

4.3 Steric Hindrance

An interesting effect of steric hindrance in surface labelling has recently been described for the pore protein PhoE of *Escherichia coli* K-12 (Voorhout et al. 1986). Using freeze-etching methods these authors detected only a very limited labelling of PhoE, similar to results obtained by immunofluorescence and whole-mount labelling. In contrast, however, an uniform dense labelling of the cell surface was observed using cryoultramicrotomy. Using mutants, it was concluded that the antigenic sites of the PhoE pore protein were not accessible in normal cells, due to steric hindrance caused by the LPS carbohydrate chains. These phenomena might occur in a variety of systems and thus emphasize the necessity to apply several, basically different approaches to study the presence and localization of particular proteins on the cell surface and inside the cell.

5 Conclusions

We have discussed the applicability of various cryotechniques in combination with immunogold labelling in the visualization and localization of the EGF receptor in A431 cells. The results demonstrate that immunogold labelling on ultrathin cryosections provides a powerful method in visualizing intracellular and cell surface-located receptors. This compromise between maintenance of antigenicity and well-preserved ultrastructure allows a detailed analysis of the metabolism of cellular proteins. The various applications of this method, furthermore, demonstrate the universal applicability.

In addition, it is shown that the application of various replica techniques in combination with immunogold labelling allows a detailed analysis of the lateral distribution of cell surface-located proteins. In general, it is concluded that modern cryotechniques in combination with immunocytochemistry allow a molecular study of the metabolism of cellular proteins. Such studies require, however, the use of well-characterized markers (antibodies and gold label) and the application of various complementary methods.

Acknowledgements. We are grateful to Drs. N. van Belzen, L. H. K. Defize, J. L. M. Leunissen, J. Leunissen-Bijvelt, W. J. Hage and S. W. de Laat for their contributions and support during the course of these studies.

References

(References marked by an asterisk are review papers or books respectively)

*Bendayan M (1984a) Protein A-gold electron microscopic immunocytochemistry: methods, applications, and limitations. J Electron Microsc Tech 1:243–270
*Bendayan M (1984b) Enzyme-gold electron microscopic cytochemistry: a new affinity approach for the ultrastructural localization of macromolecules. J Electron Microsc Tech 1:349–372
Boonstra J, Maurik P van, Defize LHK, de Laat SW, Leunissen JLM, Verkleij AJ (1985a) Visualization of epidermal growth factor receptor in cryo-sections of cultured A431 cells by immuno-gold labeling. Eur J Cell Biol 36:209–216
Boonstra J, Belzen N van, Maurik P van, Hage WJ, Blok FJ, Wiegant FAC, Verkleij AJ (1985b) Immunocytochemical demonstrations of cytoplasmic and cell-surface EGF receptors in A431 cells using cryo-ultramicrotomy, surface replication, freeze etching and label fracture. J Microsc (Oxford) 140:119–129
*Bullock GR (1984) The current status of fixation for electron microscopy: a review. J Microsc (Oxford) 133:1–15
Christensen AK, Komorowski TE, Wilson B, Ma S-F, Stevens III RW (1985) The distribution of serum albumin in the rat testis, studied by electron microscope immunocytochemistry on ultrathin frozen sections. Endocrinology 116:1983–1996
Defize LHK, Moolenaar WH, Saag PT van der, de Laat SW (1986) Dissociation of cellular responses to epidermal growth factor using antireceptor monoclonal antibodies. EMBO J 5:1187–1192
*De Meij J (1983) Colloidal gold probes in immunochemistry. In: Polak S, Noorden S van (eds) Immunochemistry. Practical applications in pathology and biology. Wright, London, pp 82–112
Fitzgerald PG, Bok D, Horwitz J (1983) Immunocytochemical localization of the main intrinsic polypeptide (MIP) in ultrathin frozen sections of rat lens. J Cell Biol 97:1491–1499
Geuze HJ, Slot JW, Strous GJAM, Lodish HF, Schwartz AL (1983a) Intracellular site of asialoglycoprotein receptor-ligand uncoupling: double label immunoelectron microscopy during receptor-mediated endocytosis. Cell 32:277–287
Geuze HJ, Slot JW, Strous GJAM, Schwartz AL (1983b) The pathway of the asialoglycoprotein-ligand during receptor-mediated endocytosis: a morphological study with colloidal gold/ligand in the human hepatoma cell line, Hep G2. Eur J Cell Biol 32:38–44
Geuze HJ, Slot JW, Strous GJAM, Peppard J, Figura K von, Hasilik A, Schwartz AL (1984) Intracellular receptor sorting during endocytosis: comparative immunoelectron microscopy of multiple receptors in rat liver. Cell 37:195–204
Haigler H, Ash JF, Singer SJ, Cohen S (1978) Visualization by fluorescence of the binding and internalization of epidermal growth factor in human carcinoma cells A-431. Proc Natl Acad Sci USA 75:3317–3321
Haigler HT, McKanna JA, Cohen S (1979) Direct visualization of the binding and internalization of a ferritin conjugate of epidermal growth factor in human carcinoma cells A-431. J Cell Biol 81:382–395
*Hayat MA (1972) Principles and techniques of electron microscopy, biological applications, vol. 2. Van Nostrand Reinhold, New York
Hohenberg H, Bohn W, Rutter G, Mannweiler K (1986) Plasma membrane antigens detected by replica techniques. In: Müller M, Becker RP, Boyde A, Wolosewick JJ (eds) The science of biological specimen preparation 1985. SEM, AMF O'Hare, IL 60666, pp 235–244
Horisberger M (1983) Colloidal gold as a tool in molecular biology. Trends Biochem Sci 7:395–397
Keller GA, Tokuyasu KT, Dutton AH, Singer SJ (1984) An improved procedure for immunoelectron microscopy: ultrathin plastic embedding of immunolabeled ultrathin frozen sections. Proc Natl Acad Sci USA 81:5744–5747
McKanna JA, Haigler HT, Cohen S (1979) Hormone receptor topology and dynamics: morphological analysis using ferritin-labeled epidermal growth factor. Proc Natl Acad Sci USA 76:5689–5693
Miller K, Beardmore J, Kanety H, Schlessinger J, Hopkins CR (1986) Localization of the epidermal growth factor (EGF) receptor within the endosome of EGF-stimulated epidermoid carcinoma (A431) cells. J Cell Biol 102:500–509

*Pinto da Silva P (1984) Freeze-fracture cytochemistry. In: Polak JM, Varndell IM (eds) Immunolabeling for electron microscopy. Elsevier, Amsterdam New York, pp 179–188
Pinto da Silva P, Kan FWK (1984) Label-fracture: a method for high resolution labeling of cell surfaces. J Cell Biol 99:1156–1161
Pinto da Silva P, Moss PS, Fudenberg HH (1973) Anionic sites on the membrane intercalated particles of human erythrocyt ghosts membrane. Freeze-etch localization. Exp Cell Res 81:127–138
Pinto de Silva P, Kachar B, Torrisi MR, Brown C, Parkinson C (1981) Freeze-fracture cytochemistry: replicas of critical point dried cells and tissues after fracture label. Science 213:230–233
Rossi G, Himmelhoch S (1985) Immunocytochemical localization of the mannose receptor on ultrathin cryosections of chicken macrophages. Eur J Cell Biol 38:280–291
Roth J, Taatjes DJ, Lucocq JM, Weinstein J, Paulson JC (1985) Demonstration of an extensive trans-tubular network continuous with the Golgi apparatus stack that may function in glycosylation. Cell 43:287–295
Schreiber AB, Lax I, Yarden Y, Schlessinger J (1983) Biological role of epidermal growth factor receptor clustering: investigation with monoclonal anti-receptor antibodies. J Biol Chem 258:846–853
*Sitte H (1984) Instruments for cryofixation, cryoultramicrotomy and cryosubstitution for biomedical TEM. Zeiss Inf MEM Mag Electron Microsc 3:25–31
Slot JW, Geuze HJ (1984) Gold markers for single and double immunolabelling of ultrathin cryosections. In: Polak JM, Varndell IM (eds) Immunolabeling for electron microscopy. Elsevier, Amsterdam New York, pp 129–142
Tillack TW, Allietta M, Moran RE, Young WW (1983) Localization of globoside and forssman glycolipids on erythrocyte membranes. Biochim Biophys Acta 733:15–24
Tokuyasu KT (1973) A technique for ultracryotomy of cell suspension and tissues. J Cell Biol 57:551–565
Tokuyasu KT (1980) Immunochemistry on ultrathin frozen sections. Histochem J 12:381–403
Van Bergen en Henegouwen PMP, Leunissen JLM (1986) Controlled growth of colloidal gold particles and implications for labelling efficiency. Histochemistry 85:81–87
Van Dongen JM, Barneveld RA, Geuze HJ, Galjaard H (1984) Immunocytochemistry of lysosomal hydrolases and their precursor forms in normal and mutant human cells. Histochem J 16:941–954
Van Ewijk W, Van Soest PL, Verkerk A, Jongkind JF (1984) Loss of antibody binding to prefixed cells: fixation parameters for immunocytochemistry. Histochem J 16:179–193
Vernooy-Gerritsen M, Leunissen JLM, Veldink GA, Vliegenthart JFG (1984) Intracellular localization of lipoxygenase-1 and -2 in germinating soybean seeds by indirect labelling with protein A-colloidal gold complexes. Plant Physiol 76:1070–1079
Voorhout WF, Leunissen-Bijvelt JJM, Leunissen JLM, Verkleij AJ (1986) Steric hindrance in immunolabelling. J Microsc (Oxford) 141:303–310
*Williams MA (1977) Autoradiography and immunocytochemistry. In: Glauert AM (ed) Practical methods in electron microscopy, vol. 6. Elsevier, Amsterdam New York
Willingham MC, Haigler HT, Fitzgerald DJP, Gallo MG, Rutherford AV, Pastan IH (1983) The morphologic pathway of binding and internalization of epidermal growth factor in cultured cells. Studies on A431, KB, and 3T3 cells, using multiple methods of labelling. Exp Cell Res 146:163–175
*Wischnitzer S (1981) Introduction to electron microscopy. Pergamon, New York
Zidovetski R, Yarden Y, Schlessinger J, Jovin TM (1981) Rotational diffusion of epidermal growth factor complexed to the cell surface receptors reflects the rapid microaggregation and endocytosis of occupied receptors. Proc Natl Acad Sci USA 78:6981–6985
Zidovetski R, Yarden Y, Schlessinger J, Jovin TM (1986) Microaggregation of hormone-occupied epidermal growth factor receptors on plasma membrane preparations. EMBO J 5:247–250

Chapter 12 Cryoultramicrotomy for Autoradiography and Enzyme Cytochemistry

PETER M. FREDERIK[1]

1 Introduction

The introduction of cryoultramicrotomy as a new tool in histochemistry and cytochemistry has led to great expectations for autoradiography and enzyme cytochemistry, the classical methods in histochemistry and cytochemistry. It is an unpleasant task to summarize the present situation and to arrive at the conclusion that these promises are only partly fulfilled. For the autoradiography of diffusible substances, there seem to be only practical limitations (work and time), however, for enzyme cytochemistry, there are also theoretical limitations to the use of thin cryosections. Autoradiography and enzyme cytochemistry have thus hardly benefitted from the development of cryoultramicrotomy and this remains in sharp contrast to immunocytochemistry where the thin cryosection has found many applications and has become a sort of "(immuno)-golden" standard (see also Boonstra et al., Chap. 11, this Vol.).

In autoradiography a photographic emulsion is placed in close contact with the (biological) material. Due to the decay of a labelled compound in the material, electrons are generated, electrons which may travel towards the photographic emulsion and induce a latent image. Under proper conditions the accumulated latent images can be turned into metallic silver (development), and after fixation and washing the silver grains can be visualized in the electron microscope, together with the underlying cell(s) or tissue. The silver grains can thus be used to locate the labelled compound and the density of grains over different cell organelles is an estimate of the (relative) concentration of the labelled compound. In the final autoradiogram, metallic silver is used to locate the compound of interest.

In enzyme cytochemistry, the other classical method of histochemistry, metal salts are employed to locate enzymes. Enzymes in cells or tissues are located, in most cases after fixation, by using the enzyme activity to convert a substrate. One of the reaction product(s) is usually immobilized at the enzyme site by precipitation ("trapping") with a heavy metal ion and the precipitate is visualized by electron microscopy together with the subcellular organelles.

1 Department of Pathology/EM Unit, University of Maastricht, P.O.Box 616, N-6200 MD Maastricht, The Netherlands

Cryotechniques in Biological Electron Microscopy
Edited by R. A. Steinbrecht and K. Zierold

Autoradiography and enzyme cytochemistry as "methods" of histochemistry have found a serious competitor in immunocytochemistry. To quote A. G. E. Pearse (1982): "Immunocytochemistry overtook enzyme cytochemistry some years ago to become the advancing edge of the discipline. With the current rapid increase in immunological knowledge, and through the resulting improved technology, it maintains its position without serious challenge". The introduction of monoclonal antibodies (Köhler and Milstein 1975) and the widespread application of this technology in lifesciences has resulted in the availability of an almost unlimited quantity and variety of antibodies, each directed against a well-defined epitope. Localization of antigens in tissue sections is done by incubating the tissue (slice or small sample) with the antibody and visualizing the antibody directly with a "tag" (gold, ferritin or peroxidase) or visualizing the antibody after interaction with a second antibody (indirect method, the last antibody is usually "tagged"). In immunocytochemistry, the antigen-antibody interaction is thus visualized. The sensitivity depends on the affinity between antigen and antibody (and eventually between subsequent antibody-antibody interactions), the precision (resolution) on stereochemical parameters (size of antibodies and distance of primary binding sites to the tag; see also Boonstra et al., Chap. 11, this Vol.).

For the application of immunocytochemistry, the antigenic site has to be preserved and be accessible to the antibodies in aqueous solutions. This last criterion can become difficult for labile compounds or substances that are diffusible in the fixatives or dehydrating agents used for cell or tissue preparation. For the localization of diffusible (organic) substances, autoradiography is still the only method available. Although well established at the light microscope level (Appleton 1964, Stumpf and Roth 1964), autoradiography of diffusible substances at the EM level is still not a routine technique and has been employed only in a few cases.

2 Cryoultramicrotomy for the Autoradiography of Diffusible Substances

In the late 1960s there was a renewed interest in cryoultramicrotomy based on the expectation that thin cryosections could be used to localize diffusible substances by autoradiography (Appleton 1969, 1974; Christensen 1969, 1971; Hodson and Marshall 1970). The idea that physical fixation (freezing) and cryopreparative methods would lead to the most faithful representation of the living state by the specimen, as observed in the electron microscope, has been playing a role since the early days of ultrastructure research on biological material. Freeze-drying and embedding has been used by Sjöstrand (1943), freeze-drying and shadowing by Wyckhoff (1946) and cryoultramicrotomy for the study of the ultrastructure of the myelin sheet by Fernández-Morán (1952). The incorporation of an ultramicrotome in a high-vacuum chamber (Moor et al. 1961) was first attempted to prepare thin cryosections in a clean environment, instead, this instrument became the cradle of freeze-fracturing. The cryopreparative procedures lost some of their attraction by the advance of chemical fixation (Sabbatini et al. 1963) and embedding (Glauert et al. 1956; Luft 1961). Only W. Bernard (e.g. Bernard and Leduc 1967) held an interest in cryoultramicrotomy and its application in histochemistry. His work inspired A.K. Christensen and T.C. Appleton to further develop the instrument with the aim to exploit cryosections for the autoradiography of diffusible substances at the EM level. The potential of this approach was based on the results of the well-established methods for autoradiography of diffusible substances at the light microscope level (Appleton 1964; Stumpf and Roth 1964; see also Rogers 1979). The same reasoning was given for the development of instruments by Dollhopf and Sitte (1969) and Hodson and Marshall (1970). With the prototype instruments thin cryosections were obtained and observed in the freeze-dried state in the electron microscope. Christensen (1971) pointed to the potential hazard of rehydration and/or melting of the frozen (dried) cryosection to the preservation of ultrastructure. Hodson and Williams (1976) propagated the use of a vacuum-transfer device to transport thin cryosections under vacuum to a coater for the subsequent application of a protective carbon layer. This strictly "dry" procedure seemed to give a faithful representation of the ultrastructure and also of the freezing damage; i.e. freezing damage that is masked upon melting or rehydration (Frederik and Busing 1981; Frederik 1982).

Although the cryoultramicrotome has been developed to produce cryosections from fresh-frozen tissue for subsequent autoradiography on such sections, the reported accounts are few. The provisions for autoradiography of diffusible substances are strict and centre around the issue such that no translocation or extraction of the labelled compound may occur up to the moment of development of the photographic emulsion. First of all, fresh-frozen tissue (no chemical prefixation or cryoprotection) has to be used to produce cryosections. These cryosections are "stabilized" by freeze-drying followed by carbon evaporation and a photographic emulsion is placed on the sections by a safe procedure. This implies that no extraction or transfer of the labelled compound may be allowed towards the photographic emulsion (usually applied in the gelled form) and vice versa that no water is allowed to travel from the photographic emulsion to the cryosection in quantities that alter the ultrastructure and/or the localization of the labelled compound. The possible exchange of a "soluble" tritiated compound between a specimen and an overlying photographic emulsion was tested in a model system (Harris and Salpeter 1983). Autoradiograms were prepared from nebulized droplets containing ^{3}H-proline (air-dryed and fixed in formaldehyde vapour), and the distribution of grains around the radioactive source was estimated (resolution test). Without a protective carbon layer between specimen and photographic emulsion, the resolution was seriously impaired. The form in which the autoradiographic emulsion is applied was also found to be critical; only with a "stripping film" and not with "Caro loops" (for the description of the emulsion-handling techniques see Rogers 1979) was high resolution attainable.

Most authors expect that a carbon coating on freeze-dried cryosections would prevent any exchange between section and emulsion (e.g. Christensen and Paavola 1972), but also (thin) polymer films have been used for the same purpose (e.g. Butvar B-98, Futaesaku and Mizuhira 1986). Experimental verification of this idea was presented by Baker and Appleton (1976), and X-ray microanalysis was used to test the protective effect of an evaporated carbon layer on the section. A freeze-dried section that had been carbon coated (20 – 30 nm) showed no gross extraction of diffusible elements, as estimated by X-ray microanalysis, when coated sections were exposed to wetted filterpaper. The same authors also presented the first account on the autoradiographic localization of a diffusible substance ($^{22}NaCl$) using thin cryosections. The conclusion from their work still holds: a way is open for studying the subcellular distribution and metabolism of materials such as lipids, peptides, sugars, electrolytes and drugs which cannot be retained by chemical fixation. No essential improvement to the principle or the technique has since been made. Magelhaes (1977) stressed the potential of the approach, illustrated by only one micrograph with a few autoradiographic grains. A grain density distribution was estimated after the absorption of labelled sugars by jejunal loops (Johnson and Bronk 1980). The ultrastructural preservation of the absorptive cells, however, was poor and hardly compatible with autoradiography at the electron microscope level.

High resolution autoradiography of diffusible substances has had a great impact on the development of cryoultramicrotomy, however, in the actual application of cryoultramicrotomy, it has only played a minor role. The reasons for the low penetration of this promising technology are probably twofold. Autoradiography is a time-consuming technique, i.e. long exposure times, data aquisition and statistical treatment (reviewed by Williams 1977 a, b) require careful planning of experiments. Localization studies on the basis of antigen-antibody interactions may require the same sort of statistical treatment of the data, but certainly lack the long exposure times of autoradiography and are, therefore, generally preferred. The second reason is probably the ultrastructural preservation upon freezing of fresh tissue. Only an outer rim of the frozen specimen is well preserved as can be demonstrated after a number of cryopreparative procedures such as freeze-substitution (Dempsey and Bullivant 1976), freeze-fracturing, cryosectioning as well as freeze-drying followed by embedding in plastic (Frederik and Busing 1981). With more sophisticated freezing methods (Elder et al. 1982; Escaig 1982), an increase of the useful volume of tissue can be obtained, in the order of 20–50 μm from the edge of the sample (see also Sitte et al., Chap. 4, this Vol.). The useful area for high resolution autoradiography on cryosections from fresh-frozen tissue is thus restricted to the few cell layers on the surface of the frozen sample. The same restriction holds for X-ray microanalysis on cryosections from fresh-frozen tissue (see Zierold and Steinbrecht, Chap. 15, this Vol.), but the statistics for autoradiography generally require many more sections to obtain reliable data. Thin cryosections from fresh-frozen tissue are no longer the privilege of a few experts and there is nothing to withhold anyone from using such sections for the autoradiographic localization of diffusible substances. Thin cryosections from fresh-frozen (no chemical fixation) tissue are the most versatile specimens for studying the localization of diffusible substances by X-ray microanalysis or by autoradiography. Cryoultramicrotomy may also play a key role in specially designed cryomethods for the localization of diffusible substances. These methods can best be described as hybrid physicochemical fixation methods. The physical fixation by rapid cooling (preservation of the ultrastructure in a split second without any extraction of soluble compounds) is combined with low temperature processing (freeze-drying, freeze-substitution, see also Steinbrecht and Müller, Chap. 7, this Vol.) in conjunction with gentle chemical fixation (vapour fixation, fixatives in the substitution fluids). In this way it is hoped to circumvent both the disadvantages of chemical fixation, which at room or body temperature is a relatively slow process accompanied by the extraction of soluble compounds, as well as the disadvantages of physical fixation, e.g. the aqueous solutions used for localization reactions or staining are incompatible with this type of fixation. Stirling and Kinter (1967) developed a freeze-drying and embedding method for the autoradiographic localization of sugars at the electron microscope level. The method was based on earlier work by Wilske and Ross (1965) for light microscope autoradiography and involves freezing of tissue samples (without chemical prefixation) and freeze-drying the samples at high vacuum. The dry samples are fixed in osmium tetroxide vapour and this step is thought to be essential in the procedure. A number of compounds (aspirin, Wilske and Ross 1965; sugars, Stirling and Kinter 1967; ouabain, Stirling 1972; steroids, Frederik and Klepper 1976) are fixed in the tissue by vapour fixation and will not be extracted during the subsequent infiltration by resin. However, flotation of thin sections on the water of a knife boat caused some extraction of the labelled compounds from the sections and, therefore, thin sections were cut on a "dry" glass knife, collected by an eyelash probe and pressed onto a Formvar-coated grid. It was found convenient to use standard embedding (Epon 812, Luft 1961; Stirling and Kinter 1967 added some silicone oil). When thin sections are prepared at room temperature on a "dry" knife, this material only yields wrinkled sections and fragments at the knife edge. On a cryoultramicrotome rather smooth thin sections were obtained by sectioning this material at $-70\,^{\circ}C$ (Frederik and Klepper 1976), sections which have been employed for the autoradiographic localization at the electron microscope level of steroids in testis (Frederik et al. 1977). Cryoultramicrotomy can thus be used to obtain thin plastic sections on a "dry" knife.

In this application the cryoultramicrotome is used to search for the optimum mechanical properties of the specimen (temperature-dependent) in relation to the thin-sectioning process. As such, the cryoultramicrotome is also used in materials research and it might be worthwile to consider the application of cryoultramicrotomy in a number of cryopreparative procedures which require "dry" sectioning. Examples could be freeze-substituted and embedded material (to be used for X-ray microanalytical or autoradiographic studies) or freeze-dried, vapour-fixed and embedded material (to be used for X-ray microanalysis of diffusible ions, e.g. Ingram et al. 1974; Wróblewski and Wróblewski 1984). This application of the cryoultramicrotome may seem to be against the philosophy of the instrument designers, but it deserves consideration when it comes to theoretical (or practical) problems with regard to the extraction and/or redistribution of soluble compounds in plastic sections floating on a water surface.

3 Thin Cryosections in Histochemistry

Thick cryosections (in the micrometer range of thickness) and tissue slices have long been used in histochemistry and cytochemistry. In combination with X-ray microanalysis to quantitate the reaction products (Bowen and Ryder 1978), a powerful technique was established with a spatial resolution compatible with sub-

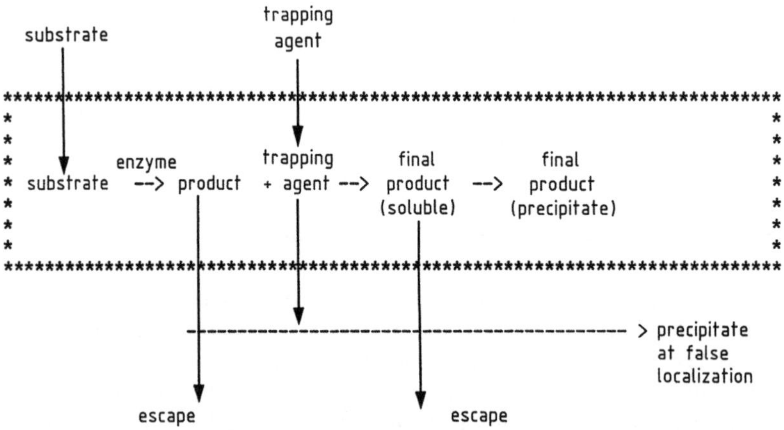

Fig. 1. The physicochemical processes involved in a cytochemical staining of an enzyme. *Solid lines* indicate diffusion processes and *broken lines* chemical processes. The *asterisks* depict the boundary of the enzyme site. (After Van Duijn and van der Ploeg 1980)

cellular localization (e.g. Bàcsy 1982; Roomans and Wróblewski 1982). Once we thought that cytochemical (precipitation) reactions could also be carried out on thin cryosections and that such a system would be ideal to localize and quantify cellular phosphatases (X-ray microanalysis to study enzyme kinetics!). The attempts to use a cerium-based trapping agent (Veenhuis et al. 1980; Robinson and Karnovski 1982; Hulstaert et al. 1983; Angermüller and Fahimi 1984) to localize phosphatases in thin cryosections have failed thus far. The reactions, when carried out on a thick (vibratome) section or on small tissue samples (a few mm^3), resulted in a fine precipitate and an "excellent" localization (references cited above). When thin sections were incubated in the same reaction mixture, no detectable precipitate was found. To come to a reasonable explanation of this negative observation, a small excursion to the fundamentals of enzyme histochemistry is required. The general type of cytochemical reaction involving the precipitation of metal salts takes the form shown in Fig. 1.

The final (insoluble) product will mark the enzyme site as long as there is no substantial diffusion of the primary product and, in addition, the primary product combines with the trapping agent at the enzyme site. Heavy metal salts, formed as the final product in a cytochemical reaction, may be sparingly soluble at thermodynamic equilibrium, but in a non-equilibrium state (the cytochemical incubation), considerable supersaturation of the final product may occur before a precipitate is formed. The capture reaction in metal-salt methods has been studied in theoretical and experimental models (reviewed by de Jong 1982), and nucleation in a (super)saturated salt solution was found to be a key event. When no exogenous nuclei are present, crystallization will only start at salt concentrations in far excess of the solubility product (homogeneous nucleation). When a few nuclei are already present, crystallization will start at the site of the nucleus (heterogeneous nucleation; precipitate not necessarily related to the site of the enzyme). Theoretical considerations on supersaturation predict a threshold activity of enzyme for the correct localization of a metal salt. In a biological matrix, heterogeneous nucleation (with possible local variations in the degree of supersaturation) must always be taken into account (de Jong et al. 1978). For a number of enzyme-cytochemical reactions, the smallest size of the enzyme site can be determined. For some reactions (azo-methods in light microscopy), enzyme sites smaller than 1 – 2 μm can be detected almost regardless of the local enzyme activity (Raap 1983). For metal-salt methods such data are scarce (Raap 1983) and not generally applicable, because the nature of the crystallization nuclei may vary from tissue to tissue. It might well be that the small amount of material, including enzyme in thin cryosections, prevents the formation of a steady state with enough primary product to yield a local precipitate (by homogeneous or heterogeneous nucleation). A thin cryosection, so much wanted for high resolution localization studies, can be a disadvantage when enzyme cytochemistry is attempted. On the basis of these considerations (and preliminary experiments), it can be argued that an enzyme localization that works perfectly on tissue samples or thick (micrometer range) sections will not be necessarily yield the same quality of results when applied to thin cryosections. Scaling down a precipitation method from the micrometer scale to the submicrometer scale has a certain risk related to (super)saturation and nucleation conditions of the precipitate to be formed. The localization methods based on the affinity between a tissue component and a probe (e.g. immunocytochemistry, lectin cytochemistry) do no have this disadvantage and this is one explanation for their leading role in cytochemistry.

4 Conclusions

Autoradiography of diffusible substances at the electron microscope level requires thin cryosections. Cryoultramicrotomy of fresh-frozen material is the most versatile way to produce "dry" thin sections for autoradiography of diffusible substances. Cryoultramicrotomy of embedded material obtained by a cryopreparative procedure (e.g. freeze-substitution, freeze-drying and vapour fixation) can also be considered. An ultrastrucltural localization study based on such a complicated methodology requires a large investment of time and effort. It is, therefore, advised to have preliminary biochemical data and/or autoradiographic studies at the light microscope level available before embarking on electron microscope autoradiography of diffusible substances.

For enzyme cytochemistry, thin cryosections are not the object of choice as long as reaction schemes are involved yielding the precipitate of a metal salt. Preliminary experiments with negative results indicate that in thin sections enzymes may escape detection by such methods. There is a theoretical explanation for this negative observation, but more experimental work is needed before a firm conclusion can be drawn.

The advantages and disadvantages of autoradiography and enzyme cytochemistry on thin cryosections have been weighed against those of other localization principles, notably the affinity principle involved in antibody or lectin interaction with tissue. Although the present situation does not seem to be in favour of autoradiography or enzyme cytochemistry, there is no reason for not changing this situation by trying harder or thinking smarter.

References

(References marked by an asterisk are review papers or books respectively)

Angermüller S, Fahimi H (1984) A new cerium-based method for cytochemical localization of thiamine pyrophosphate in the Golgi complex of rat hepatocytes. Comparison with the lead technique. Histochemistry 80:107–111

Appleton TC (1964) Autoradiography of soluble labelled compounds. J R Microsc Soc 84:277–281

Appleton TC (1969) The possibilities of locating soluble labelled compounds by electron microscope autoradiography. In: Roth LJ, Stumpf WE (eds) Autoradiography of diffusible substances. Academic Press, London New York, pp 304–319

Appleton TC (1974) A cryostat approach to ultrathin "dry" frozen sections for electron microscopy: a morphological and X-ray analytical study. J Microsc (Oxford) 100:49–74

Bàcsy E (1982) Enzymic heterogeneity of adrenocortical lysosomes: an X-ray microanalytical study. Histochem J 14:99–112

Baker JRJ, Appleton TC (1976) A technique for electron microscope autoradiography (and X-ray microanalysis) of diffusible substances using freeze-dried fresh frozen sections. J Microsc (Oxford) 108:307–315

Bernhard W, Leduc EH (1967) Ultrathin frozen sections. I. Methods and ultrastructural preservation. J Cell Biol 34:757–771

Bowen ID, Ryder IA (1978) The application of X-ray microanalysis to histochemistry. In: Erasmus DA (ed) Electron probe microanalysis in biology. Chapman & Hall, London, pp 183–211

Christensen AK (1969) A way to prepare frozen thin sections of fresh tissue for electron microscopy. In: Roth LJ, Stumpf WE (eds) Autoradiography of diffusible substances. Academic Press, London, New York, pp 349–362

Christensen AK (1971) Frozen thin sections of fresh tissue for electron microscopy with a description of pancreas and liver. J Cell Biol 51:772–804

Christensen AK, Paavola LG (1972) Frozen thin sections of fresh frozen tissue and the possibility of their use for autoradiography. J Microsc (Paris) 13:148–149

*De Jong ASH (1982) Mechanisms of metal salt methods in enzyme cytochemistry with special reference to acid phosphatase. Histochem J 14:1–33

De Jong ASH, Hak TJ,Van Duijn P, Daems WTh (1978) A new dynamic model system for the study of capture reactions for diffusable compounds in cytochemistry. I. Description of the model with special attention to phosphate capture in acid phosphatase cytochemistry. J Histochem Cytochem 26:331–339

Dempsey GP, Bullivant S (1976) A copper block method for freezing noncryoprotected tissue to produce ice-crystal free regions for electron microscopy. I. Evaluation using freeze substitution. J Microsc (Oxford) 106:251–260

Dollhopf FL, Sitte H (1969) Die Shandon-Reichert-Kühleinrichtung FC-150 zum Herstellen von Ultradünn- und Feinschnitten bei extrem niederen Temperaturen. I. Gerätetechnik. Mikroskopie (Wien) 25:17–32

Elder HY, Gray CC, Jardine AG, Chapman JN, Biddlecombe WH (1982) Optimum conditions for cryoquenching of small tissue blocks in liquid coolants. J Microsc (Oxford) 126:45–61

Escaig J (1982) New instruments which facilitate rapid freezing at 83 K and 6 K. J Microsc (Oxford) 126:221–229

Fernández-Morán H (1952) Application of the ultrathin freezing-sectioning technique to the study of cell structures with the electron microscope. Ark Fys 4:471–481

Frederik PM (1982) Cryoultramicrotomy. Recognition of artifacts. Scanning Electron Microsc 1982, II:709–721

Frederik PM, Busing WM (1981) Ice crystal damage in frozen thin sections: freezing effects and their restoration. J Microsc (Oxford) 121:191–199

Frederik PM, Klepper D (1976) The possibility of electron microscopic autoradiography of steroids after freeze drying of unfixed testes. J Microsc (Oxford) 106:209–219

Frederik PM, Van Der Molen HJ, Klepper D, Galjaard H (1977) Electronmicroscopic autoradiography of tritiated testosterone in rat testis. J Cell Sci 26:339–346

Futaesaku Y, Mizuhira V (1986) Negative staining autoradiography: a new technique for ultracryotomy utilizing an interposed film. J Histochem Cytochem 34:1085–1094

Glauert AM, Rogers GE, Glauert RH (1956) A new embedding medium for electron microscopy. Nature (London) 178:803
Harris WV, Salpeter MM (1983) Soluble compound electron microscope (EM) autoradiography: a resolution source to test the redistribution of soluble tritiated compounds during processing. J Histochem Cytochem 31:495–500
Hodson S, Marshall J (1970) Ultracryotomy: a technique for cutting ultrathin sections of unfixed biological tissues for electron microscopy. J Microsc (Oxford) 91:105–117
Hodson S, Williams L (1976) Ultracryotomy of biological tissues to preserve membrane structure. J Cell Sci 20:687–698
Hulstaert CE, Kalicharan D, Hardonk MJ (1983) Cytochemical demonstration of phosphatases in rat liver by a cerium-based method in combination with osmium tetroxide and potassium ferrocyanide postfixation. Histochemistry 78:71–79
Ingram FD, Ingram MJ, Hogben CAM (1974) An analysis of the freeze-dried, plastic embedded electron probe specimen preparation. In: Hall T, Echlin P, Kaufmann R (eds) Microprobe analysis as applied to cells and tissues. Academic Press, London New York, pp 119–146
Johnson IT, Bronk JR (1980) Distribution of actively absorbed diffusible sugars in the jejunal epithelium of the rat. J Cell Sci 45:199–210
Köhler G, Milstein C (1975) Continuous cultures of fused cells secreting antibody of predefined specificity. Nature (London) 256:495–497
Luft JH (1961) Improvement in epoxy embedding methods. J Biophys Biochem Cytol 9:409–414
Magalhaes MM (1977) Ultrastructure of the adrenal cortex in frozen thin sections. J Cell Sci 27:303–311
Moor H, Mühlethaler K, Waldner H, Frey-Wyssling A (1961) A new freezing-ultramicrotome. J Biophys Biochem Cytol 10:1–13
Pearse AGE (1982) Foreword. In: Bullock GR, Petrusz P (eds) Techniques in immunocytochemistry, vol. 1. Academic Press, London New York, p VII
Raap AK (1983) Localization mechanisms in enzyme cytochemistry. Thesis Univ Leiden, Neth
Robinson JM, Karnovsky MJ (1982) Cytochemical localization of the ectoenzyme 5′-nucleotidase by a new cytochemical method. J Histochem Cytochem 30:507 (Abstr)
*Rogers AW (1979) Techniques of autoradiography, 3rd edn. Elsevier/North Holland Biomedical Press, Amsterdam New York
Roomans GM, Wróblewski R (1982) Quantitative X-ray microanalysis of spleen lysosomes after acid phosphatase reaction. Histochemistry 75:485–491
Sabatini DD, Bensch K, Barrnett RJ (1963) Cytochemistry and electron microscopy. The preservation of cellular ultrastructure and enzymatic activity by aldehyde fixation. J Cell Biol 17:19–58
Sjöstrand FS (1943) Electron-microscopic examination of tissues. Nature (London) 151:725–726
Stirling CE (1972) Radioautographic localization of sodium pump sites in rabbit intestine. J Cell Biol 53:704–714
Stirling CE, Kinter WB (1967) High resolution radioautography of galactose-^{3}H accumulation in rings of hamster intestine. J Cell Biol 35:585–604
Stumpf WE, Roth LJ (1964) Vacuum freeze-drying of frozen sections for dry-mounting, high resolution autoradiography. Stain Technol 39:219–223
*Van Duijn P, Van der Ploeg M (1980) Microscopic cytochemistry as matrix chemistry. In: Trends in enzyme histochemistry and cytochemistry. Ciba Found Symp 73 Elsevier/North Holland Biomedical Press, Amsterdam New York, pp 209–229
Veenhuis M, Van Dijken JP, Harder W (1980) A new method for the cytochemical demonstration of phosphatase activities in yeast based on cerous ions. FEMS Microbiol Lett 9:285–291
*Williams MA (1977a) Autoradiography and immunocytochemistry. In: Glauert AM (ed) Practical methods in electron microscopy, vol. 6, pt. 1. Elsevier, Amsterdam
*Williams MA (1977b) Quantitative methods in biology. In: Glauert AM (ed) Practical methods in electron microscopy, vol. 6, pt. 2. Elsevier, Amsterdam, pp 1–234
Wilske KR, Ross R (1965) Autoradiography of lipid- and water soluble compounds: a new approach. J Histochem Cytochem 13:38–43
Wróblewski R, Wróblewski J (1984) Freeze-drying and freeze-substitution combined with low temperature embedding. Histochemistry 81:469–475
Wyckhoff RWG (1946) Frozen-dried preparations for the electron microscope. Science 104:36–37

Chapter 13 Scanning Electron Microscopy and X-Ray Microanalysis of Frozen-Hydrated Bulk Samples

ALAN T. MARSHALL[1]

1 Introduction

Living organisms normally contain a high concentration of water, thus when frozen-hydrated biological samples are examined this is the closest representation of the living state that is currently possible to obtain in the electron microscope. An exception is the "environmental microscope" (Danilatos and Postle 1982), but this has considerably more limitations in its use than the "low temperature microscope". The methods and techniques of low temperature microscopy of bulk samples evolved for both morphological purposes (Echlin et al. 1970; Echlin 1971; Nei et al. 1972; Echlin and Moreton 1973; Nei et al. 1973; Tokunaga and Tokunaga 1973; Turner and Smith 1974; Robinson 1975; Echlin and Moreton 1976; Echlin and Burgess 1977; Echlin 1978; Echlin et al. 1979) and for X-ray microanalysis (Gehring et al. 1973; Marshall and Wright 1973; Gullasch and Kaufmann 1974; Echlin and Moreton 1974; Fuchs and Lindeman 1975; Brombach 1975; Lechene et al. 1975; Marshall 1975a, b; Forrest and Marshall 1976; Zierold 1976; Marshall 1977; Yeo et al. 1977; Fuchs et al. 1978a, b; Kramer and Preston 1978; Zierold and Schäfer 1978; Echlin et al. 1980; Fuchs and Fuchs 1980; Marshall 1980a, b; Echlin et al. 1981; Marshall 1981; Marshall 1982; Marshall 1984a; Marshall and Condron 1985a, b; Marshall et al. 1985b; Echlin and Taylor 1986).

At the present time, low temperature scanning electron microscopy for morphological purposes may be considered to be well established and routine. For analytical purpose the methods, although continuing to evolve, are capable of producing valid physiological data (e.g. Echlin et al. 1982; Marshall 1983; Potts and Oates 1983; Marshall et al. 1985b; Echlin and Taylor 1986).

2 Morphology

Frozen-hydrated bulk samples prepared for scanning electron microscopy can reveal details of fluid-filled and air-filled spaces which are not readily apparent in fixed and dried specimens. Such detail can be seen in fractured samples

1 Analytical Electron Microscopy Laboratory, Department of Zoology, La Trobe University, Bundoora (Melbourne), Victoria 3083, Australia

Cryotechniques in Biological Electron Microscopy
Edited by R. A. Steinbrecht and K. Zierold

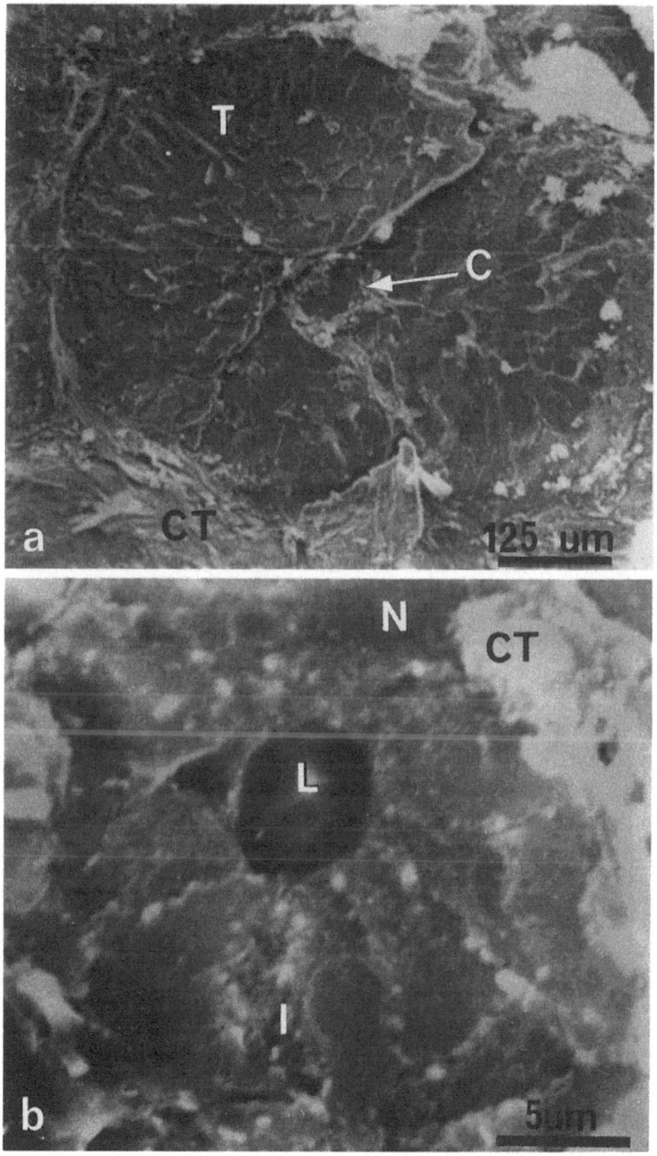

Fig. 1a, b. Fractured salt gland of duck showing: **a** a lobule containing secretory tubules (*T*), central canal (*C*) and connective tissue (*CT*); **b** a secretory tubule fractured transversely showing individual cells with nuclei (*N*), fluid-filled lumen (*L*), intercellular spaces (*I*) and connective tissue (*CT*) (Marshall et al. 1985b)

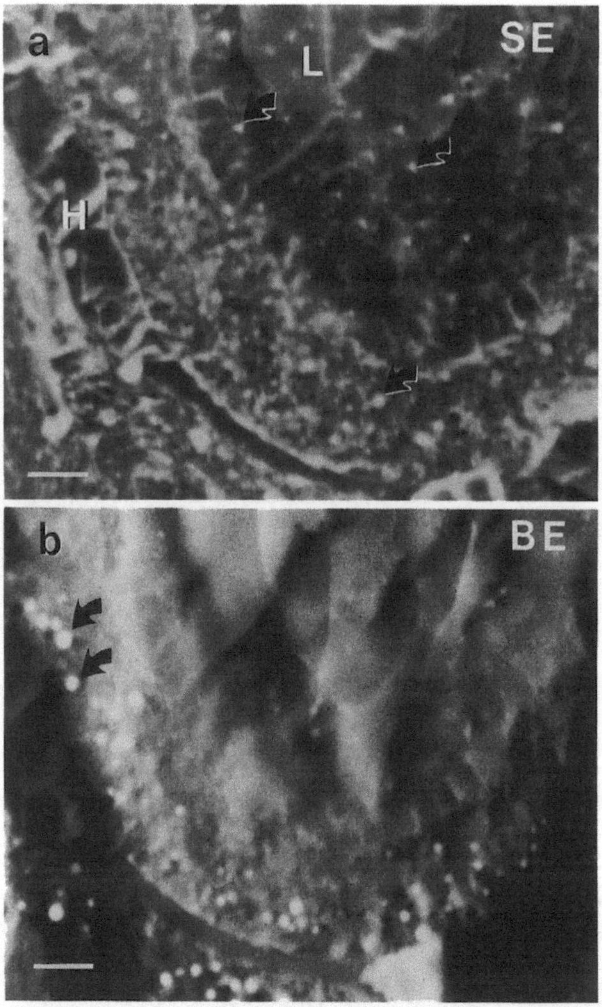

Fig. 2a, b. Fractured insect Malpighian tubule showing frozen luminal contents (*L*) and haemolymph (*H*) seen as a secondary electron image (*SE*) and a backscattered electron image (*BE*). The cellular structure, microvilli and subsurface granules (*arrows*) are readily visible in the SE mode and granules (*arrows*) on the surface are more readily seen in the BE mode (Marshall 1981). Scale bar = 5 μm

(Fig. 1) and samples which are dissected within the microscope specimen chamber (Hayes and Koch 1975). Samples of great fragility also may be better preserved in the frozen state for microscopy than when fixed and dried or freeze-dried (e.g. Tokunaga and Tokunaga 1973; Hook et al. 1980; Beckett and Porter 1982; Jones et al. 1984). Indeed, drying of any type inevitably results in some shrinkage and possibly distortion (Boyde and Franc 1981).

The type of information obtainable depends on whether secondary electron or backscattered electron images are obtained. True topographic imaging is seen in backscattered images when a low take-off angle is used. Thus, real surface detail can be seen (the two-arrowed granules in Fig. 2b are on the surface and are not readily seen in Fig. 2a). Secondary electron images appear to contain information from subsurface structures and therefore contain more detail of cell structure (Fig. 2a). Note that more subsurface luminal granules and microvilli can be seen in Fig. 2a than in Fig. 2b. The structures seen in the SE mode are possibly due in part to beam-induced conductivity contrast and edge effects.

Specimen preparation techniques and instrumentation are essentially the same as for X-ray microanalysis except that samples may be "glued" to specimen holders with solutions of, for example, methyl cellulose. Samples which are destined for X-ray microanalysis should not be brought into contact with non-physiological solutions. The specimens may be coated with a conductive coat by evaporative coating of gold or chromium or sputter coated with gold or platinum (reviewed by Robards and Sleytr 1985).

3 Analysis

For X-ray microanalysis the frozen-hydrated bulk sample offers some advantages compared to frozen sections. It is easier to prepare a frozen bulk sample and maintain it in the frozen state. The thermal capacity of bulk samples and stages is generally many times greater than that of a thin frozen-hydrated section perched precariously on a filmed grid or other filmed sample holders. Thus, the danger of desiccation is correspondingly lower. A further advantage is that considerably larger specimens can be examined as bulk samples than as sections, although it is by no means always the case that the whole surface of a bulk sample is usable for analysis.

As will become abundantly clear in later sections of this chapter, there are many difficulties in the analysis of frozen bulk samples and some disadvantages compared to the analysis of sections. Nevertheless, it can be successfully employed for the analysis of fluid-filled spaces and for intracellular analyses with some ability for discrimination between different zones of the cell. It is perhaps in the area of analysis of luminal contents of tubular organ systems and plant cell vacuoles that the technique has most potential (see also Zierold and Steinbrecht, Chap. 15, this Vol.).

4 Freezing

Specimens can be frozen in a variety of ways using fast-freezing techniques which have been developed for cryomicrotomy and freeze-etching (Plattner and Bachmann 1982; Bald 1985; Robards and Sleytr 1985, see also Sitte et al., Chap. 4, this Vol.). As previously outlined, however, specimens for scanning electron microscopy are usually larger than for microtomy and freeze-etching. This in itself will limit the freezing rates which can be achieved. It is also not desirable to contaminate the microscope column or the sample with hydrocarbon or fluorinated cryogenic fluids, especially if a windowless detector is to be used. Since good ther-

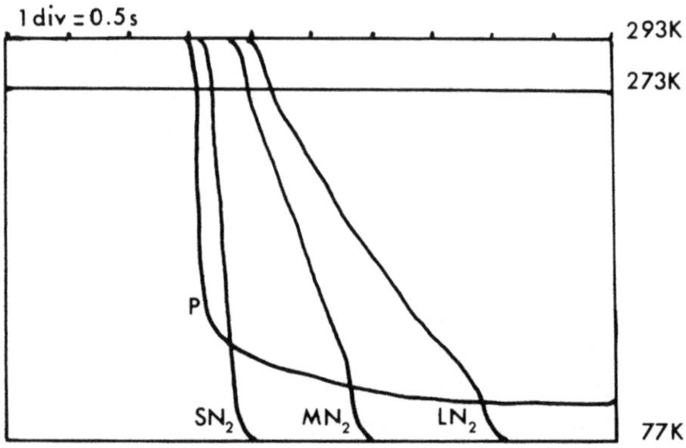

Fig. 3. Cooling curves for liquid (LN_2), melting (MN_2) and solid (SN_2) nitrogen and propane (P)

mal and electrical contact between specimen and substage is desirable, the specimen should be mounted in the substage before freezing. These points argue in favour of freezing in melting nitrogen or solid nitrogen. The present author freezes samples by mounting them in a substage and pressing them into solid nitrogen. Solid nitrogen gives much higher cooling rates than melting or liquid nitrogen (Fig. 3). The sample, protected from frost deposition, is then transferred to a fracturing device or microtome.

5 Instrumentation

5.1 Fracturing and Transfer Devices, Low Temperature Stages

A considerable amount of structural detail can be seen in a fractured sample without removing ice by sublimation. The sample is fully frozen-hydrated and thus, from a physiological point of view, in the ideal state for X-ray microanalysis. The fracturing device should have a low temperature stage in a clean, high vacuum and a means of producing a percussion fracture in the sample. A sharp blow to the sample from a cold knife should result in a fracture plane propagating through the sample. The sample must not be cut or microtomed, otherwise the surface is featureless and ice must be removed by sublimation before any structural detail can be seen. It should be noted that the latter technique has been used as a preparation method by some authors, e.g. Potts and Oates (1983). The fracturing device must have some means of coating the fractured sample with a conducting coat. Various devices have been described and some are available commercially (Echlin and Moreton 1973; Marshall 1977; Pawley and Norton 1978; Robards and Crosby 1979; Robards and Sleytr 1985). Few devices have all the necessary requirements for producing flat and reproducible fracture planes. This remains one of the limitations of the fracturing technique.

Transfer of the fractured and coated sample to the microscope must then be accomplished without frost being deposited on the sample. Again various devices have been used for this purpose, varying from the use of a simple capping device for the specimen holder and immersion in liquid nitrogen (Marshall 1977) to highly elaborate evacuated transfer chambers (Robards and Crosby 1979). The simple systems appear to work just as well as the complex ones.

Finally, the substage is positioned on the low temperature stage in the microscope column. The low temperature stage should be equipped with a thermocouple thermometer and a Faraday cup. The stage should be capable of reaching a temperature of 100 K or lower. The most efficient stage designs depend upon circulation of liquid nitrogen or nitrogen gas through the stage (Robards and Sleytr 1985). Heat transfer from stage to liquid nitrogen Dewar via copper braid is not so efficient but 100 K can be attained if the braid is attached to a cold finger protruding from the Dewar into the column and through which liquid nitrogen circulates (Taylor and Burgess 1977; Marshall et al. 1982). The higher the column vacuum, the lower the stage temperature needs to be in order to reduce the vapour pressure of ice. At temperatures lower than 143 K the sublimation rate of ice is extremely low (<0.001 nm s^{-1}), even in a perfect vacuum (Koehler 1968), and at 113 K the sublimation rate decreases by a further five orders of magnitude (Umrath 1983)!

5.2 Detectors and Vacuum Requirements

Consideration should be given to the area of the detector crystal, the solid angle of X-ray collection, detector collimation, X-ray take-off angle, window thickness and the operating voltage of the microscope (Marshall 1980c; Marshall 1982). These parameters will have a marked effect on the detection of X-rays from Na and Mg particularly. If peak to background ratios are to be used for quantitation, then it is most important to exclude extraneous X-rays from the detector.

Windowless or removable window detectors are of value in the analysis of frozen-hydrated biological samples, primarily for the determination of water content via the measurement of oxygen concentration. This is a method first suggested by Fuchs et al. (1978a), who made preliminary observations with a thin window detector, and subsequently developed by Marshall (1981, 1982, 1984a) and Marshall and Condron (1987) for windowless detectors.

The column of the microscope must be clean, which means that a reasonably high vacuum (10^{-7} torr) should be attainable (Marshall 1984b). If not, then there is a risk of specimen contamination by hydrocarbons and frost. This may be severe enough to affect absorption and the accuracy of the analysis. If a windowless detector is being used, then a clean column is essential to prevent contamination of the detector crystal (Marshall 1984a).

6 Electron Interactions

6.1 Space Charge and Conductive Coats

When frozen-hydrated bulk biological samples were first examined at low temperature in a scanning electron microscope, it was soon realized that the samples rapidly acquired a charge (Brombach 1975; Fuchs and Lindemann 1975; Marshall

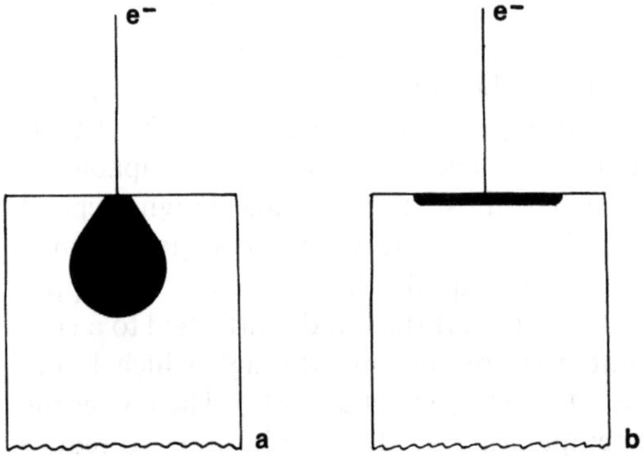

Fig. 4a, b. Electron diffusion volume: **a** normal; **b** as a result of development of an internal space charge

1975a; Marshall 1977; Fuchs et al. 1978b). This is not surprising since ice is a singularly good insulator at low temperature (Durand et al. 1967). The work of Fuchs and colleagues was done on uncoated samples and indicated that a space charge developed which resulted in the formation of a pancake-shaped electron diffusion volume below the surface of the sample (Fig. 4). With a carbon coat they found that the dissipation of charge through the coat would limit the space-charge field. Since carbon is not a good conductor, it may be that the conductivity of carbon is limiting and that with a more conductive coat, charge storage in the sample will be negligible. This seems to be borne out by the observation of Marshall and Condron (1985a) that in charging uncoated frozen-hydrated samples, the backscattered electron yield rises rapidly, whereas in coated samples it remains constant (Fig. 5).

It would seem, therefore, that if frozen-hydrated samples are well coated, no gross distortion of the electron diffusion volume occurs. This finding is consistent with estimates of X-ray excitation depth and electron penetration depth in ice (see later).

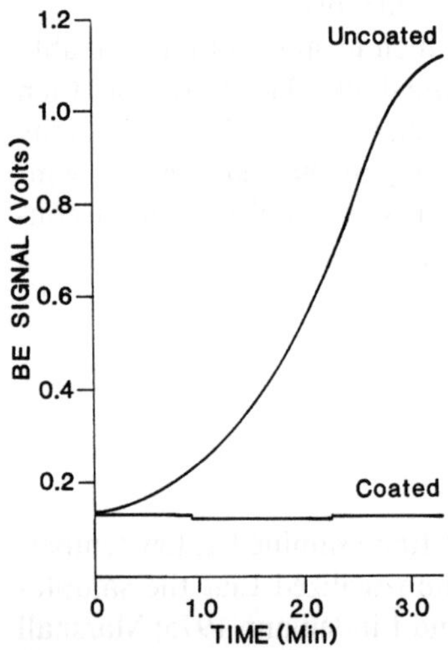

Fig. 5. Recordings of backscattered electron signal (after amplification) from coated and uncoated frozen-hydrated gelatine gels at constant beam current (Marshall and Condron 1985a)

Conductive coats, other than carbon, which are suitable for X-ray microanalysis include Be, Al and Cr; of these, Be has the best electrical and thermal conductivities (Table 1), but due to its high toxicity requires care in handling (Marshall and Carde 1983; Marshall et al. 1985a). The absorption of X-rays from low atomic number elements of biological interest may also be affected by the coat (Table 2). In this respect Be again is the best coat since it shows least absorption. A useful alternative to Be is Cr (Marshall 1977, 1980a, b), although a small correction for absorption of X-rays from low atomic number elements is advisable (Echlin and Taylor 1986).

Table 1. Thermal and electrical conductivities of coating materials

	Resistivity (μOhm cm)	Thermal conductivity (W cm^{-1} K^{-1})
	100 K	*100 K*
Be	0.09	10
Al	0.47	3
Cr	1.6	1.5
	293 K	*273 K*
C	3500	0.04

Table 2. Percentage of X-ray absorption by sample coats, assuming an X-ray emergence angle of 40° to the plane of a 20-nm-thick coat[a]

Emitter	Absorber			
	Be	C	Al	Cr
O	2.23	7.43	5.49	6.78
Na	0.30	1.22	0.89	13.97
Mg	0.17	0.73	0.54	9.07
P	0.04	0.15	1.93	2.89
S	0.03	0.10	1.33	2.00
Cl	0.02	0.08	0.51	1.34
K	0.01	0.04	0.49	0.75
Ca	0.01	0.03	0.37	0.52

[a] Calculated according to Reed (1975) using mass absorption coefficients from Henke et al. (1982).

6.2 Electron Diffusion Volume and X-Ray Depth Resolution

The possible effect of a space charge on the electron diffusion volume has already been discussed. The size of the diffusion volume and the X-ray depth distribution are of paramount importance for the determination of the analytical resolution in bulk frozen-hydrated samples.

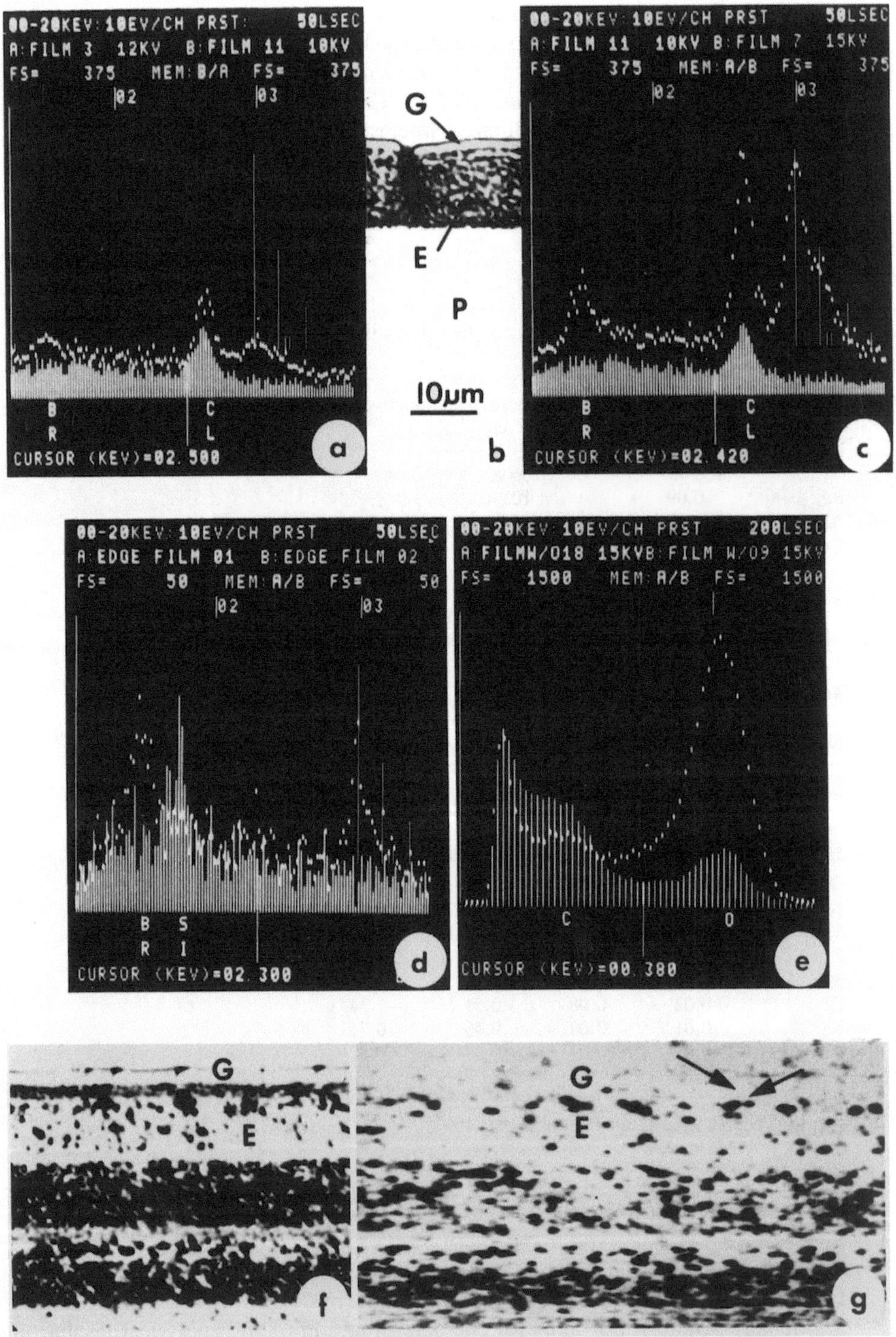
00-20KEV:10EV/CH PRST: 50LSEC
A:FILM 3 12KV B:FILM 11 10KV
FS= 375 MEM:B/A FS= 375
02
03
B
R
C
L
CURSOR (KEV)=02.500
a
G
E
P
10μm
b
00-20KEV:10EV/CH PRST 50LSEC
A:FILM 11 10KV B:FILM 7 15KV
FS= 375 MEM:A/B FS= 375
02
03
B
R
C
L
CURSOR (KEV)=02.420
c
00-20KEV:10EV/CH PRST 50LSEC
A:EDGE FILM 01 B:EDGE FILM 02
FS= 50 MEM:A/B FS= 50
02
03
B S
R I
CURSOR (KEV)=02.300
d
00-20KEV:10EV/CH PRST 200LSEC
A:FILMW/018 15KVB:FILM W/09 15KV
FS= 1500 MEM:A/B FS= 1500
C
O
CURSOR (KEV)=00.380
e
G
E
f
G
E
g

Fig. 6a–g

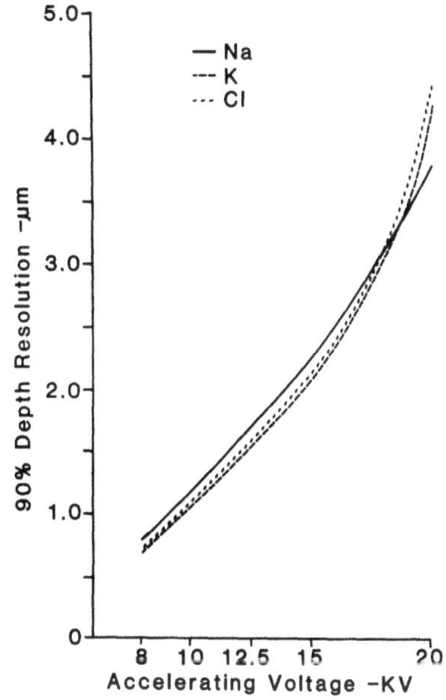

Fig. 7. Graph showing depth resolution, calculated from $\phi(\varrho z)$ curves, for Na, Cl and K as a function of accelerating voltage (Marshall 1982)

Marshall and Condron (1985a) have shown that estimates of X-ray depth distribution made from computed X-ray depth distribution [i.e. ϕ (ϱz)] curves, based on the empirical equations of Parobek and Brown (1978) correspond well to experimental determinations of electron beam penetration and X-ray emission using frozen-hydrated sections of gelatine gel and frozen-hydrated bulk samples of photographic film. In Fig. 6 it is shown that Br and Ag X-rays are not obtained from the emulsion layer, lying below a 1.5-µm gelatine layer, of frozen-hydrated photographic film at 10 kV, but small spectral peaks are seen at 12 kV and pronounced peaks at 15 kV. This agrees with the predicted resolution obtained from computer-generated $\phi(\varrho z)$ curves. These curves show that the depth resolution for Br L X-rays would be 1.2 µm at 10 kV, 1.7 µm at 12 kV and 2.4 µm at 15 kV (for details see Marshall and Condron 1985a). Results consistent with these have also been obtained from experiments on electron beam penetration in ice by Oates and Potts (1985).

Due to the fortuitous effects of absorption, all elements of biological interest have the same depth resolution at 15 kV in frozen-hydrated specimens (Fig. 7).

Fig. 6a–g. A light micrograph (**b**) of a transverse section of swollen Kodak EM film is shown with the outer gelatine layer (*G*), emulsion layer (*E*) and polyester base (*P*). Typical spectra are shown from analyses in which the electron beam was normal to the film. In **a** and **c** *bars* denote spectrum obtained at 10 kV and *dots* at 12 kV and 14.8 kV respectively. The position of the Br L and Cl K lines are indicated and the *bar markers* show the Ag L line series. A light and scanning electron micrograph in reverse contrast of Kodak Kodacolor film are shown in **f** and **g** respectively. Gelatine (*G*) and emulsion (*E*) layers are marked. The scale is as in **b**. Analysis sites are indicated by *arrow* and **d** shows a spectrum (bars) obtained from the most distant site from a silver grain (<2.0 µm) compared with a spectrum (*dots*) obtained from the closer site (<1.0 µm). The spectra in **e** represent an analysis of hydrated EM film (*dots*) compared to the same dried film (*bars*). It can be deduced that the water content of the gelatine layer was 70% (Marshall and Condron 1985a)

7 X-Ray Emission

The optimum accelerating voltage for the analysis of bulk frozen-hydrated samples is around 15 kV. Lower voltages result in a reduced X-ray yield with poor counting statistics, whereas higher voltages result in a reduction in analytical resolution. A possible source of error in analyzing this type of sample is slight charging which could cause a reduction in the electron injection voltage. This may sometimes occur due to insufficient electrical contact of the sample with the stage. A fine crack may have developed, for example, during the fracturing process. The consequence of a reduction in the effective excitation voltage can be calculated from the equation of Green and Cosslett (1961) and Green (1963):

$$I \simeq (U-1)^{1.63},$$

where I is X-ray intensity and U is the over-voltage ratio. For elements of biological interest, a 1 kV reduction at 15 kV will result in a drop in X-ray intensity of 11–15%. If peak to background ratios are being used for quantitation, there will be a fall in the ratios of 3–5%. Fortunately, a fall in the electron injection voltage can be detected by monitoring the continuum cut-off which has a value equal to the injection voltage.

If peak integrals are used for quantitation, then fluctuations in beam current may have a greater effect on X-ray intensity than fluctuations in voltage. This stems from the well-known linear relationship between beam current and X-ray intensity (Birks 1963).

The desirability of a high take-off angle has already been mentioned. The absorption of Na and Mg X-rays by the frozen matrix is very strong, but is much less for P and elements of higher atomic number (Marshall 1982); any decrease in X-ray path length through the sample is therefore beneficial in increasing the detected X-ray intensities from Na and Mg.

8 The Fracture Surface

The fracture surface poses two problems. The surface topography may actually block the direct path of emitted X-rays to the detector or the path length of emitted X-rays may be sufficiently changed to markedly affect the absorption of Na and Mg X-rays. This problem can be dealt with by using an annular backscattered electron detector on the collimator of the X-ray detector. The slope of the sample towards or away from the detector can be judged from the visual differences in grey levels or directly from the voltage output of the backscattered electron detector (Marshall 1981, 1982) (Fig. 2). A commercial version of this detector configuration is now available and has been used by Echlin and Taylor (1986).

A considerable angular deviation from the horizontal can be tolerated for the analysis of P to Ca, but rather less for Na and Mg. Tilts from the horizontal up to 10° have little effect on Na analyses and up to 20° is well within the tolerance for K (Marshall and Condron 1985b).

The local tilt angle can also be judged and measured, albeit less conveniently, by stereoscopy. With either method the local take-off angle could be adjusted by tilting the sample, however, if the stage is not mounted on a eucentric goniometer a shift in Z-position will cause inaccuracies when a highly collimated detector is used.

9 Quantitative Analysis

Two approaches have been used to obtain quantitative data from frozen bulk samples.

9.1 Peak to Background Ratios

The most straightforward method is to use peak to background ratios to reduce the problems engendered by the topography of the fracture surface. Strictly speaking, these ratios should be peak to local background, i.e. the background under the characteristic peak (Roomans 1981). Unless a complex correction procedure is used (Echlin and Taylor 1986), it is necessary to use matching standards to produce a regression of concentration on peak to background ratios. These standards are typically frozen solutions of salts and proteins or other organic compounds such as dextrans or, recently, suspensions of finely ground carbon and salts (Whitecross et al. 1982).

9.2 Corrected Intensities

The second approach to quantitation is to use peak intensities and correct these for the effects of the different matrices of sample and standards. The advantages compared to the peak to background method are that (1) peak intensities vary less than background intensities with some software packages, this, together with the fact that peak to background ratios are small numbers, improves statistical accuracy. (2) Non-matching standards which are stable and of well-defined stoichiometry can be used. The major disadvantage is that peak intensities are more susceptible to topographical influences than peak to background ratios.

The classical approach to applying matrix corrections is the ZAF (atomic number – absorption – fluorescence) procedure. This has met with little success in biological matrices using commercial software packages (Boekestein et al. 1980; Boekestein et al. 1983) due to the difficulties in defining an unknown, but unmeasurable multielement (C, H, O, N) matrix.

As an alternative, the X-ray depth distribution curve [$\phi(\varrho z)$] approach (Parobek and Brown 1978) has been applied (Marshall 1982, 1984a; Marshall and Condron 1987). In this case, the absorption and atomic number corrections are generated by microcomputer from a set of empirical equations derived by Parobek and Brown (1978) and Brown and Robinson (1979) for sample and standard. In the case of the sample, an assumed matrix is used. The variation in matrix composition over the normal cellular range has virtually no effect on the correction factors except for Na and Mg. For these elements an algorithm is used to derive correction factors for a calculated matrix composition. Differences in elemental composition between frozen-hydrated cytoplasm and nucleus are minimal so that correction factors for these two compartments are virtually identical. The corrections can, of course, also be applied to the analysis of extracellular fluids.

The fluid concentrations are rendered in mmol kg^{-1} wet weight. In order to convert these concentrations into mmol l^{-1}, it is necessary to obtain the water content of the cells.

10 Water Content or Dry Weight Fraction

Two approaches have been developed for obtaining the water content in bulk frozen-hydrated samples.

10.1 Peak to Continuum Method

Zs.-Nagy and colleagues (Zs.-Nagy et al. 1982; Lustyik and Zs.-Nagy 1985) have exploited the peak to continuum ratio method which was developed by Hall for the analysis of thin sections (Hall 1971). This was initially validated for elements $Z>11$ in bulk freeze-dried samples (Zs.-Nagy et al. 1977) and used to compare the concentrations of potassium in the frozen and dried states. This leads to an estimate of the water fraction.

10.2 Measurement of Oxygen Concentrations

If the cell cytoplasm can be approximated to a solution of protein, then by virtue of the linear relationship between O concentration and protein concentration in such a solution, a measurement of O concentration can readily be translated into water-mass fraction (Marshall 1982).

Oxygen intensities can be readily measured with a windowless detector (Marshall 1984a) (Fig. 6e). The spectral background under the O peak can be approximated by a background from clean beryllium and scaled on the electronic noise peak. After subtracting background the O counts can be integrated in a window covering a fixed energy range. This excludes some of the counts present in the pronounced low energy tail of the O peak, but avoids the necessity of stripping off the overlapping C and possibly N peaks.

If the standard used to calculate O concentration is ice then an atomic number correction is unnecessary and only the absorption correction is required. The absorption correction can be obtained from a $\phi(\varrho z)$ curve calculation. This is based again on the assumption that protein solutions adequately model cell cytoplasm. Absorption decreases in an exponential manner as protein concentration decreases and the rate of decrease is, unfortunately, greatest over the range of biologial interest (Fig. 8). Furthermore, the absorption of O X-rays is highly dependent on local take-off angle. These difficulties are dealt with by first obtain-

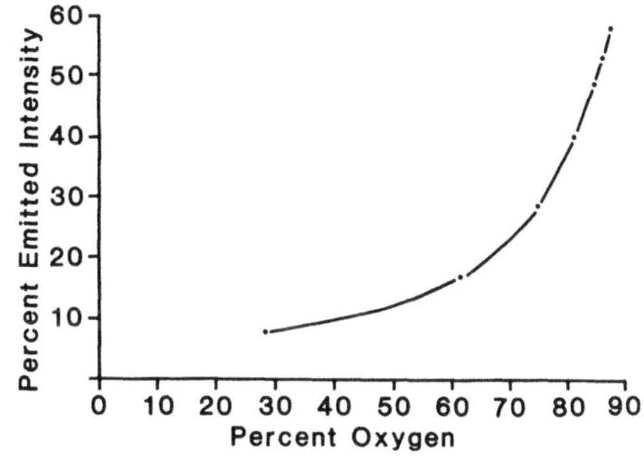

Fig. 8. Calculated emitted intensity as a percentage of generated intensity of O X-rays for increasing O concentration in model protein-water systems. The range 8–20% protein translates to 82–75% O (Marshall 1982)

ing an approximate O concentration value from a regression of protein concentration on O intensity from standards. This first estimate of O concentration is then used to calculate an absorption correction value which is, in turn, used to calculate a second estimate of O concentration. The final O concentration is translated into protein concentration to give dry weight and the water mass fraction.

In order to reduce the effect of topography on O X-ray absorption, the O intensity values are normalized by determining the ratio to the electron backscatter signal which is obtained from a backscattered electron detector mounted on the collimator of the windowless detector (Marshall and Condron 1985b). It is not feasible to use peak to background normalization since the background below the O peak is very small and is difficult to measure with any accuracy by simple means.

11 Beam Damage and Mass Loss

Beam damage, which potentially could affect quantitation, can be induced in frozen-hydrated bulk samples by high current doses 1000–4000 nC μm^{-2} especially in standards containing high (20%) protein concentrations. Biological tissues, however, appear to be somewhat more resistant to beam damage than simple protein solutions. Beam damage can be reduced substantially by reducing the current dose to approximately 40 nC μm^{-2} (e.g. a beam current of 0.2 nA for 200 s with a 1 μm^2 raster). Under these conditions the total count rate remains stable over the analysis period (Marshall 1980a, b, c, 1982). This suggests that frozen-hydrated bulk samples may be slightly less sensitive to beam damage than are 1-μm-thick frozen-hydrated sections (Hall 1986). When beam damage is incurred this is presumably not thermal damage at 100 K, since temperature rise is calculated to be relatively slight (Talmon and Thomas 1977), but due to radiolysis (Talmon 1984; see also Chap. 3, this Vol.). When static beam analyses are conducted, a cone of material is formed on the surface of the sample around the beam (Marshall 1982) if beam damage occurs. This is presumably from the deposition of the products of radiolysis. This type of damage has no apparent effect on count rate, except for O X-rays where it declines. The latter is due possibly to increased absorption because of increased X-ray path length, rather than to water loss.

12 Detection Limits

Detection limits in bulk frozen-hydrated samples have not been systematically investigated by experimentation. It is possible to detect and quantify Na at a concentration of 10 $mmol\,l^{-1}$, albeit with low precision, provided that the detector window is thin (see for example Marshall and Condron 1987). With a windowless detector Na could not be detected with a precision of three times the probable error due to counting statistics at 10 $mmol\,l^{-1}$, but could be detected at 25 $mmol\,l^{-1}$ (Marshall 1982). In order to achieve a precision of five times the probable error (Hall and Gupta 1982) with a standard detector, then the minimum concentration will be in excess of 25 $mmol\,l^{-1}$. This is somewhat higher than the detection limit for Na in thick frozen-hydrated sections (Hall and Gupta 1982). Higher atomic number elements will have correspondingly lower detection limits.

13 Conclusions

Low temperature microscopy of bulk frozen-hydrated specimens is now straightforward for morphological purposes and has particular value for resolving questions concerning the architecture of fluid- or gas-filled spaces and for examining very fragile structures. X-ray microanalysis has developed to the point where useful quantitative data can be obtained with an analytical resolution of approximately 2.0 µm. It is to be expected that further refinement of the technique will take place and that it will be more widely used as realization spreads that many investigations requiring only relatively low analytical resolution can probably be carried out more easily on frozen-hydrated bulk samples than on thick frozen-hydrated sections.

References

(References marked by an asterisk are review papers or books respectively)

*Bald WB (1985) The relative merits of various cooling methods. J Microsc (Oxford) 140:17–40

Beckett A, Porter R (1982) *Uromyces viciae-fabae* on *Vicia faba*: Scanning electron microscopy of frozen-hydrated material. Protoplasma 111:28–37

*Birks LS (1963) Electron probe microanalysis. John Wiley & Sons, New York

Boekestein A, Stols ALH, Stadhouders AM (1980) Quantitation in X-ray microanalysis of biological bulk specimens. Scanning Electron Microsc 1980/II:321–334

Boekestein A, Stadhouders AM, Stols ALH, Roomans GM (1983) Quantitative biological X-ray microanalysis of bulk specimens: an analysis of inaccuracies involved in ZAF-correction. Scanning Electron Microsc 1983/II:725–736

Boyde A, Franc F (1981) Freeze-drying shrinkage of glutaraldehyde fixed liver. J Microsc (Oxford) 122:75–86

Brombach JD (1975) Electron beam X-ray microanalysis of frozen biological bulk specimens below 130 K. 2. The electrical charging of the sample in quantitative analysis. J Microsc Biol Cell 22:233–238

Brown JD, Robinson WH (1979) Quantitative analysis by ϕ (ϱz) curves. Microbeam Anal 1979:238–240

Danilatos GD, Postle R (1982) The environmental scanning electron microscope and its applications. Scanning Electron Microsc 1982/I:1–16
Durand M, Deleplanque M, Kahane A (1967) Bulk conductivity of ice between −25° and −100 °C with ion exchange membranes. Sol Statn Commun 5:759–750
Echlin P (1971) The examination of biological material at low temperature. Scanning Electron Microsc 1971:225–232
*Echlin P (1978) Low temperature scanning electron microscopy: a review. J Microsc (Oxford) 112:47–61
Echlin P, Burgess A (1977) Cryofracturing and low temperature scanning electron microscopy of plant material. Scanning Electron Microsc 1977:491–500
Echlin P, Moreton R (1973) The preparation coating and examination of frozen biological materials in the SEM. Scanning Electron Microsc 1973:325–331
Echlin P, Moreton R (1974) The preparation of biological materials of X-ray microanalysis. In: Hall T, Echlin P, Kaufmann R (eds) Microprobe analysis as applied to cells and tissues. Academic Press, London New York, pp 159–174
Echlin P, Moreton R (1976) Low temperature techniques for scanning electron microscopy. Scanning Electron Microsc 1976:753–762
Echlin P, Taylor SE (1986) The preparation and X-ray microanalysis of bulk frozen hydrated vacuolate plant tissue. J Microsc (Oxford) 141:329–348
Echlin P, Lai C, Hayes TL (1981) The distribution and relative concentration of potassium in the root-tips of *Lemna minor* analysed using low temperature X-ray microanalysis. Scanning Electron Microsc 1981/II:489–498
Echlin P, Paden R, Dronzek B, Wayte R (1970) Scanning electron microscopy of labile biological tissues maintained under controlled conditions. Scanning Electron Microsc 1970:51–56
Echlin P, Pawley JB, Hayes TL (1979) Freeze-fracture scanning electron microscopy of *Lemna minor*. Scanning Electron Microsc 1979/III:69–76
Echlin P, Lai CE, Hayes TL, Hook G (1980) Elemental analysis of frozen-hydrated differentiating phloem parenchyma in roots of *Lemna minor* L. Scanning Electron Microsc 1980/II:383–394
Echlin P, Lai CE, Hayes TL (1982) Low temperature X-ray microanalysis of the differentiating vascular tissue in root tips of *Lemna minor* L. J Microsc (Oxford) 126:285–306
Forrest QG, Marshall AT (1976) Comparative X-ray microanalysis of frozen-hydrated and freeze-substituted specimens. In: Ben Shaul Y (ed) Electron microscopy 1976, vol 2. Proc 6th Eur Congr Electron Microsc, Israel Soc Electron Microsc, Jerusalem, pp 218–220
Fuchs W, Fuchs H (1980) The use of frozen-hydrated bulk specimens for X-ray microanalysis. Scanning Electron Microsc 1980/II:371–382
Fuchs W, Lindemann B (1975) Electron beam X-ray microanalysis of frozen biological bulk specimens below 130 K. 1. Instrumentation and specimen preparation. J Microsc Biol Cell 22:227–232
Fuchs W, Lindemann B, Brombach JD, Trosch W (1978a) Instrumentation and specimen preparation for electron beam X-ray microanalysis of frozen hydrated bulk specimens. J Microsc (Oxford) 112:75–87
Fuchs W, Brombach JD, Trosch W (1978b) Charging effect in electron-irradiated ice. J Microsc (Oxford) 112:63–74
Gehring K, Dörge A, Wunderlich P, Thurau K (1973) On the application of the electron microscope for the analysis of freeze-dried cryosections and deep frozen water containing tissue of the rat kidney. Beitr Elektronenmikrosk Direktabb 5:937–950
Green M (1963) The efficiency of production of characteristic X-ray radiation. In: Pattee HH, Cosslett VE, Engstrom A (eds) X-ray optics and X-ray microanalysis. Academic Press, London New York, pp 185–189
Green M, Cosslett VE (1961) The efficiency of production of characteristic X-radiation in thick targets of a pure element. Proc Phys Soc 78:1206–1214
*Gullasch J, Kaufmann R (1974) Energy-dispersive X-ray microanalysis in soft biological tissues: relevance and reproducibility of the results as depending on specimen preparation (air drying, cryofixation, cool-stage techniques). In: Hall T, Echlin P, Kaufmann R (eds) Microprobe analysis as applied to cells and tissues. Academic Press, London New York, pp 175–190
*Hall TA (1971) The microprobe assay of chemical elements. In: Oster E (ed) Physical techniques in biological research, vol 1A, 2nd edn. Academic Press, London New York, pp 157–275

Hall TA (1986) Properties of frozen sections relevant to quantitative microanalysis. J Microsc (Oxford) 141:319–328
*Hall TA, Gupta BL (1982) Quantification for the X-ray microanalysis of cryosections. J Microsc (Oxford) 126:333–345
Hayes TL, Koch G (1975) Some problems associated with low temperature micromanipulation in the SEM. Scanning Electron Microsc 1975/II:35–42
Henke BL, Lee P, Tanaka TJ, Shimabukuro RL, Fujikawa BK (1982) Low energy X-ray interaction coefficients: Photoabsorption, scattering and reflection. Atom Data Nucl Data Tab 27:1–44
Hook G, Lai C, Bastacky J, Hayes T (1980) Conductive coatings studied on inflated lung in the frozen hydrated and freeze-dried states. Scanning Electron Microsc 1980/IV:27–32
Jones D, McHardy WJ, Tait JM (1984) Low temperature scanning electron microscopy of biological specimens. Trans Br Mycol Soc 82:164–170
*Koehler JK (1968) The technique and application of freeze-etching in ultrastructure research. Adv Biol Med Phys 12:1–15
Kramer D, Preston J (1978) A modified method of X-ray microanalysis of bulk frozen plant tissue and its application to the problem of salt exclusion in mangroove roots. Microsc Acta Suppl 2:193–200
Lechene C, Strunk T, Warner R (1975) Perspectives in electron probe microanalysis of biological samples kept frozen. Microbeam Anal 1975:49A–49E
Lustyik G, Zs.-Nagy I (1985) Alterations of the intracellular water and ion concentrations in brain and liver cells during aging as revealed by energy dispersive X-ray microanalysis of bulk specimens. Scanning Electron Microsc 1985/I:323–337
Marshall AT (1975a) X-ray microanalysis of frozen hydrated biological specimens: the effect of charging. Micron 5:272–280
*Marshall AT (1975b) Electron probe microanalysis. In: Hayat MA (ed) Principles and techniques of scanning electron microscopy, vol. 4. Van Nostrand Reinhold, New York, pp 103–173
Marshall AT (1977) Electron probe X-ray microanalysis of frozen hydrated biological specimens. Microsc Acta 79:254–266
Marshall AT (1980a) Quanitative X-ray microanalysis of frozen-hydrated bulk biological specimens. Scanning Electron Microsc 1980/II:335–348
*Marshall (1980b) Frozen-hydrated bulk specimens. In: Hayat MA (ed) X-ray microanalysis in biology. Univ Park Press, Baltimore, pp 167–205
*Marshall (1980c) Principles and instrumentation. In: Hayat MA (ed) X-ray microanalysis in biology. Univ Park Press, Baltimore, pp 1–64
Marshall AT (1981) Simultaneous use of EDS, windowless EDS, BE and SE detectors and digital real-time line scanning for the X-ray microanalysis of frozen-hydrated biological specimens. Scanning Electron Microsc 1981/II:327–343
Marshall AT (1982) Application of $\phi(\varrho z)$ curves and a windowless detector to the quantitative X-ray microanalysis of frozen-hydrated bulk biological specimens. Scanning Electron Microsc 1982/I:243–260
Marshall AT (1983) X-ray microanalysis of the filter chamber of the cicada, *Cyclochila australasiae* Don. A water shunting epithelial complex. Cell Tissue Res 231:215–277
Marshall AT (1984a) The windowless energy dispersive X-ray detector: prospects for a role in biological X-ray microanalysis. Scanning Electron Microsc 1984/II:493–504
Marshall AT (1984b) Residual gas analysis in a SEM. J Microsc (Oxford) 133:119–120
Marshall AT, Carde D (1983) Beryllium coating for biological X-ray microanalysis. J Microsc (Oxford) 134:113–116
Marshall AT, Condron RJ (1985a) X-ray microanalytical resolution in frozen-hydrated biological bulk samples. J Microsc (Oxford) 140:109–118
Marshall AT, Condron RJ (1985b) Normalisation of light element X-ray intensities for surface topography effects in frozen-hydrated biological bulk samples. J Microsc (Oxford) 140:99–108
Marshall AT, Condron RJ (1987) A simple method of using $\phi(\varrho z)$ curves for the X-ray microanalysis of frozen-hydrated bulk biological samples. Micron Microsc Acta 18:23–26
Marshall AT, Wright A (1973) Detection of ions in insect osmoregulatory systems by electron probe X-ray microanalysis using scanning electron microscopy and a cryoscopic technique. Micron 4:31–45

Marshall AT, Carde D, Kent MJ (1982) Low temperature stages for the JSM35 SEM and the JEE4B coating unit. Micron 13:313–314

Marshall AT, Carde D, Kent MJ (1985a) Improved vacuum evaporation unit for beryllium coating for biological X-ray microanalysis. J Microsc (Oxford) 139:335–337

Marshall AT, Hyatt AD, Phillips JG, Condron RJ (1985b) Isosmotic secretion in the avian nasal salt gland: X-ray microanalysis of luminal and intracellular ion distribution. J Comp Physiol B 156:213–227

Nei T, Yotsumoto H, Hasegawa Y, Nagasawa Y (1972) Electron microscopic observation of biological specimen in their native state by employing cryogenic techniques. In: Arceneaux CJ (ed) Proc 30th Annu Electron Microsc Soc Am, Los Angeles, pp 410–411

Nei T, Yotsumoto H, Hasegawa Y, Nagasawa Y (1973) Direct observation of frozen specimens with a scanning electron microscope. J Electron Microsc 22:185–190

Oates K, Potts WTW (1985) Electron beam penetration and X-ray excitation depth in ice. Micron Microsc Acta 16:1–4

Parobek L, Brown JD (1978) The atomic number and absorption corrections in electron microprobe analysis at low electron energies. X-ray Spectrometry 7:26–30

Pawley JB, Norton JT (1978) A chamber attached to the SEM for fracturing and coating frozen biological samples. J Microsc (Oxford) 112:169–182

*Plattner H, Bachmann L (1982) Cryofixation: A tool in biological ultrastructural research. Int Rev Cytol 79:237–304

Potts WTW, Oates K (1983) The ionic concentrations in the mitochondria rich or chloride cell of *Fundulus heteroclitus*. J Exp Zool 227:349–359

*Reed SJB (1975) Electron microprobe analysis. Univ Press, Cambridge, pp 1–400

Robards AW, Crosby P (1979) A comprehensive freezing, fracturing and coating system for low temperature scanning electron microscopy. Scanning Electron Microsc 1979/II:325–344

*Robards AW, Sleytr UB (1985) Low temperature methods in biological electron microscopy. In: Glauert AM (ed) Practical methods in electron microscopy, vol 10. Elsevier, Amsterdam

Robinson VNE (1975) A simple technique for examining frozen hydrated specimens in the scanning electron microscope. J Microsc (Oxford) 101:287–292

Roomans GM (1981) Quantitative electron probe X-ray microanalysis of biological bulk specimens. Scanning Electron Microsc. 1981/II:345–356

Talmon Y (1984) Radiation damage to organic inclusions in ice. Ultramicroscopy 14:305–316

Talmon Y, Thomas EL (1977) Beam heating of a moderately thick cold stage specimen in the SEM/STEM. J Microsc (Oxford) 111:151–164

Taylor PG, Burgess A (1977) Cold stage for electron probe microanalyser. J Microsc (Oxford) 111:51–64

Tokunaga J, Tokunaga M (1973) Cryo-scanning microscopy of conidiospores formations in *Aspergillus niger*. Jeol News 11e (1):3–7

Turner RH, Smith CB (1974) A simple technique of examining fresh, frozen, biological specimens in the scanning electron microscope. J Microsc (Oxford) 102:209–214

Umrath W (1983) Berechnung der Gefriertrocknungszeiten für die elektronenmikroskopische Präparation. Mikroskopie (Wien) 40:9–37

Whitecross MI, Price GD, Preston JS (1982) Use of colloidal graphite in frozen-hydrated standard solutions for X-ray microanalysis. J Microsc (Oxford) 128:RP3–RP4

Yeo AR, Läuchli A, Kramer D, Gullasch J (1977) Ion measurements by X-ray microanalysis in unfixed, frozen, hydrated plant cells of species differing in salt tolerance. Planta 134:35–38

Zierold K (1976) X-ray microanalysis of cryo-fractured tissue specimens of the rat. In: Ben Shaul Y (ed) Electron microscopy 1976, vol 2. Proc 6th Eur Congr Electron Microsc, Israel Soc Electron Microsc, Jerusalem, pp 223–225

Zierold K, Schäfer D (1978) Quantitative X-ray microanalysis of diffusible ions in skeletal muscle bulk specimens. J Microsc (Oxford) 112:89–93

Zs.-Nagy I, Pieri C, Giuli C, Bertoni-Freddari C, Zsa.-Nagy V (1977) Energy dispersive X-ray microanalysis of the electrolytes in biological bulk specimens. I. Specimen preparation, beam penetration and quantitative analysis. J Ultrastruct Res 58:22–33

Zs.-Nagy I, Lustyik G, Bertoni-Freddari C (1982) Intracellular water and dry mass content as measured in bulk specimens by energy-dispersive X-ray microanalysis. Tissue Cell 14:47–60

Chapter 14 Cryofixation of Dynamic Processes in Cells and Organelles

GERD KNOLL[1], ARIE J. VERKLEIJ[2], and HELMUT PLATTNER[1]

1 Introduction

Electron microscope analysis of cells depends on fixation; the goal is to trap the native state of dynamic structures and distribution patterns of components (Plattner and Bachmann 1982). This intention of fixation is, of course, not restricted to applications in electron microscopy. Attempts to literally "freeze" the living structure by cryofixation can be traced back to the last century, and were extended to electron microscopy in its very early days (compiled by Rash 1983). The main problem with simply freezing a biological specimen in order to fix it, is destruction caused by ice crystals. This freezing damage hampers further analysis, in particular at the high resolution level obtained with the electron microscope. For this reason, chemical fixation procedures have routinely been used, although their limitations and risks have been well recognized (Fernandez-Moran 1964; Rebhun 1965). On the other hand, the drawbacks of chemical treatments have stimulated the development of rapid freezing techniques which reduce the formation of ice crystals (Van Harreveld and Crowell 1964).

Twenty years have passed since the first significant results were obtained using rapid freezing with freeze-substitution as a follow-up procedure (Van Harreveld et al. 1965). A more general interest in rapid freezing has arisen since freeze-fracturing was recognized as a unique tool to study membrane dynamics (e.g. Verkleij and Ververgaert 1975). Important considerations in this application are that specimens are kept at low temperatures till the final replication, and that the replicas are not sensitive to beam damage. Therefore, several hazards of other techniques are avoided.

Rapid freezing techniques are now capable of suppressing freezing damage, at least in small samples, such as single cells and superficial tissue layers (see Sitte et al., Chap. 4, this Vol.). This allows for the complete avoidance of any chemical pretreatments and increases the potential of trapping native structures and distribution patterns, even in fast, transient stages. Both, maintenance of the natural situation and the trapping of rapid events, are possible due to the high time resolution achieved using rapid freezing. Principles and consequences of "time resolution" will be discussed in the first part of this chapter. In the second part,

1 Fakultät für Biologie, Universität Konstanz, D-7750 Konstanz, FRG
2 Department of Molecular Cell Biology, University of Utrecht, The Netherlands

Cryotechniques in Biological Electron Microscopy
Edited by R. A. Steinbrecht and K. Zierold

the progress obtained by rapid freezing versus chemical fixation in preserving native cellular structures will be exemplified. The third part will show the need for rapid freezing to arrest membrane constituents and the potential and limitations of this technique in trapping rapid events (i.e. here membrane fusion).

2 Cooling Rate, Freezing Time, Time Resolution

The "quality" of cryofixation is judged in most cases by the size of the visible segregation phenomena caused by ice-crystal growth. Despite a strong relationship between ice-crystal size and cooling rates (Kopstad and Elgsaeter 1982a; Bald 1986), several other factors, such as sample geometry and solute concentrations, considerably affect the size of ice-crystal-induced segregation patterns in specimens frozen with the same velocity (Plattner and Bachmann 1982). In the analysis of dynamic events, however, it is also important to know the time resolution of the cryofixation in comparison with the dynamic features under investigation. This time resolution will depend both on the rate of cooling and the temperature required for the final arrest of mobile components (both parameters define the freezing time) in comparison with the mobility of constituents or the lifetime of events to be analyzed.

The most straightforward approach for the determination of cooling rates may seem to measure the temperature drop by means of miniature thermocouples within the samples. However, such measurements are only valuable for comparison of individual parameters affecting cooling rates, because in these microscopic systems the introduction of the measuring system inevitably changes the measured system itself (Bald 1983). Whereas the highest values reported, $5.5 \times 10^4\ \mathrm{K\,s^{-1}}$ measured at 40 µm distance from the area of contact with the quenching medium (Escaig 1982), may give an idea of the minimal cooling rates achievable in that zone, a theoretical analysis of cooling rates, assuming ideal conditions, predicts values in excess of $10^6\ \mathrm{K\,s^{-1}}$ within up to 10 µm distance from the "freezing contact" (Kopstad and Elgsaeter 1982b). However, calculations of this kind appear to be rather problematic (see Bachmann and Mayer, Chap. 1, this Vol.).

The freezing point of aqueous solutions during rapid freezing depends on the cooling rate and this in turn depends on the progression of the ice front (Kopstad and Elgsaeter 1982a). Because exact values for these interdependent variables are not available, the time necessary for freezing is not known. Measurements of dielectric changes (assumed to be caused by water-ice transitions) in the specimens during freezing indicate 2–4 ms as necessary for the freezing of 10 µm (Heuser et al. 1979; Van Harreveld and Trubatch 1979). It is uncertain, however, whether this value results from inadequate interpretation of the signal (Bald 1983; Jones 1984) or whether it reflects the actual freezing time prolonged by the experimental, not ideal, freezing conditions (Kopstad and Elgsaeter 1982a; Bald 1983). Calculations based on simplified models consistently point to ca. 0.1 ms to freeze a specimen with adequate quality to a depth of 10 µm (Kopstad and Elgsaeter 1982b; Bald 1983; Jones 1984). The freezing time for any partial layer within these 10 µm will be considerably shorter.

Is the “freezing time”, defined as the interval between the start of cooling and the final solidification of water, an appropriate parameter for the estimation of “time resolution”? Alternatively, Jones (1984) assumed as freezing time relevant for the time resolution to trap morphological changes, the time needed for the temperature drop during the actual solidification process of the aqueous phase. This temperature interval was determined to be $< 2-3$ K (Kopstad and Elgsaeter 1982a). Assuming a cooling rate of $10^5\ \mathrm{K\,s^{-1}}$, this would result in a time resolution of 30 μs. Taking the solidification range as 10 K and assuming cooling rates in excess of $4 \times 10^4\ \mathrm{K\,s^{-1}}$, Jones (1984) estimated the resolution time to be 250 μs. For membranes, however, freezing times and, consequently, time resolution are defined by the solidification of the lipids rather than that of the water. Lipid phase transitions and segregation phenomena may occur at or only slightly below physiological temperatures. High cooling rates have, therefore, to be applied right at the beginning of the freezing process. High time resolution as a prerequisite to preserve the native situation has to be based on the freezing time in the whole range between these transition/segregation phenomena and the freezing of the water compartment. This actual freezing time has to be smaller than the time needed for the lipid transitions and eventually resulting segregation phenomena (see Sect. 4). We assume that both parameters are relevant, because phase separations in the aqueous phase might influence the behaviour of membrane components.

In conclusion, we see that time resolution has to be defined in relation to the kinetic properties of the subject under investigation. As stated above, exact data for cooling rates and freezing times are not available today, but it appears reasonable to assume that the overall freezing time and the resulting time resolution potential of the best cryofixation methods would not be more than ca. 1 ms. A more detailed knowledge would be very helpful for using rapid freezing as a tool to understand more precisely the dynamic properties of cellular constituents (see Sect. 4).

3 Cellular Structures: Rapid Freezing Versus Chemical Fixation

Principal drawbacks of chemical fixation include its slow action (seconds to minutes), its selectivity for cellular components (mainly proteins by aldehyde fixatives), the breakdown of selective ion permeability and the occurrence of structural changes (for review see Plattner and Zingsheim 1983).

A straightforward approach to evaluate the potential of a fixation procedure to preserve dynamic cellular structures in the native state is the direct observation of the living cell during the action of the fixative under the light microscope. For plant cells, all chemical fixation procedures commonly used for electron microscopy failed to prevent vesiculation of intracellular membranes following the arrest of intracellular movement (Mersey and McCully 1978). Plant cells are especially difficult to fix; however, similar results have been obtained for monolayers of cultured chicken cells (Buckley 1973).

Even the overall *shape of cells* is not reliably preserved by chemical fixation methods. A significant example is the apparent paucity of extracellular space in central nervous tissue which conflicts with the physiological data on extracellular ions. These data could only be correlated with ultrastructural data obtained by using rapid freezing and freeze-substitution (Van Harreveld et al. 1965).

There are several examples for the sensitivity of intracellular membrane structures to chemical fixation. *Lysosomes* are shown to change their shape during chemical fixation (Robinson et al. 1986). Specific *lamellar structures*, which may play a role in the controversial mechanism of membrane insertion into nervous growth cones, are destroyed by chemical fixation (Cheng and Reese 1985). The slow action of chemical fixation due to the low permeability of plant cells led to the "accepted view at present" (Fernandez and Staehelin 1985) that some secretory proteins in plants bypass the *Golgi apparatus*. Rapid freezing does prevent the vesiculation of this organelle (Fernandez and Staehelin 1985) and may lead to a revision of the "accepted view".

Analysis of rapidly frozen native cells after freeze-substitution as whole mounts or thin sections reveals the fine structure of the *cytoskeletal meshwork* embedded in a granular *cytoplasmic matrix* (Bridgman and Reese 1984). The results obtained differ from the interpretations of cytoplasmic fine structure derived from other preparative procedures, including chemical fixation and/or extraction steps. Combining the EM analysis of rapidly frozen cells with video-enhanced differential interference contrast light microscopy of the living cells, the relationship between the cytoskeletal meshwork and organelle movement has been studied (Bridgman et al. 1986). The results show that the movement of organelles along microtubules is continuous or saltatory, depending on the local mesh size of the cytoskeleton.

The structure of rapidly frozen *tight junctions* suggests a lipidic nature of this junction type (Kachar and Reese 1982). This model predicts restricted mobility of membrane constituents in the outer membrane monolayer of an individual cell, but a continuity of that layer between adjacent cells. These predictions have been tested by investigating the mobility of fluorescently labelled lipids. For these lipids the tight junction hinders diffusion in the outer layer of the plasma membrane between the apical and lateral membrane domains, but also between adjacent cells. Therefore, the model has been rejected and a proteinaceous nature for the tight junctions proposed (van Meer et al. 1986).

Special contact sites between the *enveloping membrane systems* of mitochondria, chloroplasts and gram-negative bacteria have been observed by conventional techniques (Hackenbrock 1968; Bayer 1979; Carde et al. 1982). In contrast, low temperature methods, including rapid freezing and freeze-substitution, have shown for *bacteria* a defined gel matrix without any contacts (Hobot et al. 1984). On the other hand, for mitochondria and chloroplasts the existence of such structures could be confirmed. For *chloroplasts* special "contact particles" have been described (Cline et al., 1985). Since 1965 *mitochondria* have been analyzed in more detail by rapid freezing techniques (Malhotra and Van Harreveld 1965). In contrast to the well acquainted intermembrane space, as visible after chemical fixation, after rapid freezing the membranes are closely apposed in situ (pentalaminar appearance, Malhotra and van Harreveld 1965) (Fig. 1 a), whereas after isolation a clear space becomes visible, exhibiting defined contacts (Knoll 1984) (Fig. 1 b). The close apposition can be restored by metabolic activation of the organ-

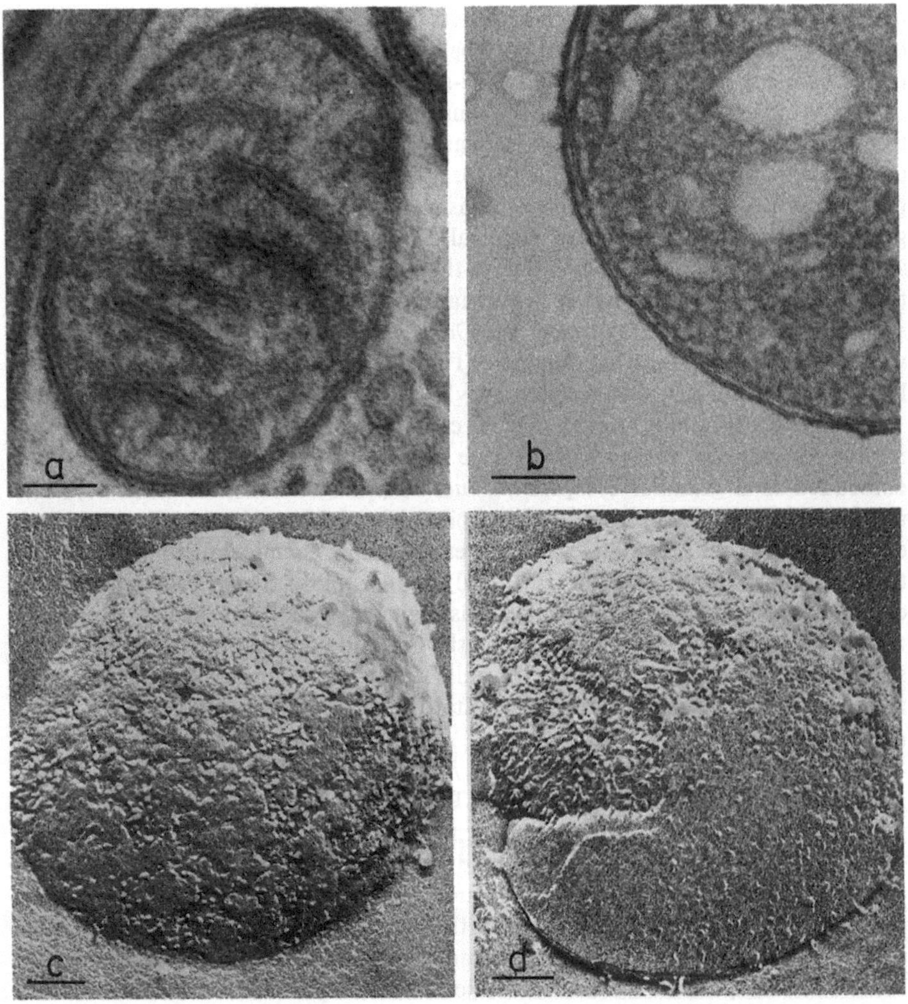

Fig. 1a–d. Enveloping membranes of mitochondria. **a** In situ the membranes are closely apposed and of pentalaminar appearance. **b** In isolated mitochondria the membranes separate with only few contacts remaining. **c** Close apposition of the membranes is reflected by frequent deflections of the fracture plane between the inner and outer membranes (mitochondria in situ or well-coupled, isolated mitochondria during oxidative phosphorylation). **d** Uncoupling of the mitochondria abolishes these frequent deflections. *Bars*: **a** 50 nm, **b–d** 100 nm (**a** Malhotra and Van Harreveld 1965; **b** Knoll 1984; **c, d** Klug et al. 1984)

elles (Fig. 1c, d). This led to the suggestion that membrane-to-membrane contacts are dynamic and functionally regulated structural features (Knoll and Brdiczka 1983). The significance of these findings is emphasized by the observation that in tumour cells the close apposition is abolished (Denis-Pouxviel et al. 1987), and that in hepatocytes the degree of contacts seems to be under hormonal control (Brdiczka et al. 1986). Contact sites are also postulated for the import of proteins (Schleyer and Neupert 1985). Non-bilayer lipids seem to be involved in the establishment of the contacts (Nicolay 1984; van Venetie and Verkleij 1982). In a mechanistic interpretation, a semifusion of the membranes is proposed (van Venetie and Verkleij 1982).

All these examples demonstrate clearly that chemical pretreatments may result in significantly different morphological features as compared to rapidly frozen native specimens. Hence, elaborate morphological investigations by conventional methods lose much of their convincing power. Such studies should always, if possible, be corroborated by the investigation of some cryofixed controls. The rationale of this argumentation is the reasonable assumption that a method that operates in the millisecond range can exert less side effects than one that requires seconds or minutes.

4 Membrane Dynamics

The potential of cryofixation became most obvious in studies of membrane dynamic events, such as lateral phase segregations, membrane fusion and lipid polymorphism.

4.1 Lateral Mobility

When cooling mixed lipid systems with different phase-transition temperatures of the components (see below), a phase separation may occur (Verkleij and Ververgaert 1975). The degree of this separation will depend on the cooling rate, the difference in the transition temperatures and the lateral mobility of the lipids. The lateral diffusion coefficient of fluorescent lipid analogues has been measured to be as high as $3 \times 10^{-7}\,cm^2\,s^{-1}$ (Gaub et al. 1984). This would allow a lateral displacement of > 300 nm within 1 ms by diffusion (as root-mean-square displacement, Saffman and Delbrück 1975). The significance of this effect for the analysis of biological membranes is illustrated in Fig. 2: in many biological membranes (depending on their lipid composition), intramembrane particles (IMPs) are squeezed out of growing smooth patches of lipids in the lamellar gel state during quenching from above the transition temperature. Rapid freezing under the same conditions leads to the homogeneous distribution of the IMPs.

Segregation of proteins can be prevented by glutaraldehyde fixation above the transition temperature (e.g. Kleemann and McConnell 1974). However, specific artefacts have been shown to be induced by glutaraldehyde and/or glycerol impregnation also as far as heterogeneities of IMP distributions are concerned (reviewed by Plattner and Bachmann 1982). Results claiming non-thermotropic, but physiological heterogeneities in IMP distributions (see compilation by Verkleij and Ververgaert 1978) should be obtained or confirmed by rapid freezing (e.g. modulations during the cell cycle, de Laat et al. 1983; polarized distribution patterns, Friend 1982; aggregation patterns, Landis and Reese 1981). The structure of gap junctions has been claimed to depend on their functional state (Peracchia 1980). Rapid freezing under exactly controlled conditions (including avoidance of mechanical stress, oxygen depletion, etc.) showed no significant dif-

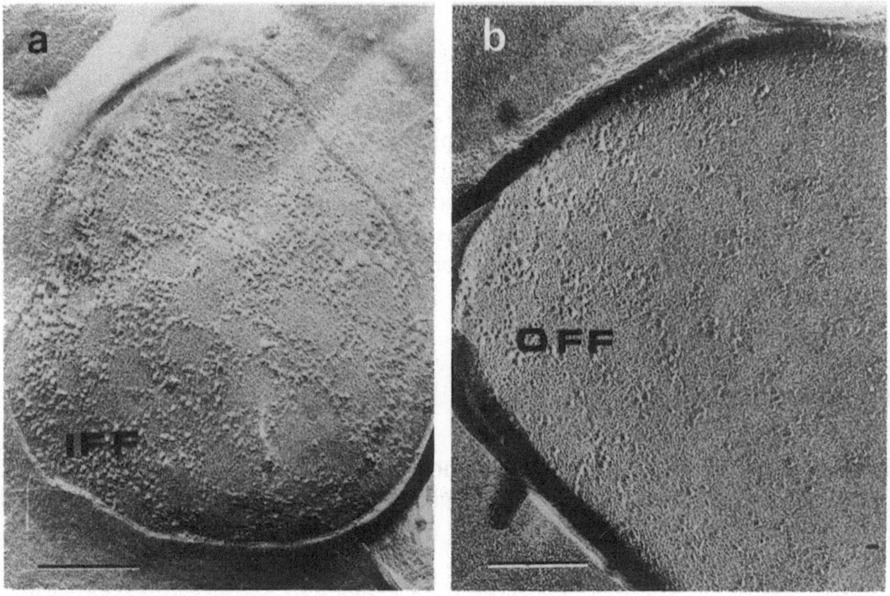

Fig. 2a, b. Fracture faces of *Acholeplasma laidlawii* frozen with a moderate cooling rate from above the transition temperature of the membrane lipids. Growing patches of lipids in the lamellar gel phase squeeze out IMPs due to insufficient cooling rates. **a** Inner fracture face (*IFF*); **b** outer fracture face (*OFF*). *Bars*: 100 nm (Verkleij et al. 1972)

ferences, except for chemically fixed preparations, when different tissues were analyzed (Green and Severs 1984; Miller and Goodenough 1985).

The analysis of distribution patterns of specific lipids in membranes is valid only after rapid freezing. The "universal tendency to cluster" of gangliosides (Grant and Peters 1984) is not observed after rapid freezing (Thompson et al. 1985) and, therefore, is likely to be a freezing artefact of thermotropic phase separations.

4.2 Membrane Fusion

The investigation of exocytotic quantal transmitter release (Heuser et al. 1979) had great impact on the perspectives for the application of rapid freezing to study rapid physiological processes (Shotton 1980; Rash 1983). This concerns not only the correlation of the well-characterized transmitter release with the event of exocytosis on a millisecond timescale, but also the mechanism of membrane fusion in general. In a careful re-investigation using freeze-substitution, even a correlation discrepancy of only 3 ms could be corrected (Torri-Tarelli et al. 1985). This illustrates the potential of rapid freezing for the analysis of dynamic processes.

Analysis of nerve terminals has the outstanding advantage of a very high degree of synchronization of the events triggered at exactly defined time points in correlation to the freezing process. This advantage is not given in most other

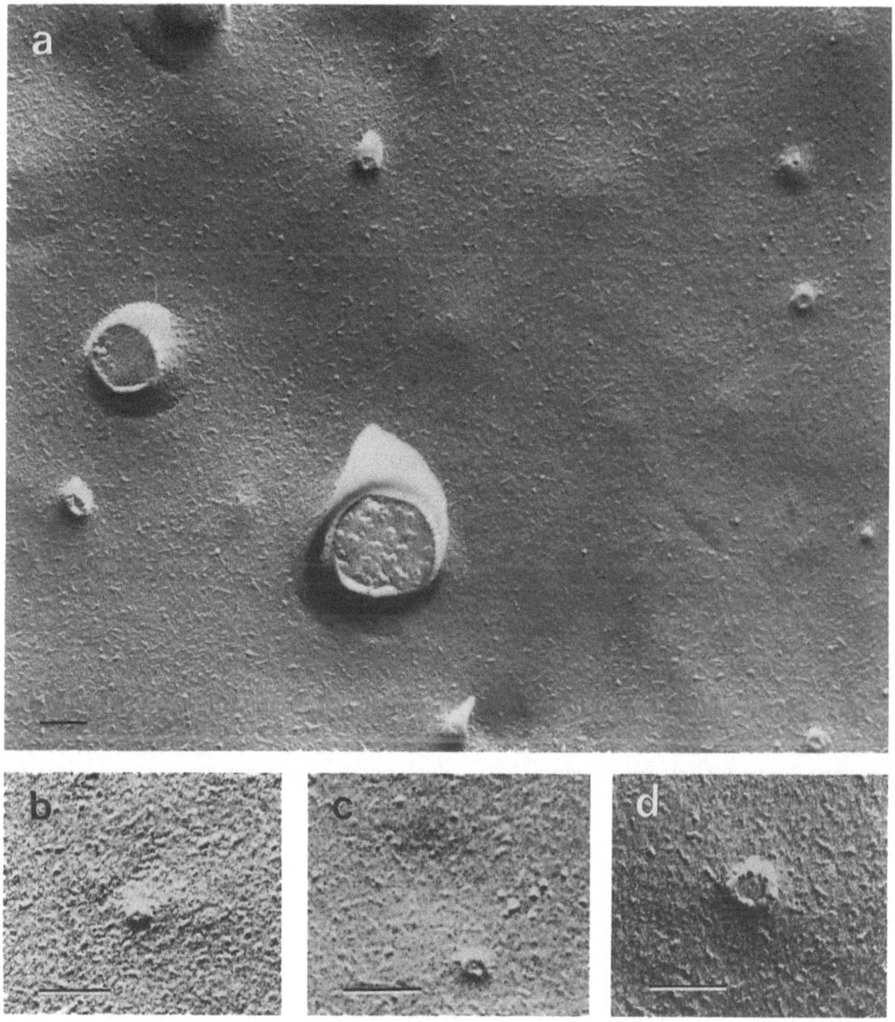

Fig. 3a–d. Membrane events in adrenal chromaffin cells during exocytosis. **a** Membrane survey of the exoplasmic fracture face exhibits exo-endocytotic openings of varying sizes without any IMP clearing. **b–d** Suggested sequence of stages. *Bars*: 100 nm (Schmidt et al. 1983)

systems used for the investigation of biological membrane fusion (see discussion in Chandler and Heuser 1979; Pinto da Silva and Kachar 1980). Therefore, statistical analysis (Schmidt et al. 1983) is necessary in most cases (Fig. 3).

A common result of all these studies concerning biological membrane fusion events is the absence of an IMP clearing or of diaphragms which have been commonly observed after chemical pretreatments (see compilations and discussions in Plattner 1981; Chandler 1984; Düzgünes 1985). An explanation for the specificity of these features – though they are not found in vivo – is given by Chandler (1984), assuming a very local increase of membrane fluidity just before the fusion event.

As it seems clear now that membrane fusion is a local event which leads to a narrow pore (which then rapidly enlarges) at the contact point of the fusing membranes (Plattner 1981, see also Fig. 3), nothing is known about the intermediate structure just at the moment between apposition and fission of the bilayers.

To circumvent the heterogeneity of structural features and the low frequency of fusion events observed with biological membranes, fusion of lipid models under defined conditions was analyzed. The initial events during fusion of phosphatidyl serine vesicles after the administration of Ca^{2+} (Miller and Dahl 1982; Rand et al. 1985) showed a first strong apposition of the vesicles which finally led to rupture. Only ruptures occurring by chance in the contact areas resulted in fusion. Therefore, Rand et al. (1985) came to the conclusion that the phosphatidyl serine model is not representative for the local and non-leaky biological fusion.

Alternatively, the local involvement of H_{II}-phase-preferring lipids via inverted micelles as fusion intermediates has been proposed (Verkleij 1984), and lipidic particles as a structural equivalent of these inverted micelles have been observed in freeze-fracture preparations of artificial lipid fusion systems (Verkleij et al. 1979). This was possible only after several rounds of fusion where joining stages may have a longer lifetime. The failure to visualize these structures during initial fusion rounds (Bearer et al. 1982; Düzgünes et al. 1985) is likely due to the fact that fusion proceeds so fast that intermediates escape detection. Calculations point to a very short lifetime (submillisecond half-life) of such intermediates (Siegel 1984), and the query arises whether rapid freezing is capable of capturing structures with such a short lifetime (Verkleij 1984). It also remains open whether the very same structures (inverted micelles) occur during membrane fusion in vivo.

4.3 Lipid Phases and Transition Intermediates

The limits of physical fixation by freezing became especially apparent from the study of membrane systems, where results can be compared with those obtained by different methodologies, like NMR, X-ray diffraction, differential scanning calorimetry, etc. Biological phospholipids undergo thermotropic phase transitions from the lamellar liquid crystal to the lamellar gel phase upon cooling. Pure lipid systems exhibit a sharp temperature range for the transition (1 K for phosphatidyl ethanolamine; Caffrey 1985). In order to arrest the lipids in the liquid crystalline phase, the freezing time between the transition temperature and the freezing temperature of water (see Sect. 2) must be smaller than the time needed for the transition. Real-time X-ray diffraction studies indicate a transition time of < 1 s for phosphatidyl ethanolamine. The limited time resolution of this approach hampers the measurement of smaller time periods (Caffrey 1985). Freeze-fracture and X-ray diffraction before and after freezing of several lipid systems show that conventional freezing procedures are often too slow to prevent the transition and rapid freezing is necessary (Ververgaert et al. 1973; Gulik-Krzywicki and Costello 1978; Melchior et al. 1982).

Many naturally derived lipids exhibit an additional thermotropic phase transition, from the hexagonal inverted phase (H_{II}) to the lamellar liquid crystal phase. This transition has a broader temperature range (5 – 6 K for phosphatidyl ethanolamine), but again is too rapid (< 1 s) for the time-resolving capacity of real-time X-ray diffraction (Caffrey 1985). The H_{II}-phases of phosphatidyl ethanolamine systems with relatively low (283 K) transition temperatures are easily trapped, whereas for higher transition temperatures fast freezing is absolutely required and may even fail if the transition temperature is higher than 313 K (Verkleij 1984). Systems rapidly frozen from above a transition temperature of 303 K show H_{II}-phases, lipidic particles and lamellar phases (van Venetie and Verkleij 1981). The lipidic particles observed in this case are interpreted as intermediates captured during the transition, which could not be prevented completely by rapid freezing. However, it is not known whether this thermotropic transition intermediate has kinetics comparable to fusion intermediates. A second difference concerns the problem of frequency and synchrony: the temporal existence of the thermotropic intermediate is outstandingly well synchronous and coincides with the moment of freezing which induces its existence. As discussed above, however, a comparable perfect synchronization can only be expected, if at all, in exceptional cases of membrane fusion in vivo.

5 Conclusions and Outlook

1. Faithfully, cellular structures and ion distributions are preserved when cells are rapidly frozen.
2. Membrane dynamics will require rapid freezing for a faithful preservation of structure. Nevertheless, in some cases the preservation of lipids in the native situation may be a problem.
3. It is possible to trap rapid events (e.g. membrane fusion intermediates), but the difficulties increase considerably with decreasing lifetime of the events and of the intermediate structures.

The direct way to analyze cellular dynamics is, of course, the observation of living cells with the light microscope. The potential, e.g. for the analysis of organelle movement (Bridgman et al. 1986), ion changes (Kruskal et al. 1986) and mobility of membrane constituents (Metcalf et al. 1986a, b) has increased considerably over the last years by the introduction of video-processing techniques, the development of fluorescent indicators and tracers, the possibility of localized excitation, etc. A correlation of these studies with electron microscopy after rapid freezing would allow an extension of resolution with high fidelity. This approach has resulted already in a significant gain in knowledge of intracellular movements (for a correlation with the interference of the cytoskeleton, see Bridgman et al. 1986) and of membrane fusion between liposomes (Rand et al. 1985).

Labelling techniques applied to dynamic structures rapidly frozen in the native state are only at the beginning. The first results achieved are promising (Allred and Staehelin 1986). The reliability obtained with regard to dynamics and the ultrastructural integrity of the components under investigation can be ascertained only when rapid freezing techniques are applied. Future applications will include, e.g. the localization of soluble, reversibly structure-bound ligands. Even combined observation of the same label by light and electron microscopy seems now possible (De Brabander et al. 1986). This opens up new perspectives for the use of cryofixation as a tool for the ultrastructural analysis of dynamic processes in cells and organelles.

Acknowledgements. Part of the experimental work of G. K. and H. P. referred to in this chapter was supported by the Deutsche Forschungsgemeinschaft (SFB 138+156). G. K. was supported by an EMBO fellowship during the preparation of the manuscript.

References

(References marked by an asterisk are review papers or books respectively)

Allred DR, Staehelin LA (1986) Spatial organization of the cytochome b_6-f complex within chloroplast thylakoid membranes. Biochim Biophys Acta 849:94–103

Bald WB (1983) Optimizing the cooling block for the quick freeze method. J Microsc (Oxford) 131:11–23

Bald WB (1986) On crystal size and cooling rate. J Microsc (Oxford) 143:89–102

*Bayer ME (1979) The fusion sites between outer membrane and cytoplasmic membrane of bacteria: their role in membrane assembly and virus infection. In: Inouye M (ed) Bacterial outer membranes. Biogenesis and functions. John Wiley & Sons, New York, pp 167–202

Bearer EL, Düzgünes N, Friend DS, Papahadjopoulos D (1982) Fusion of phospholipid vesicles arrested by quick-freezing. The question of lipidic particles as intermediates in membrane fusion. Biochim Biophys Acta 693:93–98

Brdiczka D, Riesinger I, Adams V, Bremm G (1986) Functions in metabolic regulation of contact sites between mitochondrial boundary membranes. In: 4th EBEC short reports (1986-200). Univ Press, Cambridge, pp 200

Bridgman PC, Reese TS (1984) The structure of cytoplasm in directly frozen cultured cells. I. Filamentous meshworks and the cytoplasmic ground substance. J Cell Biol 99:1655–1668

Bridgman PC, Kachar B, Reese TS (1986) The structure of cytoplasm in directly frozen cultured cells. II. Cytoplasmic domains associated with organelle movements. J Cell Biol 102:1510–1521

Buckley IK (1973) Studies in fixation for electron microscopy using cultured cells. Lab Invest 29:398–410

Caffrey M (1985) Kinetics and mechanism of the lamellar gel/lamellar liquid-crystal and lamellar/inverted hexagonal phase transition in phosphatidyl ethanolamine: a real-time X-ray diffraction study using synchrotron radiation. Biochemistry 24:4826–4844

*Carde J-P, Joyard J, Douce R (1982) Electron microscopic studies of envelope membranes from spinach plastids. Biol Cell 44:315–324

Chandler DE (1984) Comparison of quick-frozen and chemically fixed sea urchin eggs: Structural evidence that cortical granule exocytosis is preceded by a local increase in membrane mobility. J Cell Sci 72:23–36

Chandler DE, Heuser J (1979) Membrane fusion during secretion. Cortical granule exocytosis in sea urchin eggs as studied by quick-freezing and freeze-fracture. J Cell Biol 83:91–108

Cheng TPO, Reese TS (1985) Polarized compartmentalization of organelles in growth cones from developing optic tectum. J Cell Biol 101:1473–1480

Cline K, Keegstra K, Staehelin LA (1985) Freeze-fracture electron microscopic analysis of ultrarapidly frozen envelope membranes on intact chloroplasts and after purification. Protoplasma 125:111–123

De Brabander M, Nuydens R, Geuens G, Moeremans M, De Mey Y (1986) The use of submicroscopic gold particles combined with video contrast enhancement as a simple molecular probe for the living cell. Cell Motil 6:105–113

De Laat SW, Tertoolen LGJ, Van der Saag PT, Bluemink JG (1983) Quantitative analysis of modulations in numerical and lateral distribution of intramembrane particles during the cell cycle of neuroblastoma cells. J Cell Biol 96:1047–1055
Denis-Pouxviel C, Riesinger I, Bühler C, Bremm G, Brdiczka D, Murat J-C (1987) Regulation of mitochondrial hexokinase in cultured HT29 human cancer cells. Biochim Biophys Acta (in press)
*Düzgünes N (1985) Membrane fusion. In: Roodyn DB (ed) Subcellular biochemistry, vol 11. Plenum Press, New York, pp 195–286
Düzgünes N, Straubinger RM, Baldwin PA, Friend DS, Papahadjopoulos D (1985) Proton-induced fusion of oleic acid-phosphatidyl ethanolamine liposomes. Biochemistry 24: 3091–3098
Escaig J (1982) New instruments which facilitate rapid freezing at 83 K and 6 K. J Microsc (Oxford) 126:221–229
Fernandez DE, Staehelin LA (1985) Structural organization of ultrarapidly frozen barley aleurone cells actively involved in protein secretion. Planta 165:455–468
*Fernandez-Moran H (1964) New approaches in correlative studies on biological ultrastructure by high-resolution electron microscopy. J R Microsc Soc 83:183–195
*Friend DS (1982) Plasma-membrane diversity in a highly polarized cell. J Cell Biol 93:243–249
Gaub H, Sackmann E, Büschl R, Ringsdorf H (1984) Lateral diffusion and phase separation in two-dimensional solutions of polymerized butadiene lipids in dimyristoylphosphatidylcholine bilayers. Biophys J 45:725–731
*Grant CWM, Peters MW (1984) Lectin-membrane interactions. Information from model systems. Biochim Biophys Acta 779:403–422
Green CR, Severs NJ (1984) Gap junction connexon configuration in rapidly frozen myocardium and isolated intercalated disks. J Cell Biol 99:453–463
Gulik-Krzywicki T, Costello MJ (1978) The use of low temperature X-ray diffraction to evaluate freezing methods used in freeze-fracture electron microscopy. J. Microsc (Oxford) 112: 103–113
Hackenbrock CR (1968) Chemical and physical fixation of isolated mitochondria in low-energy and high-energy states. Proc Natl Acad Sci USA 61:598–605
Heuser JE, Reese TS, Dennis MJ, Jan Y, Jan L, Evans L (1979) Synaptic vesicle exocytosis captured by quick freezing and correlated with quantal transmitter release. J Cell Biol 81:275–300
Hobot JA, Carlemalm E, Villinger W, Kellenberger E (1984) Periplasmic gel: New concept resulting from the reinvestigation of bacterial cell envelope ultrastructure by new methods. J Bacteriol 160:143–152
Jones GJ (1984) On estimating freezing times during tissue rapid freezing. J Microsc (Oxford) 136:349–360
Kachar B, Reese TS (1982) Evidence for the lipidic nature of tight junction strands. Nature (London) 296:464–466
Kleemann W, McConnell HM (1974) Lateral phase separations in *Escherichia coli* membranes. Biochim Biophys Acta 345:220–230
Klug GA, Krause J, Östlund A-K, Knoll G, Brdiczka D (1984) Alterations in liver mitochondrial function as a result of fasting and exhaustive exercise. Biochim Biophys Acta 764:272–282
Knoll G (1984) Dynamische Interaktionen der mitochondrialen Hüllmembranen. Eine ultrastrukturelle Analyse auf der Basis schneller Kryofixation. Diss, Univ Konstanz
Knoll G, Brdiczka D (1983) Changes in freeze-fractured mitochondrial membranes correlated to their energetic state. Dynamic interactions of the boundary membranes. Biochim Biophys Acta 733:102–110
Kopstad G, Elgsaeter A (1982a) Theoretical analysis of the ice crystal size distribution in frozen aqueous specimens. Biophys J 40:155–161
Kopstad G, Elgsaeter A (1982b) Theoretical analysis of specimen cooling rate during impact freezing and liquid-jet freezing of freeze-etch specimens. Biophys J 40:163–170
Kruskal BA, Shak S, Maxfield FR (1986) Spreading of human neutrophils is immediately preceded by a large increase in cytoplasmic free calcium. Proc Natl Acad Sci USA 83:2919–2923
Landis DMD, Reese TS (1981) Astrocyte membrane structure: Changes after circulatory arrest. J Cell Biol 88:660–663

Malhotra SK, Van Harreveld A (1965) Some structural features of mitochondria in tissues prepared by freeze-substitution. J Ultrastruct Res 12:473–487
Melchior DL, Bruggemann EP, Steim JM (1982) The physical state of quick-frozen membranes and lipids. Biochim Biophys Acta 690:81–88
Mersey B, McCully ME (1978) Monitoring the course of fixation of plant cells. J Microsc (Oxford) 114:49–76
Metcalf III TN, Villanueva MA, Schindler M, Wang JL (1986a) Monoclonal antibodies directed against protoplasts of soybean cells: analysis of the lateral mobility of plasma membrane-bound antibody MVS-1. J Cell Biol 102:1350–1357
Metcalf III TN, Wang JL, Schindler M (1986b) Lateral diffusion of phospholipids in the plasma membrane of soybean protoplasts: Evidence for membrane lipid domains. Proc Natl Acad Sci USA 83:95–99
Miller DC, Dahl GP (1982) Early events in calcium-induced liposome fusion. Biochim Biophys Acta 689:165–169
Miller TM, Goodenough DA (1985) Gap junction structures after experimental alteration of junctional channel conductance. J Cell Biol 101:1741–1748
Nicolay K, Timmers RJM, Spoelstra E, Van der Neut R, Fok JJ, Huigen YM, Verkleij AJ, De Kruijff B (1984) The interaction of adriamycin with cardiolipin in model and rat liver mitochondrial membranes. Biochim Biophys Acta 778:359–371
*Peracchia C (1980) Structural correlates of gap junction permeation. Int Rev Cytol 66:81–146
*Pinto da Silva P, Kachar B (1980) Quick freezing vs. chemical fixation: capture and identification of membrane fusion intermediates. Cell Biol Int Rep 4:625–640
*Plattner H (1981) Membrane behaviour during exocytosis. Cell Biol Int Rep 5:435–459
*Plattner H, Bachmann L (1982) Cryofixation: A tool in biological ultrastructural research. Int Rev Cytol 79:237–304
*Plattner H, Zingsheim HP (1983) Electron microscopic methods in cellular and molecular biology. In: Roodyn DB (ed) Subcellular biochemistry, vol 9. Plenum Press, New York, pp 1–236
Rand RP, Kachar B, Reese TS (1985) Dynamic morphology of calcium-induced interactions between phosphatidylserine vesicles. Biophys J 47:483–489
*Rash JE (1983) The rapid-freeze technique in neurobiology. TINS 6:208–212
*Rebhun LI (1965) Freeze-substitution: fine structure as a function of water concentration in cells. Fed Proc 24 (Suppl 15):217–233
Robinson JM, Okada T, Castellot JJ, Karnovsky MJ (1986) Unusual lysosomes in aortic smooth muscle cells: presence in living and rapidly frozen cells. J Cell Biol 102:1615–1622
Saffman PG, Delbrück M (1975) Brownian motion in biological membranes. Proc Natl Acad Sci USA 72:3111–3113
Schleyer M, Neupert W (1985) Transport of proteins into mitochondria: translocation intermediates spanning contact sites between outer and inner membranes. Cell 43:339–350
Schmidt W, Patzak A, Lingg G, Winkler H, Plattner H (1983) Membrane events in adrenal chromaffin cells during exocytosis: a freeze-etching analysis after rapid cryofixation. Eur J Cell Biol 32:31–37
*Shotton D (1980) Quick freezing – the new frontier in freeze-fracture. Nature (London) 283:12–14
Siegel DP (1984) Inverted micellar structures in bilayer membranes. Formation rates and half-lifes. Biophys J 45:399–420
Thompson TE, Allietta M, Brown RE, Johnson ML, Tillack TW (1985) Organization of ganglioside GM_1 in phosphatidylcholine bilayers. Biochim Biophys Acta 817:229–237
Torri-Tarelli F, Grohovaz F, Fesce R, Ceccarelli B (1985) Temporal coincidence between synaptic vesicle fusion and quantal secretion of acetylcholine. J Cell Biol 101:1386–1399
Van Harreveld A, Crowell J (1964) Electron microscopy after rapid freezing on a metal surface and substitution fixation. Anat Rec 149:381–386
Van Harreveld A, Crowell J, Malhotra SK (1965) A study of extracellular space in central nervous tissue by freeze-substitution. J Cell Biol 25:117–137
Van Harreveld A, Trubatch J (1979) Progression of fusion during rapid freezing for electron microscopy. J Microsc (Oxford) 115:243–256
Van Meer G, Gumbiner B, Simons K (1986) The tight junction does not allow lipid molecules to diffuse from one epithelial cell to the next. Nature (London) 322:639–641

Van Venetie R, Verkleij AJ (1981) Analysis of the hexagonal II phase and its relation to lipidic particles and the lamellar phase. A freeze fracture study. Biochim Biophys Acta 645:262–269

Van Venetie R, Verkleij AJ (1982) Possible role of non-bilayer lipids in the structure of mitochondria. A freeze-fracture electron microscopic study. Biochim Biophys Acta 692:379–405

*Verkleij AJ (1984) Lipidic intramembranous particles. Biochim Biophys Acta 779:43–63

*Verkleij AJ, Ververgaert PHJT (1975) Architecture of biological and artificial membranes, as visualized by freeze-etching. Annu Rev Phys Chem 26:101–122

*Verkleij AJ, Ververgaert PHJT (1978) Freeze-fracture morphology of biological membranes. Biochim Biophys Acta 515:303–327

Verkleij AJ, Ververgaert PHJT, Van Deenen LLM, Elbers PF (1972) Phase transitions of phospholipid bilayers and membranes of *Acholeplasma laidlawii* B. visualized by freeze-fracturing electron microscopy. Biochim Biophys Acta 288:326–332

Verkleij AJ, Mombers C, Gerritsen WJ, Leunissen-Bijvelt J, Cullis PR (1979) Fusion of phospholipid vesicles in association with the appearance of lipidic particles as visualized by freeze-fracturing. Biochim Biophys Acta 555:358–361

Ververgaert PHJT, Verkleij AJ, Verhoeven JJ, Elbers PF (1973) Spray-freezing of liposomes. Biochim Biophys Acta 311:651–654

Chapter 15 Cryofixation of Diffusible Elements in Cells and Tissues for Electron Probe Microanalysis

KARL ZIEROLD[1] and RUDOLF ALEXANDER STEINBRECHT[2]

1 Introduction

The interaction of the electron beam with the specimen not only provides information on its ultrastructure, but also on its elemental composition. In principle, the irradiating electron beam (probe) is focussed on the area of interest in the specimen, while the secondary radiation (e.g. X-rays) generated by the interaction with the specimen is processed through a spectrometer to provide information on the kind and amount of the elements present. Today, electron probe X-ray microanalysis (EPXMA) is the most widely used method in biological microanalysis. Also, electron energy loss spectrometry (EELS) acquires more and more attention. It would be beyond the scope of this book to explain the physical principles and the methodology of these analytical methods here, in particular, since there are numerous, well-written reviews on this subject (e.g. Chandler 1977; Hren et al. 1979; Hall and Gupta 1983; Morgan 1985). Rather, we want to discuss the particular importance of cryotechniques in this field and to point out the still existing problems of specimen preparation. Its crucial steps are not only relevant for EPXMA and EELS, but also for other microanalytical techniques such as proton probe X-ray microanalysis and laser probe mass spectrometry.

Indeed, the history of EPXMA surprisingly resembles that of electron microscopy in general, the development of instrumentation having been far in advance of that of specimen preparation. Being well established in the material sciences since the 1960s, biological specimens only gave meaningful results in those cases where strongly bound elements were analyzed, e.g. mineralized tissue (Hall and Höhling 1969) or metal-intoxicated tissue (Yarom et al. 1973). However, most elements of biological interest, such as electrolytes, are quantitatively extracted from the tissue by conventional specimen preparation, in particular, by chemical fixation, dehydration and sectioning on a water surface. Attempts to precipitate ions during chemical fixation only occasionally were successful and the results are qualitative at best.

The introduction of cryofixation changed the situation, as documented by the increasing number of laboratories producing meaningful results. Rapid freezing not only is the best approach to preserve the original composition of a specimen,

1 Max-Planck-Institut für Systemphysiologie, Rheinlanddamm 201, D-4600 Dortmund, FRG
2 Max-Planck-Institut für Verhaltensphysiologie, D-8131 Seewiesen, FRG

Cryotechniques in Biological Electron Microscopy
Edited by R.A. Steinbrecht and K. Zierold

it also allows one to arrest physiological processes under defined conditions with high time resolution (see Knoll et al., Chap. 14, this Vol.). Although cryofixation may preserve the original distribution of diffusible substances, this is by no means guaranteed, as it depends on the cooling rate like the preservation of structure (see Bachmann and Mayer, Chap. 1, this Vol.). Moreover, severe artefacts may be produced during sampling of the specimen before it is frozen, and likewise later during the various procedures following cryofixation. Even the processing of the raw analytical data is not an easy task and has many pitfalls. For the latter aspect, the reader again is referred to the reviews on electron probe microanalysis as given above (see also Hall and Gupta 1982).

In this chapter we want to compare the different cryotechniques in use and discuss their relative advantages and disadvantages for the in situ analysis of mobile elements. We will also give brief reference to some papers, which highlight the potential of microanalysis in giving physiologically relevant data on the intra- and intercellular concentration of ions, which, with no other method so far, can be obtained with comparatively high accuracy of localization.

2 Specimen Preparation

Cryopreparation pathways as used for the quantitative analysis of diffusible ions and water in cells and tissues by electron microprobe methods are schematically outlined in Fig. 1. The preparation starts by specimen sampling, followed by cryofixation. Then there exist essentially three different ways of processing, which yield three different types of electron microscopic specimens:

1. Dehydration by freeze-drying or freeze-substitution with subsequent embedding in resin (see Steinbrecht and Müller, Chap. 7, this Vol.). The result is an ultrathin plastic section, which can be studied by STEM (scanning transmission electron microscopy) or TEM (transmission electron microscopy).

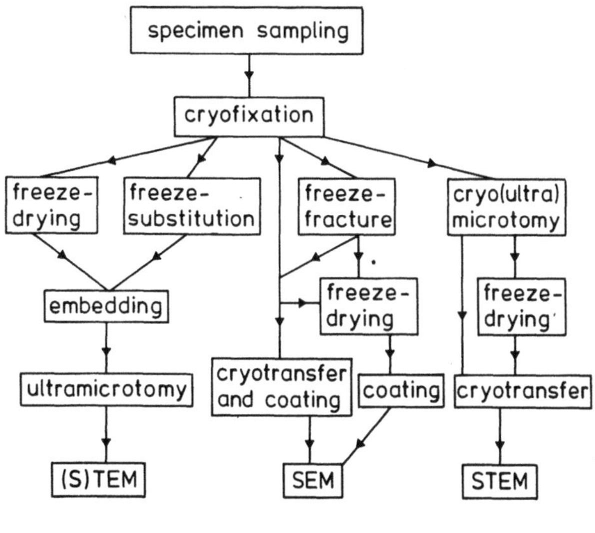

Fig. 1. Pathways of cryopreparation as used for electron probe microanalysis of biological specimens

2. Preparation of bulk specimens for SEM (scanning electron microscopy). The cryofixed sample is cryotransferred to the SEM either directly or after freeze-fracturing and may be studied in the frozen-hydrated state (see Marshall, Chap. 13, this Vol.). The specimens can also be analyzed after sublimation of water by freeze-drying. In both cases, the specimen requires coating by an electrically conductive layer, e.g. carbon, before microanalysis in the SEM is carried out.
3. Cryosections are cut from the cryofixed specimen by cryo(ultra)-microtomy and are transferred either directly into the STEM or freeze-dried before microanalysis (see Zierold, Chap. 6, this Vol.). It is also possible to observe and analyze a section first in the frozen-hydrated state and then a second time after freeze-drying.

In the following, the advantages and limitations of these different approaches are discussed.

2.1 Specimen Sampling

Reasonable cryofixation of biological specimens for electron microscope purposes is restricted to small specimens ranging up to 40 µm in size due to physical and technical reasons (Bachmann and Mayer, Chap. 1, this Vol.; Sitte et al., Chap. 4, this Vol.), unless high-pressure freezing is employed, which may extend the specimen size to about 1 mm (Moor, Chap. 8, this Vol.). Thus, tissue specimens often need to be excised from an intact organ or removed from the natural environment in a defined physiological state. It is conceivable that disconnecting a specimen from its blood supply or severing its innervation may disturb the physiological state of the cells and, consequently, the distribution of ions to such an extent that further processing of such specimens for microanalysis becomes useless. Therefore, meaningful microanalytical data were first obtained from tissue specimens which could be prepared as small samples appropriate for cryofixation without losing their physiological characteristics, e.g. frog skin (Dörge et al. 1974), insect Malpighian tubules (Gupta et al. 1976) and single muscle fibres (Somlyo et al. 1977). Still today, muscle fibres and epithelia are preferred objects of investigation. Single cells and tissue cultures could become the specimens of choice for electron probe microanalysis for the same reason (Warley et al. 1983; Zierold et al. 1984; Schmitz et al. 1985). The development of appropriate sampling or in situ freezing techniques (e.g. von Zglinicki et al. 1986), which provide good cryofixation of a defined physiological state, is the main challenge for the application of analytical electron microscopy to more complex biological specimens.

2.2 Cryofixation

With respect to ion localization, cryofixation aims towards the immobilization of all diffusible substances. Two crucial questions have to be answered: (1) How much redistribution of mobile substances occurs during cooling and freezing?

(2) Are the electrolyte ions really immobile in a cryofixed specimen? A conclusive answer from physics is not yet available. We know that the viscosity of water increases considerably during supercooling; this slows down diffusion. During crystallization of the water, the solutes are swept into the intercrystalline space by the growing ice-crystal dendrites. Thus, the displacement of ions during freezing is given by the mean size of the segregation compartments. Further displacement during storage at LN_2 temperature is certainly smaller, but may become critical at elevated temperatures as used for further preparation. Fortunately, practical experience so far has not given any indication for redistribution of elements in cryofixed specimens even after storage for several years in liquid nitrogen. EPXMA results from freeze-dried cryosections show that the characteristic gradients of electrolyte concentrations across cell membranes remain unchanged (R. A. Steinbrecht and K. Zierold, unpubl.). Moreover, no ion shifts depending on the direction of ice-crystal growth were so far observed by microanalysis. Thus, the size of the segregation compartments due to ice-crystal growth is the main limit to the analytical spatial resolution in EPXMA and related methods.

2.3 Plastic Embedding Methods

These methods start with the removal of the water either by freeze-drying or by freeze-substitution. From the microanalytical point of view, the question arises: Where do the ions remain or go after their original solvent is exchanged by vacuum, as is the case in freeze-drying, or by an organic solvent as in freeze-substitution? Most of the elements of interest such as Na, Mg, P, S, Cl, K and Ca are not volatile even in the high vacuum of the electron microscope. Exceptions are chlorine when it reacts to Cl_2 and some compounds of sulphur, e.g. SO_2. After sublimation of the ice during freeze-drying, most ions are assumed to react with counter-ions or bind to the closest organic molecule. These processes are facilitated by electrostatic forces between electric charges, which are increased 80 times due to the drop of the dielectric constant from 80 for ice to 1 for vacuum.

In cells containing enough organic matrix (dry weight content >10%), this redistribution of ions to stable compounds is assumed to take place within a few nanometers, i.e. within the size of the ice-crystal segregation zones, which should be smaller than the attempted spatial microanalytical resolution. However, in compartments rich in water, such as vacuoles or extracellular spaces, the redistribution will be much greater. These compartments are not suitable for quantitative microanalysis after freeze-drying (see also Sect. 2.5). Generally, it is recommended to dry for a long time (days and weeks) at low temperature (below 200 K) in order to minimize ion redistribution and tissue shrinkage (Edelmann 1986).

The use of freeze-substitution for microanalytical purposes is controversial, as can be seen from the literature (Marshall 1980a; Harvey 1982; Zierold and Schäfer 1983; Meyer et al. 1985; Roos and Barnard 1985; Edelmann 1986). The extent to which ions are lost from the cells or redistributed into different intracellular compartments may differ considerably depending on the biological object, the substitution medium and the exact protocol of the process. However, it should be mentioned that ion retention is reported preferentially after use of diethylether as substitution liquid (dielectric constant = 4.3), whereas much ion loss is found after freeze-substitution in ethanol, methanol, or acetone with the higher dielectric constants 25.1, 33.5, 21.4, respectively. This observation agrees with the notion that ion retention within the specimen requires electrostatic binding when the water is removed.

The embedding of the freeze-dried or freeze-substituted biological object into plastic is done in order to achieve a specimen feasible for ultramicrotomy. Sections have to be cut with a dry knife in order to prevent washing out of diffusible substances. No trough liquids have been found so far which are safe for microanalytical purposes. Cutting ultrathin dry plastic sections is much easier at low temperature in a cryoultramicrotome than at room temperature.

Plastic sections can also be used for the determination of the local water content. As in this method water in the original biological specimen is replaced by the plastic, labeling of the plastic will represent the original local water content. Ingram and Ingram (1983) have used bromine as a plastic marker in order to determine the water distribution in muscle tissue. Thus they were able to determine element concentrations per kg wet weight from freeze-dried, plastic-embedded sections by microanalysis.

It is still controversial whether and how far ions are redistributed during plastic infiltration. Recent studies show remarkably lower ion gradients after freeze-drying and embedding in Spurrs low viscosity resin as compared to data obtained from freeze-dried cryosections (Roos and Barnard 1985). Wróblewski and Wróblewski (1986) have compared X-ray microanalytical data obtained from muscle after different cryopreparation procedures including low temperature plastic embedding. These data indicate, for example, a higher potassium retention (as shown by higher K/P-ratios) in cryosections than in plastic sections after freeze-drying or freeze-substitution. It remains open whether longer drying times or lower embedding temperatures (e.g. 190 K with Lowicryl K11M) will improve the situation. At present, caution is still advised with microanalytical data obtained from plastic sections unless they are corroborated by data from cryosections.

2.4 Bulk Specimens

The microanalysis of frozen-hydrated bulk specimens is described by Marshall (Chap. 13, this Vol.). As cryofixed cells or tissue blocks often exhibit ice contamination on the surface, cryofracturing under vacuum is recommended in order to obtain a clean uncontaminated surface for X-ray microanalysis in the SEM. The original ion distribution as achieved after cryofixation and cryotransfer can be assumed to remain preserved. The drawbacks of this method are the relatively low spatial analytical resolution of about 2 μm and the invisible X-ray excitation volume below the specimen surface. Comparison of the X-ray continuum in the frozen-hydrated state and after subsequent freeze-drying allows the determination of the water content (see Sect. 2.5). For the interpretation of results from freeze-dried bulk specimens the same arguments concerning freeze-drying have to be considered as mentioned in Section 2.3.

2.5 Cryosections

From the microanalytical point of view the increase of the spatial resolution is the main reason for the use of cryosections. Of course, cryosections have to be cut dry at low temperature in order to preserve the same element distribution as in the frozen bulk specimen. A cutting temperature below 180 K is accepted to guarantee the localization of elements as cryofixed. Although the physics of the cryosectioning process is not yet completely understood, there is evidence that no through-section melting of the ice occurs which would allow diffusible elements to redistribute. Additionally, there is no indication from microanalytical results for ion displacement along the surface of the cryosections (smearing) as could be expected in the case of superficial melting (see Zierold, Chap. 6, this Vol.).

Frozen-hydrated cryosections are expected to provide the most reliable data, but localization of elements may be difficult due to inherent low contrast. Furthermore, full use of the high spatial resolution obtainable with ultrathin sections cannot be made because of radiation damage. EPXMA and EELS require such high electron doses for recording reliable results that the mass lost by radiation damage does not allow the use of ultrathin sections (Zierold 1983, 1986a; see also Talmon, Chap. 3, this Vol.; Zierold, Chap. 6, this Vol.). Therefore, analysis of frozen-hydrated specimens is done successfully only in cryosections of at least 1 μm thickness (Hall and Gupta 1983) or in bulk specimens (Marshall 1980b). For example, an etched depth of 100 nm caused by electron irradiation would destroy the whole ultrathin section, but is still tolerable in a 1-μm-thick section and even negligible in bulk specimens. However, the disadvantage of increasing specimen thickness is the loss of spatial resolution, which is about 30 nm in 0.1-μm-thick sections, about 300 nm in 1-μm-thick sections and 2 μm in bulk specimens, depending on the instrument operation parameters (Hall and Gupta 1983).

The problem of mass loss in frozen-hydrated specimens can be overcome by freeze-drying as will be described below. Hall and Gupta have developed an elegant method to determine the local water content in 1-μm-thick cryosections from the difference in continuum measured before and after freeze-drying (for detailed accounts see Hall and Gupta 1979, 1982). Sasaki et al. (1983) have applied this method to 100 nm thick cryosections. However, the authors do not report how they have mastered the problem of radiation damage they must have encountered in frozen-hydrated sections according to the described experimental conditions. An additional advantage of frozen-hydrated sections is that ion concentrations in water-rich compartments can be measured. The risk of ion redistribution and precipitation by freeze-drying was already discussed in Section 2.3.

Ultrathin freeze-dried cryosections have proved to be most feasible for measurements of the element distribution in compartments of cells and tissues with a water content of less than 90% (e.g. A. V. Somlyo et al. 1977, 1985; Hagler and Buja 1984; Wendt-Gallitelli and Wolburg 1984; A.P. Somlyo et al. 1985; Zierold 1986a). Freeze-dried cryosections are stable in the electron beam and the high electron optical contrast allows good ultrastructural localization of the measured element distributions. The achieved detection limits are in the range of 10 mmol kg^{-1} dry weight depending on the geometrical arrangement and type of

X-ray analyzer. Some authors have achieved detection limits of less than 1 mmol kg^{-1} dry weight for calcium (A.P. Somlyo et al. 1985). In comparison with sections of plastic-embedded material, freeze-dried cryosections exhibit considerably higher image contrast. Also, the analytical detection limit is better by a factor of three to five due to the mass thickness of the embedding material. The spatial analytical resolution depends on specimen thickness. In 100-nm-thick freeze-dried sections, a spatial analytical resolution of 30 nm can be achieved. Thus, less than 1000 characteristic atoms can be localized if they are present in such a small volume. The current limitations in the quantitative microanalysis of cryosections were reviewed recently by Hall and Gupta (1984) and Hall (1986a, b).

About 100-nm-thick cryosections are thin enough to give rise to a STEM darkfield signal which is linearly related to the local mass thickness. Thus, the local dry weight content and hence, the local water content can be determined in different cellular compartments of freeze-dried cryosections (Zierold 1986b).

3 Some Biological Applications and Results

It is impossible in this context to enumerate the various applications in biological specimens. Moreton (1981), in his extensive review, gives over 150 references and the number of studies has increased ever since!

Intracellular concentrations of electrolyte elements are now available for a wide variety of tissues, confirming the general notion of low Na, high K, low Cl content in cytoplasm and nucleus (see Table 3 in Gupta and Hall 1982). Earlier ideas of a high sodium content in cell nuclei comparable to that in the extracellular space, therefore, have been abandoned (see discussion in Jones et al. 1979). Another example where EPXMA has substantially contributed to a general physiological concept is the role of mitochondria in calcium storage. Wherever mitochondria were investigated in intact cells after proper cryofixation, their calcium content was extremely low. A.P. Somlyo et al. (1985), for example, measured in mitochondria of parenchymal liver cells, a mean calcium concentration of 0.8 ± 0.1 mmol kg^{-1} dry weight. Only damaged cells accumulate larger amounts of Ca in their mitochondria (see e.g. Buja et al. 1985).

Unfortunately, with a detection limit in the range of 1 mmol Ca kg^{-1} (Gupta and Hall 1978), most of the physiologically relevant cytoplasmic calcium concentrations within cells are beyond the instrumental possibilities. Functionally important accumulations of Ca have been extensively studied in muscle cells (Somlyo et al. 1977; Wendt-Gallitelli and Wolburg 1984). A.V. Somlyo and co-workers (1985) were able to follow the release and uptake of calcium in the terminal cisternae of the sarcoplasmic reticulum in situ during tetanic contraction of single muscle fibres. Extraordinarily high calcium concentrations prevail in the luminal fluid of nematocysts (stinging organelles) of sea anemones. The changes in ion concentration as observed by EPXMA during the discharge of these nematocysts have led to a novel concept of the discharge mechanism (Lubbock et al. 1981; Gupta and Hall 1984).

Quite understandably, the electron probe microanalysis of cryofixed tissue has had particular impact on the investigation of the transport of ions and water across cells and tissues. It provides direct data on the electrolyte concentrations in the cells as well as in narrow extracellular lumina. In almost all ion-transporting epithelia studied, irrespective of whether secreting or absorbing, increased electrolyte concentrations have been measured in local intercellular spaces (see Table 2 in Gupta and Hall 1982). These data have substantially supported current hypotheses on coupled transport of ions and water (reviewed by Gupta 1984). Following the application of stimulating and inhibiting agents, the changes of intra- and extracellular electrolyte concentrations could be well monitored by EPXMA and are consistent with electrophysiological and other independent data (e.g. Gupta and Hall 1983; Dow et al. 1984; Beck et al. 1985; Dörge et al. 1985; Marshall et al. 1985; Larsson et al. 1986).

4 Conclusion and Outlook

As has been shown in the preceding section, cryopreparation is suitable for the study of mobile elements by electron probe microanalysis. In particular, cryosections have proved to provide reliable data for cell biological and physiological research, which at the highest spatial resolution so far cannot be obtained or confirmed by any other means. When comparing results obtained by EPXMA with other microanalytical methods, one has to keep in mind that EPXMA measures the total concentration of an element, whereas for example ion-selective microelectrodes measure the ion activity of this element. Other specific differences have to be considered in the evaluation of data from other techniques, such as flame photometry, radioactive tracers or specific fluorescent indicator dyes. Where such comparisons have been made, the data often are in good consistency (e.g. Gupta et al. 1978; Acker et al. 1985; Rick et al. 1985). Thus, microanalytical data obtained with different methods are supplementary rather than competitive.

The study of diffusible elements in defined physiological states of the cell would be further improved by progress in the following directions:

1. Specimen sampling and cryofixation have to be optimally adapted to the particular object of investigation so that the physiological state is maintained and freezing damage is minimized.
2. Thinner cryosections with uniform thickness are desired to make full use of electron energy loss spectrometry.
3. Remedies are required to reduce the radiation damage in ultrathin frozen-hydrated cryosections so that artefacts caused by freeze-drying are avoided and the physiologically important water content can be directly determined with high spatial resolution.

References

(References marked by an asterisk are review papers or books respectively)

Acker H, Pietruschka F, Zierold K (1985) Comparative measurements of postassium and chloride with ion-sensitive microelectrodes and X-ray microanalysis in cultured skeletal muscle fibers. In Vitro Cell Dev Biol 21:45–48

Beck F, Dörge A, Rick R, Thurau K (1985) Osmoregulation of renal papillary cells. Pflügers Arch 405 (Suppl 1):s28–s32

Buja LM, Hagler HK, Parsons D, Chien K, Reynolds RC, Willerson JT (1985) Alterations of ultrastructure and elemental composition in cultured neonatal rat cardiac myocytes after metabolic inhibition with iodoacetic acid. Lab Invest 53:397–412

*Chandler JA (1977) X-ray microanalysis in the electron microscope. Elsevier/North-Holland Biomedical Press, Amsterdam

Dörge A, Gehring K, Nagel W, Thurau K (1974) Localization of sodium in frog skin by electron microprobe analysis. Naunyn-Schmiedeberg's Arch Pharmacol 281:271–280

Dörge A, Rick R, Beck F, Thurau K (1985) Cl transport across the basolateral membrane in frog skin epithelium. Pflügers Arch 405 (Suppl 1):s8–s11

Dow JAT, Gupta BL, Hall TA, Harvey WR (1984) X-ray microanalysis of elements in frozen-hydrated sections of an electrogenic K^+ transport system: The posterior midgut of tobacco hornworm (*Manduca sexta*) in vivo and in vitro. J Membr Biol 77:223–241

*Edelmann L (1986) Freeze-dried embedded specimens for biological microanalysis. Scanning Electron Microsc 1986/IV:1337–1356

*Gupta BL (1984) Models of salt and water flow across epithelia: an evaluation by electron probe X-ray microanalysis. In: Pequeux A, Gilles R, Bolis L (eds) Osmoregulation in estuarine and marine animals. Springer, Berlin Heidelberg New York, pp 191–211

*Gupta BL, Hall TA (1978) Electron microprobe X-ray analysis of calcium. Ann N Y Acad Sci 307:28–51

*Gupta BL, Hall TA (1982) Electron probe X-ray microanalysis. In: Baker PF (ed) Techniques in cellular physiology, pt 2, P128. Elsevier/North-Holland Biomedical Press, Amsterdam, pp 1–52

Gupta BL, Hall TA (1983) Ionic distribution in dopamine-stimulated NaCl fluid-secreting cockroach salivary glands. Am J Physiol 244:R176–R186

Gupta BL, Hall TA (1984) Role of high concentrations of Ca, Cu, and Zn in the maturation and discharge in situ of sea anemone nematocysts as shown by X-ray microanalysis of cryosections. In: Bolis L, Zadunaisky J, Gilles R (eds) Toxins, drugs, and pollutants in marine animals. Springer, Berlin Heidelberg New York, pp 77–95

Gupta BL, Hall TA, Maddrell SHP, Moreton RB (1976) Distribution of ions in a fluid-transporting epithelium determined by electron-probe X-ray microanalysis. Nature (London) 264:284–287

Gupta BL, Berridge MJ, Hall TA, Moreton RB (1978) Electron microprobe and ion-selective microelectrode studies of fluid secretion in the salivary glands of *Calliphora.* J Exp Biol 72:261–284

Hagler HK, Buja LM (1984) New techniques for the preparation of thin freeze dried cryosections for X-ray microanalysis. In: Revel J-P, Barnard T, Haggis GH (eds) The science of biological specimen preparation. SEM, AMF O'Hare, IL 60666, pp 161–166

Hall TA (1986a) Properties of frozen sections relevant to quantitative microanalysis. J Microsc (Oxford) 141:319–328

*Hall TA (1986b) The history and the current status of biological electron-probe X-ray microanalysis. Micron Microsc Acta 17:91–100

Hall TA, Gupta BL (1979) EDS quantitation and application to biology. In: Hren JJ, Goldstein JI, Joy DC (eds) Introduction to analytical electron microscopy. Plenum, New York, pp 169–197

*Hall TA, Gupta BL (1982) Quantification for the X-ray microanalysis of cryosections. J Microsc (Oxford) 126:333–345

*Hall TA, Gupta BL (1983) The localization and assay of chemical elements by microprobe methods. Q Rev Biophys 16:279–339

*Hall TA, Gupta BL (1984) The application of EDXS to the biological sciences. J Microsc (Oxford) 136:193–208

Hall TA, Höhling HJ (1969) The application of microprobe analysis to biology. In: Möllenstedt G, Gaukler KH (eds) X-ray optics and microanalysis. Springer, Heidelberg Berlin New York, pp 582–591

*Harvey DMR (1982) Freeze substitution. J Microsc (Oxford) 127:209–221

*Hren JJ, Goldstein JI, Joy DC (1979) Introduction to analytical electron microscopy. Plenum, New York

Ingram MJ, Ingram FD (1983) Electron microprobe calibration for measurement of intracellular water. Scanning Electron Microsc 1983/III:1249–1254

Jones RT, Johnson RT, Gupta BL, Hall TA (1979) The quantitative measurement of electrolyte elements in nuclei of maturing erythrocytes of chick embryo using electron-probe X-ray microanalysis. J Cell Sci 35:67–85

Larsson L, Aperia A, Lechene C (1986) Ionic transport in individual renal epithelial cells from adult and young rats. Acta Physiol Scand 126:321–332

Lubbock R, Gupta BL, Hall TA (1981) Novel role of calcium in exocytosis: Mechanism of nematocyst discharge as shown by X-ray microanalysis. Proc Natl Acad Sci USA 78:3624–3628

*Marshall AT (1980a) Freeze-substitution as a preparation technique for biological X-ray microanalysis. Scanning Electron Microsc 1980/II:395–408

*Marshall AT (1980b) Frozen-hydrated bulk specimens. In: Hayat MA (ed) X-ray microanalysis in biology. Univ Park Press, Baltimore, pp 167–196

Marshall AT, Hyatt AD, Phillips JG, Condron RJ (1985) Isosmotic secretion in the avian nasal salt gland: X-ray microanalysis of luminal and intracellular ion distributions. J Comp Physiol B 156:213–227

Meyer R, Schmitz M, Zierold K (1985) The influence of different cryopreparations on the distribution of ions in bullfrog myocard cells. Scanning Electron Microsc 1985/I:419–431

*Moreton RB (1981) Electron-probe X-ray microanalysis: Techniques and recent applications in biology. Biol Rev 56:409–461

*Morgan AJ (1985) X-ray microanalysis in electron microscopy for biologists. Univ Press, Oxford

Rick R, Beck FX, Dörge A, Thurau K (1985) Electron microprobe analysis of chloride secretion in the frog cornea. Current Eye Res 4:377–384

Roos N, Barnard T (1985) A comparison of subcellular element concentrations in frozen-dried, plastic-embedded, dry-cut sections and frozen-dried cryosections. Ultramicroscopy 17:335–344

Sasaki S, Nakagaki I, Mori H, Imai Y (1983) Intracellular calcium store and transport of elements in acinar cells of the salivary gland determined by electron probe X-ray microanalysis. Jap J Physiol 33:69–83

Schmitz M, Meyer R, Zierold K (1985) X-ray microanalysis in cryosections of natively frozen *Paramecium caudatum* with regard to ion distribution in ciliates. Scanning Electron Microsc 1985/I:433–445

Somlyo AP, Bond M, Somlyo AV (1985) Calcium content of mitochondria and endoplasmic reticulum in liver frozen rapidly in vivo. Nature (London) 314:622–625

Somlyo AV, Shuman H, Somlyo AP (1977) Elemental distribution in striated muscle and the effects of hypertonicity – Electron probe analysis of cryo sections. J Cell Biol 74:828–857

Somlyo AV, McClellan G, Gonzalez-Serratos H, Somlyo AP (1985) Electron probe X-ray microanalysis of post-tetanic Ca^{2+} and Mg^{2+} movements across the sarcoplasmic reticulum in situ. J Biol Chem 260:6801–6807

Warley A, Stephen J, Hockaday A, Appleton TC (1983) X-ray microanalysis of HeLa S3 cells. II. Analysis of elemental levels during the cell cycle. J Cell Sci 62:339–350

*Wendt-Gallitelli M-F, Wolburg H (1984) Rapid freezing, cryosectioning, and X-ray microanalysis on cardiac muscle preparations in defined functional states. J Electron Microsc Tech 1:151–174

Wróblewski J, Wróblewski R (1986) Why low temperature embedding for X-ray microanalytical investigations? A comparison of recently used preparation methods. J Microsc (Oxford) 142:351–362

Yarom R, Hall TA, Stein H, Robin GC, Makin M (1973) Identification and localization of intraarticular colloid gold: ultrastructural and electron probe examinations of human biopsies. Virchows Arch Abt B Cell Pathol 15:11–22

Zglinicki T von, Bimmler M, Purz H-J (1986) Fast cryofixation technique for X-ray microanalysis. J Microsc (Oxford) 141:79–90

*Zierold K (1983) X-ray microanalysis of frozen-hydrated specimens. Scanning Electron Microsc 1983/II:809–826

*Zierold K (1986a) Preparation of cryosections for biological microanalysis. In: Müller M, Becker RP, Boyde A, Wolosewick JJ (eds) The science of biological specimen preparation 1985. SEM, AMF O'Hare, IL 60666, pp 119–127

Zierold K (1986b) The determination of wet weight concentrations of elements in freeze-dried cryosections from biological cells. Scanning Electron Microsc 1986/II:713–724

Zierold K, Schäfer D (1983) Röntgenmikroanalyse diffundibler Substanzen im Ultradünngefrierschnitt. Acta Histochem Suppl 28:63–72

Zierold K, Schäfer D, Pietruschka F (1984) The element distribution in ultrathin cryosections of cultivated fibroblast cells. Histochemistry 80:333–337

Part IV Appendix

Chapter 16 Safety Rules for Cryopreparation

HELLMUTH SITTE, KLAUS NEUMANN, and LUDWIG EDELMANN[1]

1 Introduction

As is often the case in every day life, the main risk in the routine use of cryopreparation methods is frequently due to inadequate knowledge or underestimation of the hazards. Vapourization of larger quantities of nitrogen in badly ventilated, small rooms or cool rooms may be lethal (Sect. 2). The same applies to explosive propane/air mixtures (Sect. 3). Eyes and skin may be severely damaged by secondary cryogen splashing, e.g. liquid propane, ethane or halogenated hydrocarbons (Sect. 4). The problems presented by primary cryogen splashing, e.g. liquid nitrogen (LN_2) or liquid helium (LHe) are completely different (Sect. 5). Finally, precautions should be taken when working with inflammable, secondary cryogens and when using glass containers or vessels of materials which become brittle at cryogenic temperatures (Sects. 6 and 7). Transport and removal of cryogens require particular safety measures (Sect. 8).

Instructions have been kept as short and precise as possible: only the main sources of danger are dealt with and the discussion of less important problems is deliberately omitted.

2 Asphyxiation with Gaseous Nitrogen

One litre of LN_2 expands to almost 700 litres or 0.7 m^3 gaseous nitrogen (GN_2) as it changes from liquid to gas. The resulting hazard is usually underestimated as GN_2 is inert and therefore non-toxic. Normally, in addition to approximately 78% GN_2, air contains 21% oxygen. If the oxygen content in the air falls below a level of 18%, asphyxia suddenly occurs followed by deep unconsciousness without any warning symptoms (e.g. dizziness or drowsiness). Before the affected person loses consciousness he is unable to call for help or to draw attention to himself. If somebody remains unconscious in a low oxygen environment, death quickly occurs.

Cryoultramicrotomes are often installed in small, badly ventilated rooms and this presents a particularly acute hazard. If between 2 and 7 litres $LN_2\,h^{-1}$ are

1 Fachbereich 3.5, Medizinische Biologie, Universität des Saarlandes, D-6650 Homburg-Saar, FRG

Cryotechniques in Biological Electron Microscopy
Edited by R.A. Steinbrecht and K. Zierold

evaporated, approximately 1.4 to 5 m^3 GN_2 are released hourly. This causes the oxygen content in the air to decrease so drastically that lethal accidents in small rooms (e.g. height×length×depth less than 3×3×3 m^3 or 27 m^3) cannot be excluded. Adequate safety is in these cases only to be achieved with an efficient ventilation, thus providing ample air exchange. Work carried out in cool rooms where GN_2 is released is particularly critical because generally no ventilation is provided and the cool room doors are usually made comparatively airtight. A "dead man's system" appears to be a more effective warning device than an oxygen indicator; with the former system the operator regularly (e.g. every 2 min) presses a button. If the button is not pressed within the set time, an alarm is set off indicating the accident. The unconscious person should be transferred immediately to a well-ventilated environment and, if necessary, artificial respiration should be given.

3 Gaseous Propane Explosions

Gaseous propane (0.0020 g cm^{-3}) has a higher density than air (0.0013 g cm^{-3}). A highly explosive mixture results if the propane content of the air is between 2.1 and 9.5% (vol/vol). Storing and releasing propane in cellar rooms, which is strictly prohibited, is therefore very dangerous. Due to the higher density, propane gas settles to the floor and cannot flow off. In rooms at ground level or in the upper floors of a building, however, this gas can be used almost without hazard. Ignition can be caused by an electrical spark or open flame and often the building concerned is completely demolished. This type of accident may be caused by very small leaks in the valve system or the storage vessel. Explosions may also occur as a result of liquid propane being poured into a drain, if the propane/air mixture forming in the drain is ignited, e.g. if a glowing cigarette end is thrown into the drain. However, the handling of propane is hazard free if it is never used in cellars and if care is taken that larger quantities of propane are never evaporated into the atmosphere of a confined space (see also Sect. 8).

4 Burns Caused by Secondary Cryogen Splashing

The secondary cryogens used almost exclusively for cryofixation (e.g. liquid propane, ethane or halogenated hydrocarbons; see Sitte et al.; Chap. 4, this Vol.) are particularly dangerous because their boiling point is well above the temperature at which they are used. Cooling (cryofixation!) is, therefore, particularly efficient. For work during which splashing may occur (e.g. high-speed immersion cryofixation), a protective shield must be used. This particularly applies to the eyes. The most effective protection is provided by a face mask, as illustrated in Fig. 1. Should the cryogen come into contact with the skin, the affected part – especially if this is an eye – should immediately be rinsed with running water. It is often recommended that luke warm water should be used, but this advice is not very helpful in the case of an emergency as no time should be lost regulating mixer

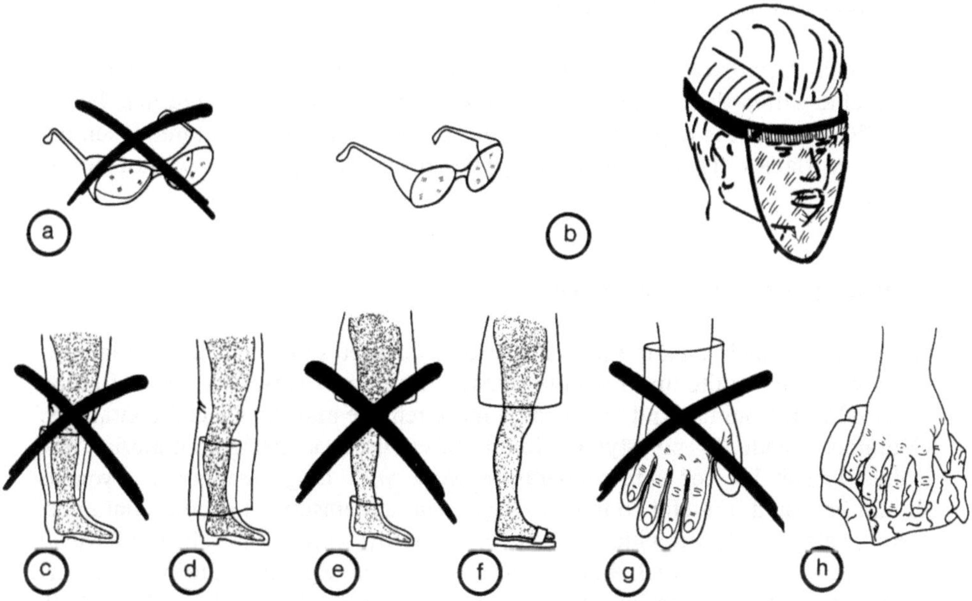

Fig. 1a–h. Safety measures during handling of liquid cryogens (see Sects. 4 and 5). Open spectacles or better still a protective shield tightly closed at the forehead **(b)** are preferable to completely closed protective spectacles **(a)**; protective shields of the type illustrated optimally protect the whole face from splashing cryogen. Boots **(c)** and shoes **(e)** open at the top are dangerous as cryogen may collect inside. Trousers should be pulled over boots **(d)**; open sandals **(f)** are the safest footwear during cryowork. Protective gloves **(g)**, often recommended, should never be worn for the same reason: if cryogen splashing cannot be excluded, a dry paper tissue **(h)** should be used to prevent hand burns on cold solid surfaces

taps. The injury should then be treated like a burn, preferably by a dermatologist or ophthalmologist.

5 Burns Caused by Primary Cryogen Splashing

In contrast to secondary cryogens, primary cryogen splashing (LN_2 or LHe) does not present any direct danger if the skin or even an eye is affected: the high vapour pressure of these continually boiling liquids leads to the buildup of an insulating gaseous layer ("Leidenfrost phenomenon") on contact with a warm surface. In contrast to secondary cryogens (see Sect. 4), a real hazard only exists if smaller quantities of cryogen accumulate in a confined space. The cryogen evaporates and causes extreme cooling in this restricted area. This may occur, for example, if LN_2 splashes into shoes, into gloves or even into special protective glasses, closed at the sides (Fig. 1). The low viscosity of primary cryogens favours accidents of this type and it is usually not possible to remove the shoes, gloves or glasses so quickly that burns are avoided. Therefore, the instructions illustrated in Fig. 1 should be followed. Here, too, the eyes are best protected by the face mask. Clothing open at the top and closed at the bottom should generally not be worn. The procedure set out in Section 4 for treating burns should also be applied here.

6 Ignition of Combustible Secondary Cryogens

Many secondary cryogens (e.g. propane, ethane) are combustible and highly inflammable. Often the comparatively low vapour pressure at temperatures below $-100\,°C$ is all that is required for ignition. Extreme caution is required when using naked flames. This particularly applies if cryogen splashing cannot be excluded.

7 Bursting of Cryogen Containers

Glass Dewar vessels used in the past repeatedly caused severe accidents in laboratories when they quite unexpectedly burst. This type of vessel should, therefore, be withdrawn from use and replaced by more reliable and safer metal containers. Containers made of polystyrene foam, often used as packing material, are eminently suitable for safe preparation work with LN_2. Secondary cryogens should be stored in metal containers. Under no circumstances should containers made of glass or plastic, which become brittle at low temperatures, be used. Metal vessels filled with LN_2 or LHe must never be completely sealed during transport or storage as the evaporation of this primary cryogen will quickly lead to a high pressure buildup which may cause the container to explode. In this context, see also Section 8 below.

8 Transport and Disposal of Cryogens

Primary cryogens (LN_2, LHe) or liquefied, inflammable, secondary cryogens (propane, ethane) should never be transported in a lift: if the lift becomes stuck between two floors, the bad ventilation and the confined space become a serious safety risk as the cryogen evaporates. Primary cryogen vessels for LN_2 and particularly for LHe should never be completely closed during transport, otherwise evaporation would lead very quickly to the explosion of the vessel. Small quantities of primary cryogens may only be disposed of outside confined areas: to this end, they should be poured into metal vessels filled with large pebbles. The same applies to liquid, secondary cryogens. If propane or ethane waste is flared off (see Fig. 3 in Sitte 1984), there should never be LN_2 below the secondary cryogen. As the density of LN_2 is higher than the densities of liquefied propane or ethane, this is possible by inappropriate manipulation.

9 Concluding Remarks

The competent safety officer should always be consulted when equipping laboratories or when installing apparatus for cryopreparation at temperatures below $-110\,°C$. Without his explicit approval no laboratories or systems of this sort should start operation, especially if developed and produced by the operator himself. It is also in the interests of the staff carrying out cryopreparation and of a responsible superior that all co-workers involved in cryopreparation are urged to study this article and/or the publications given in the references below. The individual should be required to confirm in writing that he is acquainted with these instructions.

References

(References marked by an asterisk are review papers or books respectively)

Cryogenics safety manual – a guide to good practice (1985) Safety panel. Obtainable from British Cryogenics Council, c/o Institution of Chemical Engineers, 165–171 Railway Terrace, Rugby CV 21 3HQ, England

Edwards high vaccum publication 59-K 100-00-880 (1979) Recommended safety precautions for handling cryogenic liquids. Obtainable from B. O. C. Cryo-Products, Manor Royal, Crawley, West Sussex RH 10 2LW, England

Reichert-Jung (1982 ff) Precautions to be observed during cryopreparation. Manual and leaflet included in the manuals for Reichert-Jung cryosystems FC 4 D, KF 80 and CS-auto. Obtainable from Reichert-Jung, Optische Werke AG, Product Management EM, Hernalser Hauptstr. 219, A-1170 Wien, Austria, or the local representatives

*Robards AW, Sleytr UB (1985) Low temperature methods in biological electron microscopy. In: Glauert AM (ed) Practical methods in electron microscopy, vol 10 (See appendix 1, Safety) Elsevier, Amsterdam, pp 507–511

Sitte H (1984) Equipment for cryofixation, cryo-ultramicrotomy and cryosubstitution in biomedical TEM routine. Zeiss Inf. MEM Mag Electron Microsc 3:25–31

Glossary

See also entries in Subject Index

Cryofixation	Solidification of a biological specimen by cooling with the aim of minimal displacement of its components.
Cryopreparation	Preparation technique for the electron microscopy of biological specimens which utilizes the solid state of water by means of low temperature.
Cryoprotectant	Agent for chemical pretreatment of biological specimens in order to reduce the formation of segregation compartments (→). Possible adverse side effects of c. have, however, to be taken into account.
Cryotransfer	Transfer of a cryofixed specimen (e.g. between different instruments) at low temperature without changing the state of its water.
Freeze-drying	Dehydration of a cryofixed specimen by sublimation of the ice under vacuum.
Freeze-etching	Superficial freeze-drying (→) of a cryofixed and usually freeze-fractured (→) specimen.
Freeze-fracturing	Cleavage of a cryofixed specimen in order to expose internal structures.
Freeze-substitution	Dehydration of a cryofixed specimen by exchange of ice against an organic solvent.
Frozen-hydrated	Solid state of a cryofixed specimen containing its full water content (in the vitreous or crystalline form).
Hydration shell	Layer of surface-modified water molecules at an interface, e.g. between macromolecules and bulk water.
Radiation damage	Alterations in an electron microscopic specimen caused by the electron beam.
Recrystallization	Formation or growth of more stable ice crystals out of a less stable state.

Segregation compartment	Structure caused by the segregation of solvent and solute during crystallization of the solvent. The formation of segregation compartments (segregation pattern) is the most obvious artefact of cryofixation (often referred to as freezing damage or ice-crystal damage).
Vitrification	Glass-like solidification without formation of crystals. Vitrified specimens may devitrify (crystallize) by energy uptake (e.g. heat or electron irradiation). Identification of the vitrified state, e.g. of solid water, in the electron microscope is done best by diffraction.

Subject Index

Numbers in *italics* refer to figures

acetone 152, 156, 166, 276
acrolein 156f., 166, 217, 221, 226
activation energy 3
adsorption 50ff., *51*, *58*, *59*, 202
aggregation of macromolecules 40ff., *40*, *42*, 48
albumin 200, *200*, 219
aldehyde fixation 56, 154, 217
animal tissue, see tissue
antibody 216, 226f., 232
antigenicity 166, 216f., 226
artefacts 108f., 160ff.
–, adsorption 50ff., *51*, 202
–, chemical fixation 40ff., *42*, 161, *162*, 166f., *167*, 226, 234f., 260ff., 272
–, cryoultramicrotomy 120ff., 140ff., *142*, *144*
–, drying *51*, 139, 151, 160, 242
–, flattening 50ff., *51*, *53*, 54ff.
–, freeze-drying 46, 152, 160ff.
–, freeze-substitution 47, 152, 160ff.
–, freezing 4f., 28f., *94*, *102*, 108f., 135, 139, 161, 176ff., 189f., 260ff., 275
–, rehydration 137, 157ff.
autoradiography 231ff.

bacteria *59*, *167*
bare grid technique, see cryofixation, thin films
beam heating, see electron beam heating
beam/specimen interaction, see electron beam/specimen interaction
beryllium 247
Biot number 24
blood *94*, *100*, *102*, *104*
boiling point, see temperature, boiling
bond breakage, see polymer scission or radiation damage, mechanisms
bouncing phenomenon 98, 102f.
bubble formation, see radiation damage, cryospecimens
bulk specimens 159, 240ff., *241*, *242*, *248*, *249*, 274, 276
butylbenzene 134

cage effect 79
calcium 166, 278
capillarity 54ff.
cell junctions 261, 263
– surface 222, *223*, *224*, *225*, *227*
– suspensions *94*, 95, *103*, 107f.
cellular plasm, see cytoplasm
chain scission, see polymer scission
chemical fixation 40ff., *40*, *42*, 105, 133, 134, 154f., 159ff., *162*, *167*, 217, 221, 226, 234f., 260ff., 272
chloroplasts 261
chromatin 41ff., *100*, 108, 115
cilia *138*, *162*
coating, carbon 135, 205, 233, 246
–, metal 247
cohesion force 48
cold block, see cryofixation, impact
– stage, see cryoelectron microscopy
collapse *51*, *55*, 57ff., *58*, *59*, 151, 209
colloidal gold 217
contamination 65, 137, 211, 244f.
contrast 124ff., 137ff., 140, 214, 277
convection, see cryogen convection
cooling efficiency 24ff., 89ff., 98f., 106, 177ff.
– rate 11, 15, 24ff., *26*, *27*, 88, 93, *93*, 98f., 105f., 109, 118, 176ff., *178*, 181, *185*, *244*, 259f.
– –, measurement 28
copolymerization 45
copper block 98
critical point drying 55, 159
crosslinking 40ff., *40*, *42*, 67
cryobiology 4, 116
cryoelectron microscopy 4, 64ff., 114ff., *121*, *123*, *125*, 137ff., 244f.
cryofixation 3, 22, 25ff., 41, 87ff., 243f., 259f., 274f.
–, emulsions 194
–, evaluation 108, 160ff., 189f., 234f., 259f.
–, high pressure 14, 109f., 175ff., *182*, *184*, *185*, *186*, 194
–, immersion 89ff., *90*, *94*, 107, *163*

cryofixation, impact 25, 95ff., *96*, *100*, *101*, *102*, *103*, *104*, 107, *163*
–, instrumentation *90*, *96*, 182ff., *182*, *184*
–, jet freezing *90*, 93ff., 107, *164*, 193f.
–, pliers 101
–, specimen size 24ff., 88, 94f., 103, 106f., 117f., 187f., 190, 193
–, specimen-jet freezing 193f.
–, spray freezing *90*, 95f., 193f., *193*, *197*, *200*
–, suspension 92
–, thin films *90*, 107, 122ff., *123*
–, time resolution 29f., 105, 190, 259f.
–, tissue 106f., 187f., 243f.
cryofracturing, see freeze-fracturing
cryogen 88ff., 106, 285ff.
– container 288
– convection 27, 92
– handling 285ff., *287*
– transport 288
cryoglueing 133f., *134*
cryopreparation 35, 45ff., 120f., *132*, *150*, 217ff., *222*, 273ff., *273*
–, safety 285ff., *287*
cryoprotectants 19, 116ff., 134, 152, 176, 187, 217
cryoprotection factor 68f.
cryosectioning, see cryoultramicrotomy
cryosections 120ff., *121*, 132ff., *138*, *142*, 217ff., *218*, *227*, 235f., 277f.
–, freeze-dried 137ff., *138*, *142*, 233, 274, 276f.
–, frozen-hydrated 120ff., *121*, 137, *138*, *142*, 274, 277f.
–, transfer 135, *136*
cryosubstitution, see freeze-substitution
cryotransfer 137ff., 233, 244f., 274
cryoultramicrotomy 105, 108, 120ff., 132ff., *144*, 217ff., 232ff., 235, 285f.
cutting speed 135, 141
cytochemistry 166f., 216ff., 231ff.
cytomembranes, see membranes, biological
cytoplasm 38ff., *100*, *101*, *104*, *121*, *138*, *162*, *163*, *164*, *186*, *218*
cytoskeleton 115, 261, 267

danger, see cryopreparation, safety
decoration 211ff.
dehydration 41ff., 47, 150ff., 242
denaturation 44, 52, 180
detection limit 237, 254, 277f.
deuterium oxide 207
devitrification 15, 92, 99, 102, 118f., 176
Dewar vessel 288
dextran 118
dielectric constants 275
diethylether 152, 157, 166, 276
differential scanning calorimetry 11f., 15, 28, 266
diffusible substances 165f., 232ff., 243, 254, 272ff.
dimethylsulfoxide 118, 134
dispersions 192ff.
DNA, see nucleic acids
double-labelling 220
drift 127
dynamic processes 258ff., 273, 278

electrical charging 68, 127, 135, 245ff., *246*
electrolyte 20ff., 234, *249*, 275f., 278
electron beam heating 67f., 71
– beam/specimen interaction 65ff., 245ff., *246*, 277
– diffraction 68, 119, 127
– dose 65f., *72*, 74, 127
– energy loss 65, *66*
– – – spectroscopy, see microanalysis
– exposure, see electron dose
– probe X-ray microanalysis, see microanalysis
– spin resonance 29
element composition 272f.
embedding for cryoultramicrotomy, see cryoglueing
–, low temperature, see low temperature embedding
–, room temperature 47, 55f., 156f., 158, 275f.
emulsions 192ff., *193*
enzyme cytochemistry 166, 231ff., *236*
etching depth, see ice, sublimation rate
ethane 89, 106, 285ff.
ethanol *7*, 152, 157, 166, 276
eutectic temperature, see temperature, eutectic
exocytosis 264f., *265*, 278

ferritin *197*
fixation, chemical, see chemical fixation
flattening artefact, see artefacts, flattening
formaldehyde 154, *167*, 217, 221, 226
free radicals, see radicals
freeze concentration 17, 22ff., 24
freeze-drying 45f., 137f., 149, *150*, 152ff., 159, 205, 207ff., *208*, 235, 242, 275f.
–, applications *58*, *59*, 160ff., *163*, *209*, *210*, *212*, *214*
–, equipment 157f., 207, 211
–, procedures 158f., 207ff.
–, rate 153f., 157f., 207ff.
–, temperature 152ff., 207ff., *208*
freeze-etching *162*, 189, 192ff., *193*, *195*, *197*, *200*, 222f., *222*, *223*, *264*

freeze-fracturing *186*, 189, *201*, 220ff., 240ff., *262*, *265*, 276
freeze-substitution 41ff., 46, 149ff., *150*, 160ff., 275f.
–, applications *42*, *94*, *100*, *101*, *102*, *104*, 160ff., *162*, *163*, *164*, *167*, 261, *262*, 264, 275f.
–, equipment 155f.
–, media 152, *153*, 156f.
–, procedures 156f., 159
–, rate 152ff., *153*, 156f.
–, temperature 152ff.
freezing damage, see artefacts, freezing
– point, see temperature, melting
– time, see cryofixation, time resolution
freon 89, 285ff.
frozen-hydrated specimens 52, *53*, 74f., 108, 114ff., *121*, *123*, *125*, 137, 214, 240ff., *241*, *242*, *248*, *249*

G-value 66, 70
gel 38f.
gelatine 217, 226
glues for cryosectioning, see cryoglueing
glutaraldehyde 156, 226, 263
glycerol 119, 176, 194, 263
gold, see colloidal gold
Golgi apparatus 220, 261

heat capacity 5, *13*, *20*, 92, 97
– conductance 5, 24ff.
– transfer 5, 24ff.
helium, liquid 88, 98f., 106, 285ff.
heptane 134
high-pressure freezing, see cryofixation, high pressure
histochemistry, see cytochemistry
HPI-layer, see periodic specimens
humid chamber *96*, 103, 106
hydration shell 35ff., *36*, 38, 45ff., 57, 151, 199f., 208f., 214
hydrophilicity 36f., 45, 50ff.
hydrophobicity 37f., 45, 50ff., 223

ice 6ff., 70ff., 108f., 180
–, amorphous, see water, vitreous
–, crystallization 11ff., 28, 176
–, crystallization enthalpy 5
–, cubic 11, 122
–, heat capacity 5
–, hexagonal 71, 122
–, mass loss 71, *71*, *72*
–, mechanical properties 6, *7*, 122, 144f.
–, physics 3ff.
–, recrystallization 182, see also devitrification
–, structure *9*, 9
–, sublimation 45f., 71, 136, 137, 194f.
–, sublimation rate 7, *7*, 197f., *208*
–, thermal conductivity 5, 71
–, vapour pressure *7*
–, vitreous, see water, vitreous
ice-crystal damage, see artefacts, freezing
– growth 22f., 28, 92, 152, 176, 179
image processing 205f., *209*, *212*, 213
imaging 124ff., *125*, 137, 242f., 277f.
immersion velocity 92, *93*
immunocytochemistry 133, 135, *136*, 166, *167*, 216ff., 232
immunogold labelling 216ff, *218*, *222*, *223*, *224*, *225*, *227*
immunolabelling, see immunocytochemistry
impact velocity 102f.
interface 47ff., *49*, 52, 122ff., 201f.
–, ice/organic material *72*, 75ff., *78*
intramembrane particles 263, *264*
ion activity 279
– association 22, 165f.
distribution 139, 141, 275f.
– localization, see microanalysis
– selective electrodes 279
– transport 278f.
ionization 65, 76
–, water 70ff.
isopentane 89, 134

jet freezing, see cryofixation, jet freezing

knife, diamond 135
–, glass 135
– marks, see artefacts, cryoultramicrotomy
–, steel 135

label efficiency 225ff., *227*
label-fracturing *222*, 223ff., *224*, *225*
labelling, see autoradiography or immunocytochemistry
–, post-section 217, 219
laser probe mass spectrometry, see microanalysis
latent dose effect 68
latex *72*, *78*
Leidenfrost phenomenon 89, 106, 287
lipids 56, 119, 165, 180, 234, 260f., 266
liposomes 165, 200f., 267
liver *101*, *138*, *186*
low dose imaging 127
– temperature embedding *42*, 43, 60, 155, 158f., *164*, 166, *167*, 276
Lowicryl 155, 158f., 276
lysosomes 220, 261

macromolecules 35ff., 66, 118, 192ff., *197*, *200*
–, size and shape 198f.

mass density 140
– loss 69, 71, *71*, *72*, 77, *78*, 126f., *138*, 253, 277
melamine embedding 56
melting point, see temperature, melting
membrane contacts 261, 263
– dynamics 263ff.
– fusion *104*, 180, 264ff.
membranes, biological 56, 161, *218*, 219f., 260f., *265*
metal-mirror freezing, see cryofixation, impact
methane 38
methanol 152, 156, 165f., 276
micelles 200, 266
microanalysis 137ff., 165f., 233ff., 240ff., *248*, *249*, 272ff.
microtubules *104*, 115, *162*, 261
mitochondria *101*, *138*, *164*, 261, *262*, 278
molecular weight determination 196ff.
monolayer 95, 211
muscle *100*, 115, 119, *163*, 278

Nanoplast 56
nitrogen, gas 136, 285ff.
–, liquid 88, 98f., 106, 183, *244*, 285ff.
– slush, see nitrogen, solid
–, solid 89, 244, *244*
nuclear magnetic resonance 14, 16f., 21, 38, 266
nucleation 11f., 118, 177, 179, 236
nucleic acids 41ff., 115
Nusselt number 24

optical diffraction *210*, *212*
osmium tetroxyde 43, 56, 139, 154, 156, 165, 220, 226, 235

particle density *195*, 196ff.
periodic specimens 119, 206f., *209*, *210*, *212*, 213f., *214*
phase contrast 124ff., *125*
– diagram *10*, *18*, *23*
– separation 48f.
– transition 118f., 180, 266f.
physiological state 261ff., 274, 278f.
plant tissue 108f., 161, 166, 180
plasma membrane, see membranes, biological
plasmon excitation 65
plastic section 160, 166, 220, 235, 275f.
plunging, see cryofixation, immersion
polybutylacrylate *72*, *78*
polymer crosslinking 67, 76
– scission 66, 76
polymerization, heat 157
–, UV 159
polymers 66f.
polymethylmethacrylate *72*, *78*
polysaccharides 44, 234
polystyrene *72*, *78*
precipitation 237, 272
propane 89, 106, *244*, 285ff.
protein A-gold 219, 226f.
proton probe X-ray microanalysis, see microanalysis

rad 66
radiation damage 64ff., *72*, *78*, 126ff.
– –, applications 80
– –, cryospecimens *72*, 73ff., *78*, 126ff., 137ff., *138*, 253, 277
– –, definition 65
– –, detection methods 67f.
– –, ice 69ff., 79
– –, low temperature 67ff., 128
– –, mechanisms 70ff., 77ff.
– –, organic material 67ff., *72*, *78*
– –, polymers 66f.
– –, support film 76
– –, vitreous water 73f., 77, *78*, 127
radicals 67, 70f., 77ff.
radioactive tracers, see autoradiography
radiolysis 70ff., 74ff.
rapid freezing, see cryofixation
receptor, EGF- *218*, 219, *223*, *224*, *225*, *227*
recrystallization 24, 92, 139, 141, 152, 182
rehydration artefacts, see artefacts, rehydration
relaxation time *12*, 14, 16f., *16*, *17*, *19*, 29f., 118ff.
replication 199, 205ff., 210ff.
repolymerization 79
resolution, spatial 126f., 140, 213f., 234, 247ff., 254, 275, 278
–, time, see cryofixation, time resolution

safety, see cryopreparation, safety
sandwich specimen *90*, 92, 95, 106, 108, 184f., *185*, 194
segregation 5, 18, 139, 152, 194, 260, 263
–compartment 22, *94*, 108f., 140, *142*, 177f., *193*, 259, 275f.
self-diffusion coefficient 7
SEM *142*, 159, 240ff., *241*, *242*, *248*, 274, 276
shadowing, see replication
shrinkage *55*, 57ff., 139, 151, 159, 165
signal-to-noise-ratio 124
silver mirror 98
single cells *94*, *100*, *102*, 107f., *121*, *138*, 274
slamming, see cryofixation, impact

sol 38f.
solubility 152, 236
solutions, heat capacity *20*
–, kinetics 19ff., *19*, 118f.
–, melting temperature 18
–, phase diagram 18, *18*, *23*
–, supercooled 19ff.
–, thermodynamics 17ff.
–, vitrification 21f., 114ff.
solvents, effects on macromolecules 38ff.
space charge 245ff.
specimen carriers *90*, *96*, *103*, 103, 109, 134, 184f., *185*, 187f.
specimen excision 106, 187, 274
spray freezing, see cryofixation, spray freezing
STEM 137ff., *138*, 274
steric hindrance 228
sublimation, see ice, sublimation
substitution media, see freeze-substitution, media
sucrose 118, 135, *136*, 176, 217
superconducting lens 68
supercooling, see water, supercooled or solutions, supercooled
support film 50ff., *51*, 76, 122ff., 135f., 213
surface energy 49, *49*, 54ff., *55*, 213
– –, data 53
– relief reconstruction *209*, *212*, 213f.
– tension, see surface energy
suspensions 92, *103*, 103, 107, 120, 122ff., 134

temperature, boiling 11, 89
–, crystallization 10
–, devitrification 15, *23*, 177
–, eutectic 18, *18*
–, glass transition 10, *23*
–, homogeneous nucleation 10, *23*
–, indicator dye 29
–, measurement 28, 92f., *93*, *185*, 187, *244*
–, melting 10, *23*, 89, 152
–, recrystallization *23*, 133, 152
thawing *132*, 219, 226
thermal conductivity 5, *6*, 92, 97
– energy 5
thermocouples, see temperature, measurement
time resolution, see cryofixation, time resolution
tissue *100*, *101*, 106f., *138*, *162*, *163*, *186*, *241*, *242*
tissue culture 95, 217, 274
– slicer *96*, 103
toluene 134, 141

ultra high vacuum 207, 211
ultramicrotomy 143, 160, 276
ultrastructure 8, 40f., 44f., 108, 115, 137ff., 160ff., 189f., 220, 234, 260ff.
umbelliferone 29
uranyl acetate 156f., 165, 219

vapour fixation 139, 154, 235
vesicles 115, 119, 200f., *201*, 220, 261
virus *42*, *53*, *58*, *123*, *125*
viscosity *7*
vitreous, see water, vitreous
vitrification 15, 88, 108, 114ff., 177f.
vitrified specimens 116ff., *121*, *123*, *125*, 141

water, bound, see water, surface-modified
–, bulk 8, 29, *30*
– content 4, 139f., 176, 252, 276ff.
–, crystallization enthalpy 5
–, density *13*, 119
–, dielectric constant 275
–, enthalpy 5
–, heat capacity 5, *13*
–, kinetics 11ff., *12*, 118f.
–, mobility 17, 21, 118ff.
–, molecular orientation 36ff., *36*
–, nucleation 11f., 118, 177, 179, 236
–, phase diagram *10*, 179
–, physics 3ff.
–, relaxation 16f., *16*, *17*
–, structure *9*, 9
–, supercooled 11, 13, 119, 177
–, surface-modified 8, 15f., 29, *30*, 36ff., 75, 151, 154, 208f.
–, thermal conductivity 5
–, thermodynamics 13
–, viscosity 11
–, vitreous 7f., 15, 28, 73f., 77, 114ff., 135
wetting angle *49*

X-ray absorption 247ff.
– detection 245, 250ff.
– diffraction 15, 28, 118f., 266f.
– excitation 65, 250, *253*
– microanalysis, see microanalysis

